PERIODIC TABLE OF THE ELEMENTS

IA 1	IIA 2	IIIB 3	IVB 4	VB 5	VIB 6	VIIB 7	VIIIB 8	VIIIB 9	VIIIB 10	IB 11	IIB 12	IIIA 13	IVA 14	VA 15	VIA 16	VIIA 17	18
1 **H** 1.008																	2 **He** 4.003
3 **Li** 6.941	4 **Be** 9.012											5 **B** 10.81	6 **C** 12.01	7 **N** 14.01	8 **O** 16.00	9 **F** 19.00	10 **Ne** 20.18
11 **Na** 22.99	12 **Mg** 24.31											13 **Al** 26.98	14 **Si** 28.09	15 **P** 30.97	16 **S** 32.06	17 **Cl** 35.45	18 **Ar** 39.95
19 **K** 39.10	20 **Ca** 40.08	21 **Sc** 44.96	22 **Ti** 47.88	23 **V** 50.94	24 **Cr** 52.00	25 **Mn** 54.94	26 **Fe** 55.85	27 **Co** 58.93	28 **Ni** 58.70	29 **Cu** 63.55	30 **Zn** 65.38	31 **Ga** 69.72	32 **Ge** 72.59	33 **As** 74.92	34 **Se** 78.96	35 **Br** 79.90	36 **Kr** 83.80
37 **Rb** 85.47	38 **Sr** 87.62	39 **Y** 88.91	40 **Zr** 91.22	41 **Nb** 92.91	42 **Mo** 95.94	43 **Tc** (98)	44 **Ru** 101.1	45 **Rh** 102.9	46 **Pd** 106.4	47 **Ag** 107.9	48 **Cd** 112.4	49 **In** 114.8	50 **Sn** 118.7	51 **Sb** 121.8	52 **Te** 127.6	53 **I** 126.9	54 **Xe** 131.3
55 **Cs** 132.9	56 **Ba** 137.3	57 **La*** 138.9	72 **Hf** 178.5	73 **Ta** 180.9	74 **W** 183.9	75 **Re** 186.2	76 **Os** 190.2	77 **Ir** 192.2	78 **Pt** 195.1	79 **Au** 197.0	80 **Hg** 200.6	81 **Tl** 204.4	82 **Pb** 207.2	83 **Bi** 209.0	84 **Po** (209)	85 **At** (210)	86 **Rn** (222)
87 **Fr** (223)	88 **Ra** (226.0)	89 **Ac**** (227)	104 (261)	105 (262)	106 (263)	107	108	109									

Atomic number — **H** (Symbol) — 1.00794 (Atomic weight)

*Lanthanides

58 **Ce** 140.1	59 **Pr** 140.9	60 **Nd** 144.2	61 **Pm** (145)	62 **Sm** 150.4	63 **Eu** 152.0	64 **Gd** 157.3	65 **Tb** 158.9	66 **Dy** 162.5	67 **Ho** 164.9	68 **Er** 167.3	69 **Tm** 168.9	70 **Yb** 173.0	71 **Lu** 175.0

**Actinides

90 **Th** 232.0	91 **Pa** (231)	92 **U** 238.0	93 **Np** (237)	94 **Pu** (244)	95 **Am** (243)	96 **Cm** (247)	97 **Bk** (247)	98 **Cf** (251)	99 **Es** (252)	100 **Fm** (257)	101 **Md** (258)	102 **No** (259)	103 **Lr** (260)

Study Guide to Accompany
UNIVERSITY CHEMISTRY
Fourth Edition
Mahan / Myers

Barbara A. Sawrey
University of California
San Diego

The Benjamin/Cummings Publishing Company, Inc.
Menlo Park, California • Reading, Massachusetts
Don Mills, Ontario • Wokingham, U.K. • Amsterdam • Sydney
Singapore • Tokyo • Madrid • Bogota • Santiago • San Juan

Copyright © 1987 by The Benjamin/Cummings Publishing Company, Inc.

All rights reserved. No part of this publication may be reproduced, stored in a retrieval system, or transmitted, in any form or by any means, electronic, mechanical, photocopying, recording, or otherwise, without the prior written permission of the publisher. Printed in the United States of America. Published simultaneously in Canada.

ISBN 0-201-05835-9

ABCDEFGHIJ-AL-8987

The Benjamin/Cummings Publishing Company, Inc.
2727 Sand Hill Road
Menlo Park, California 94025

TABLE OF CONTENTS

Fundamentals		1
Chapter 1	Stoichiometry and the Basis of Atomic Theory	10
Interchapter A	Dimensional Analysis	23
Chapter 2	The Properties of Gases	27
Chapter 3	Liquids and Solutions	40
Interchapter B	Dimensional Analysis Revisited	59
Chapter 4	Chemical Equilibrium	60
Chapter 5	Ionic Equilibria in Aqueous Solutions	75
Chapter 6	Valence and the Chemical Bond	104
Chapter 7	Oxidation-Reduction Reactions	118
Chapter 8	Chemical Thermodynamics	140
Chapter 9	Chemical Kinetics	163
Chapter 10	The Electronic Structure of Atoms	181
Chapter 11	The Chemical Bond	192
Chapter 12	Molecular Orbitals	200
Chapter 13	Periodic Properties	210
Chapter 14	The Representative Elements: Groups I-IV	217
Chapter 15	The Nonmetallic Elements	225
Chapter 16	The Transition Metals	234
Chapter 17	Organic Chemistry	248
Chapter 18	Biochemistry	260
Chapter 19	The Nucleus	267
Chapter 20	The Properties of Solids	279
Appendix A	Mathematical Review	284
Appendix B	Calculus Review	287
Appendix C	Problem Answers	295

INTRODUCTION

To the Student and the Instructor:

Learning is an individual process that each student must perform in his or her own mind. Knowledge cannot simply be transplanted from the instructor to text to the student. Active participation on the student's part is required. Resources such as the text, instructor and study guide are meant to help the student who actively participates in this learning process.

This study guide is designed to be used in conjunction with *University Chemistry* by B. Mahan and R. Myers. Except for the background chapter, each chapter in the guide correlates directly with the chapters in the text. In order to use this guide most efficiently, read the background chapter entitled Fundamentals in the guide first, then read the textbook material. You should read the textbook before attempting to use the rest of the guide.

A brief summary of each section is given and sample problems are worked out in the guide. There are also problem sets at the end of each chapter to give the student further practice. All answers to these problems are provided in Appendix C.

Since problem solving is the key to learning introductory chemistry, there is a large variety of questions available in both the text and study guide. Questions range in difficulty, but the emphasis in this guide is on easy and moderate problems.

Appendixes A and B provide a review of scientific notation, logarithms and some simple calculus that the student may wish to use for reference.

By taking advantage of the resources available--the instructor, textbook and study guide--student can meet and enjoy the challenge of *University Chemistry*.

FUNDAMENTALS

This chapter serves as an introduction to *University Chemistry*. Students entering a general chemistry course have diverse backgrounds and diverse goals, but it is assumed that all students using the text have had a high school chemistry preparatory course. This preliminary chapter, therefore, provides a "refresher course" covering some underlying principles of chemistry and some of the vocabulary. After all, you can't expect to speak the language of chemistry without knowing some basic vocabulary. As your knowledge increases, so will your repertoire of words and phrases.

Those students who wish to review the basic mathematical principles needed for this course, such as working with scientific notation and some calculus operations, are referred to Appendices A and B located at the end of this study guide.

What Is Chemistry?

Chemistry is the branch of science that deals with the composition and properties of all matter, and how matter can be changed from one form to another. A *chemical* then is nothing more than a substance, any substance. The word *chemical* is much maligned in present society, but all substances are chemicals. The air we breathe, the water we drink, the food we eat, and our bodies themselves are nothing more than organized collections of chemicals.

Only a few years ago we could say that chemistry was divided into four branches: analytical, inorganic, organic, and physical. At the present time there are so many divisions--biophysical, bio-inorganic, organometallic, polymer, nuclear, etc.--as to be confusing to the neophyte. Most of these divisions are designed to show that chemistry serves as a bridge between fields once considered to be quite separate. In fact, most students in your chemistry class will not become classical chemists but will use chemistry as applied to biology, physics, geology, or engineering. Chemistry is commonly known as the central science, as it clearly relates to all these other fields.

Scientific Method

Aristotle (384-322 BC), the Greek teacher and philosopher, is usually considered to be the first proponent of what we call a scientific method. His own scientific work was mainly biological, and he was an outstanding observer and recorder of biology.

The techniques and methods of logic were very important to Greek philosophers, but because of their fascination with theory coupled with rudimentary methods of measurement, *experimentation* was the missing element of Aristotle's scientific method.

It was not until the seventeenth century, nearly two thousand years after Aristotle, that experimentation became a respectable pursuit of a scientist or natural philosopher. Sir Francis Bacon (1561-1626), although not himself an experimentalist, succeeded in gaining the acceptance of experimental science through his scholarly writings. Bacon's contemporaries--Galileo, William Harvey, and William Gilbert--experimentalists all, were then seen to be engaged in a fashionable occupation.

Modern scientific method hinges on this experimentation. Though no specific set of rules exists for following the scientific method, the approach generally involves three steps, which may overlap somewhat. The first step is the *observation of phenomena*, from which a *hypothesis* or *theory* is formulated. Then *experiments* are devised to test the hypothesis. It is not strictly possible to prove a theory, only to show it is not possible to disprove it. This means that scientists' theories are always subject to modification or refinement. We will see an excellent example of how a theory is modified with new experimental evidence when we study atomic theory.

Elements

An *element* is a substance that cannot be broken down by physical or chemical means to a simpler material. An example is copper. It can be melted, even vaporized but it cannot be converted to something simpler than copper. The elements are the most fundamental chemicals.

Each element, when it is identified, is given a name and a symbol of one or two letters. Some of the earliest elements to be identified were gold (Au), silver (Ag), sulfur (S), carbon (C) and iron (Fe). In most instances the symbol is simply an abbreviation of the modern element name. Other abbreviations are based on the Latin name of the element, such as aurum (gold) or argentum (silver). A complete list of known elements, 106 to date, and their symbols can be found inside the back cover of the text. A different organization of the elements is found in the *Periodic Table* inside the front cover of the text. There we see the elements ranked from one through 106 in a pattern that seems at first glance to be arbitrary. In fact the vertical and horizontal grouping of the elements in the Periodic Table is very significant.

The number associated with each element (for instance element 20 is calcium) is also very important. It is called the *atomic number* and is explained in more detail in the next section.

Atoms

Think of a piece of copper and imagine cutting the piece in half to make two smaller pieces of copper. If you were to continue cutting each new smaller piece into half again and again, we realize that we still have a sample of the element copper, only smaller pieces than the original material. Let's extend this process to pieces so tiny that we can no longer see them, much less cut them in half. At what point in this process can we no longer subdivide the piece of copper? In other words, what is the smallest possible piece of copper that still behaves like copper?

The smallest bit of a element is called an *atom*, a very old Greek word that means uncut or indivisible, and a word used as long ago as 400 BC by the Greek philosopher Democritus to describe the fundamental unit of matter. Democritus' ideas were not popular in his life-time since atoms could not be seen or measured. It was not until 1803 that John Dalton revived the notion of atoms. Dalton's atomic theory was not immediately popular either but, eventually, overwhelming evidence convinced the skeptics. Atomic theory as proposed by Dalton has undergone many revisions, major and minor, over the last century.

The details of atomic theory will be developed in subsequent chapters but some of the basic concepts will be discussed here.

Our model of the atom shows it to consist of a dense center called the *nucleus* which is surrounded by a cloud-like collection of *electrons*. The nucleus is positively charged while the electrons are negatively charged. The attraction of the positive and negative charges for each other holds the atom together. The word atom generally refers to the neutral species where positive and negative charges are exactly balanced. If the atom has a net negative charge because the neutral species has gained one or more extra electrons, then it is called an *anion* or negative ion. If the atom has a net positive charge because there are fewer electrons than in the neutral state, then it is called a *cation* or positive ion.

Fundamentals

Although the nucleus has the positive charge, it can be further subdivided into two main components; protons and neutrons. Each proton has a charge defined as +1 and a neutron has no charge. *The number of protons defines the element.* The number of protons, given by the symbol Z, is also the atomic number. Therefore we say that a lithium (Li) atom has 3 protons, or Z=3, or has an atomic number of 3. Any neutral atom must have an atomic number equal to the number of electrons since an electron has a charge defined as -1. So a neutral Li atom has 3 protons and 3 electrons, a Li^+ cation has 3 protons and 2 electrons, while a Li^- anion has 3 protons and 4 electrons.

The number of neutrons in a nucleus can be variable. Continuing with Li as an example, approximately 92% of all Li atoms that naturally occur on earth have 4 neutrons. The remaining 8% of Li atoms have only 3 neutrons in the nucleus. Nuclei that vary only by the number of neutrons are called *isotopes*. Isotopes can be indicated in the following way. The sum of the protons and neutrons in a nucleus is called the *mass number*, which is written as left superscript on the elemental symbol. The two Li isotopes would be represented as ^6Li and ^7Li.

The chemistry of an element would be expected to depend on its atomic number and so be only independent of its atomic mass (and therefore number of neutrons). However, very careful measurements of chemical reactions show that they are very slightly different for different isotopes. By doing the same chemical reaction time and time again, one can eventually separate isotopes due to this slight difference. For most chemical purposes we will assume that all isotopes of the same element will have the same chemistry.

Table 0.1 presents some of the pertinent physical data for the proton, neutron and electron. While protons and neutrons have very similar weights, the electron is only about 1/2000 of a proton by weight. Protons, neutrons and electrons, while not the smallest sub-atomic particles known today, are considered to be the fundamental building blocks of matter.

Table 0.1 Protons, Neutrons and Electrons

Name	Symbol	Charge Number	Actual Charge	Mass
proton	p	+1	1.602189×10^{-19} coulombs	1.672649×10^{-24} grams
neutron	n	0	0	1.674954×10^{-24} grams
electron	e^-	-1	$-1.602189 \times 10^{-19}$ coulombs	9.109534×10^{-28} grams

An interesting observation about nuclei is that their masses are less than the mass sum of protons and neutrons which form them. This is a direct result of the large amounts of energy which are given - off when nuclei form. This follows the mass-energy relationship discussed in Chapter 19 of the text.

Since all isotopes of an element have essentially the same chemistry, the existence of isotopes can usually be ignored. Most of the fundamental aspects of chemistry were well established before the first isotope, ^2H, was even isolated in 1931 by Harold Urey. If all the isotopes of a given element are taken together as a group, the average mass of this group of isotopes is used as the mass of the element. This is discussed later, in Chapter 1.

Molecules

When atoms are combined in such a way that they are held fast to one another by what we call chemical bonds, we have *molecules*. Molecules have a specific ratio of elements from which they are composed and the atoms are configured in a specific fashion. The simplest molecule is formed by two hydrogen atoms. This molecule is symbolized by the formula H_2. Oxygen atoms also combine to form a molecule, O_2. Water is the combination of two hydrogen atoms and one oxygen atom. The proper way to write the formula of water is H_2O. The alternate forms, OH_2 or HOH are not as desirable. A molecule is the smallest unit of a *compound* just as an atom is the smallest unit of an element.

The subscripts in a formula indicate the number of individual atoms found in each molecule. For example, every molecule of C_2H_4 has two carbon atoms and four hydrogen atoms while every C_4H_8 molecule contains four carbon atoms and eight hydrogen atoms. The C : H ratio is the same for both compounds but the formulas represent two different molecules.

Once we know the formula of a molecule, the next question is: "What does it look like?" Many different kinds of models have been used to answer this question, two of the most successful models are: valence-shell-electron-pair-repulsion theory (VSEPR) and orbital hybridization theory. They will be covered in detail in chapters 6 and 11 respectively.

The properties of molecules are determined both by the geometric arrangement of the nuclei and by the distribution of the electron cloud around these nuclei. One of the goals in the study of chemistry is to understand how molecular size, shape, and electronic structure determine all the various chemical properties of a molecule.

Atomic and Molecular Size

We can't see an atom but chemists have found that it can be adequately represented as a sphere, with a small nucleus at the center and a diffuse electron cloud whose limits represent the outer edge of the sphere. Individual atoms may range in size from less than 1.0 Å (1 Angstrom = 10^{-10}m) in radius for H or He, to more than 2.0 Å in radius for Rb or Cs. The size units currently in vogue also include the picometer (1 pm = 10^{-12}m) and the nanometer (1 nm = 10^{-9}m).

The size of a molecule is not just the sum of the sizes of the component atoms. An H_2 molecule, for example, is more compact than two H atoms. Molecular sizes vary greatly, from the smallest, H_2, to huge polymers and proteins that might be hundreds of angstroms in length.

Chemical Nomenclature

A big part of learning the language of chemistry is learning how to identify a formula by its chemical name. Many of the names, such as water (H_2O), ammonia, (NH_3) and sodium chloride (NaCl) will be familiar to you. However, for most names you will need to know some rules of nomenclature. Don't be overwhelmed by the number of rules nor by the number of exceptions. Many chemicals are referred to by a historical name, not a systematic name.

Prior to listing some of the rules of nomenclature we must first decide if a formula represents a molecule in which each of the atoms is essentially neutral *or* is a neutral collection of cations and anions. We can get an indication of which category applies to a formula by looking at a periodic table.

Fundamentals

Elements located on the far right-hand side of the periodic table and H are *non-metals*. Non-metals do not conduct electricity. The middle and left-hand side of the table contain the *metals*, which do conduct electricity. Between the metals and non-metals lie a few *semi-metals*

A binary compound is one formed from only two different elements, such as H_2O, NH_3 or NaCl. If the two elements are both non-metals or semi-metals, there is a strong tendency for the atoms to share their outer electrons in the molecule, called *covalent bonding*. If in a binary formula the elements are a metal and a non-metal there is likely to be a complete or partial transfer of one or more electrons from the metal to the non-metal resulting in anions and cations, and called *salts*. The ions present in salts are held together by electrostatic forces and are said to exhibit *ionic bonding*. A great number of salts contain *polyatomic ions*, many of which can be found in Table 0.3.

Salts do not actually contain molecules although their formulas appear molecular in nature. For example, solid NaCl is composed of many Na^+ ions each surrounded by many Cl^- ions. *There are no distinct NaCl molecules in the solid or liquid phase.* The formula of a salt represents the ratio of cations to anions in the compound but does not represent a true molecule. Nonetheless chemists frequently refer to the "molecular" formulas for salts, understanding the invalidity of the term.

The distinction between covalent and ionic bonding as presented here is far too general for any purpose other than nomenclature. Actually, pure covalent or pure ionic bonds are not the norm. Most bonds lie intermediate between these two extremes. Later chapters in the text (6 and 11) will be more explanatory on this topic.

The *nomenclature of compounds containing only non-metals or semi- metals* follows a couple of simple rules:

1) the element listed first in the formula is named first using its simple elemental name, followed by the name of second element which has a -ide suffix attached.

2) Greek prefixes are usually used to indicate the number of atoms for each of the elements in the formula. The prefixes are:

$$\text{mono} = 1$$
$$\text{di} = 2$$
$$\text{tri} = 3$$
$$\text{tetra} = 4$$
$$\text{penta} = 5$$
$$\text{hexa} = 6$$

The *mono* prefix is normally omitted. Table 0.2 contains some examples of molecules named by these rules, as well as some common exceptions.

Fundamentals

Table 0.2 Some Names for Molecules

Water	H_2O
Hydrogen sulfide	H_2S
Hydrogen peroxide	H_2O_2
Ammonia	NH_3
Hydrazine	N_2H_4
Dichlorine oxide	ClO_2
Hydrogen chloride	HCl
Nitric oxide	NO
Nitrous oxide	N_2O
Dinitrogen Pentoxide	N_2O_5
Carbon monoxide	CO
Carbon dioxide	CO_2
Sulfur dioxide	SO_2
Carbon tetrachloride	CCl_4
Arsenic trichloride	$AsCl_3$

The *nomenclature of salts* is even more straightforward: the cation name is followed by the anion name. Since the cation appears first in the formula, this method is quite easy. No prefixes are needed, provided that you know the names of the ions! Table 0.3 is a fairly comprehensive list of common ions. Each ion has a specific charge associated with it, so there should be no ambiguity concerning how many of each ion are needed. *The salt must have an overall charge of zero.*

Example 1. Na_2S is named sodium sulfide. Since Na has a +1 and S a -2 charge there is only one possible neutral combination of the ions and the di prefix for sodium would be redundant.

Example 2. $FeSO_4$ is named iron (II) sulfate or ferrous sulfate. The Roman numerals indicating the charge on the iron ion are essential because Fe^{2+} and Fe^{3+} are both common. This is often the case for transition metals - those metals in the center of the periodic table.

Fundamentals

Example 3. The compound aluminum sulfate must have the formula $Al_2(SO_4)_3$ since Al^{3+} and SO_4^{2-} could not form a neutral compound of any other formula.

Many of the anions can be combined with H^+ in a formula that represents an *acid*. The simplest working definition of an acid is a compound that releases H^+ when dissolved in water. This turns out to be too narrow and restrictive a definition in many instances but is an easily understood classical statement of acid behavior.

The *nomenclature of inorganic acids* is derived from the name of the anion. A few acids are binary in nature, H^+ is combined with a monoatomic anion. These acids, when in water, have the prefix hydro- and the suffix -ic followed by acid.

Example 4. HCl in the gas phase is named hydrogen chloride, while dissolved in water it is named hydrochloric acid.

Example 5. H_2S in the gas phase is named hydrogen sulfide while it is named hydrosulfuric acid when dissolved in water.

When the acid contains a polyatomic anion there are the following possibilities:

1) If the anion ends in -ite, simply replace the ending with -ic acid, and
2) If the anion ends in -ate, replace the ending with -ous acid.

Example 6. H_2SO_4 is the acid of the sulfate ion and is named sulfuric acid.

Example 7. H_2SO_3 is the acid of the sulfite ion and is named sulfurous acid.

Fundamentals

Table 0.3 Some Names for Common Ions

	Cations		Anions
*Hydrogen	H^+	Hydroxide	OH^-
		Hydride	H^-
Ammonium	NH_4^+	Acetate (abv. OAc^-)	CH_3COO^-
Lithium	Li^+	Fluoride	F^-
Potassium	K^+	Chloride	Cl^-
Silver	Ag^+	Bromide	Br^-
Sodium	Na^+	Iodide	I^-
Copper(I) or cuprous	Cu^+	Nitrate	NO_3^-
Copper(II) or cupric	Cu^{2+}	Nitrite	NO_2^-
Mercury(I) or mercurous	Hg_2^{2+}	Carbonate	CO_3^{2-}
Mercury(II) or mercuric	Hg^{2+}	Hydrogen carbonate or bicarbonate (poor choice)	HCO_3^-
		Cyanide	CN^-
Barium	Ba^{2+}	Chromate	CrO_4^{2-}
Cadmium	Cd^{2+}		
Calcium	Ca^{2+}	Oxalate	$C_2O_4^{2-}$
Lead(II) or plumbous	Pb^{2+}	Sulfate	SO_4^{2-}
Magnesium	Mg^{2+}	Hydrogen sulfate or bisulfate (poor choice)	HSO_4^-
		Sulfite	SO_3^{2-}
Zinc	Zn^{2+}	Hydrogen sulfite or bisulfite (poor choice)	HSO_3^-
Manganese(II)	Mn^{2+}	Sulfide	S^{2-}
		Hydrogen sulfide	HS^-
Iron(II) or ferrous	Fe^{2+}	Thiosulfate	$S_2O_3^{2-}$
Iron(III) or ferric	Fe^{3+}		
Tin(II) or stannous	Sn^{2+}	Oxide	O^{2-}
Tin(IV) or stannic	Sn^{4+}	Peroxide	O_2^{2-}
Aluminum	Al^{3+}		
		Permanganate	MnO_4^-
*In water it is usual to write as H_3O^+		Hypochlorite	ClO^-
		Chlorite	ClO_2^-
		Chlorate	ClO_3^-
		Perchlorate	ClO_4^-

Fundamentals

Problems

1. Name the following:

   a) P_4O_{10} d) N_2O_3

   b) OF_2 e) PF_5

   c) SF_6 f) HBr (pure gas)

2. Write the formula for each of the following compounds:

   a) silicon tetrachloride   d) sulfur trioxide
   b) ammonia   e) chlorine trifluoride
   c) dinitrogen tetroxide   f) tetraphosphorous hexiodide

3. Name the following:

   a) HCN (pure gas)   f) Ag_2CrO_4

   b) $NaHCO_3$ g) $Ca_3(PO_4)_2$

   c) LiH   h) NaClO

   d) NH_4NO_3 i) ZnO

   e) $KMnO_4$ j) FeS

4. Write the formula for each of the following compounds:

   a) magnesium sulfate   e) mercury (I) chloride
   b) ammonium acetate   f) sodium chlorite
   c) potassium peroxide   g) tin (IV) sulfide
   d) mercury (II) chloride   h) aluminum sulfide

5. Name the following as acids.

   a) $HClO_4$ d) HCN

   b) HNO_2 e) H_2CrO_4

   c) CH_3COOH f) HI

6. Write the formula for each of the following compounds.

   a) chlorous acid
   b) phosphoric acid
   c) carbonic acid
   d) permanganic acid
   e) oxalic acid

CHAPTER 1
STOICHIOMETRY AND THE BASIS OF ATOMIC THEORY

1.1. The Origins of the Atomic Theory

The French chemist Joseph Proust showed in 1799 that copper carbonate always contained the same proportions of copper, oxygen and carbon by weight, no matter how it was prepared or what the source. After Proust showed that this was generally true of compounds; their proportions by weight were invariant, this came to be called Proust's law or the law of definite proportions.

John Dalton went on to show that simple whole number ratios not only applied to the weight of elements in a single compound but also applied to the same elements in a second possible compound, such as CO and CO_2 or CH_4 and C_2H_4. For example, Dalton found that methane (CH_4) always had a C:H weight ratio of 3:1 while ethane (C_2H_4) always had a C:H weight ratio of 6:1. Therefore the weight ratio of C in C_2H_4 to C in CH_4 was always 2:1. This is called the law of multiple proportions.

Based on these laws and other chemical weight data collected by Gay-Lussac, Dalton began to formulate modern atomic theory; that there were small particles, atoms, of each element and all the atoms of an element were alike while the atoms of different elements were not alike. Substances could then be viewed as specific combinations of atoms. Any different combination of these atoms would necessarily result in a different substance.

1.2. Determination of Atomic Weights and Molecular Formulas

Rather than work with the ratio of elemental weights in a compound, it makes more sense to establish some fixed scale of weights from which the ratios may easily be calculated if needed. It certainly would be more convenient to have characteristic weights associated with each type of atom. Chemists traditionally used oxygen atoms as their standard and determined all other atomic weights relative to oxygen. However, since 1961, all scientists have used the isotope carbon-12 (^{12}C has 6 protons, 6 neutrons and 6 electrons) as the weight standard. One mole (6.0221×10^{23} atoms) of ^{12}C weights *exactly* 12.00000 grams and

1 atom of ^{12}C weighs exactly 12.00000 atomic mass units,

where 1 amu = 1.66054×10^{-24} g.

If you look at any modern periodic table you will see that the atomic weight of carbon is not listed as 12.00000 g but rather as 12.0111 g. You might well wonder why this is so. ^{12}C remains our weight standard but the atomic weights reported on the periodic chart reflect the natural occurrence of all isotopes of an element. In other words, not all carbon is isotope ^{12}C; a small amount is ^{13}C. Similarly not all copper is ^{63}Cu. Some copper atoms are ^{65}Cu with two extra neutrons. Using the relative abundance of the two isotopes we can show how the listed atomic weight can be calculated.

isotope	natural abundance	atomic weight
^{63}Cu	69.090%	62.9298
^{65}Cu	30.910%	64.9278

Stoichiometry and the Basis of Atomic Theory   Chapter 1

The observed atomic weight of naturally occurring copper is simply the weighted average of all the isotopes, as shown below:

(fraction of natural abundance) x (isotope weight) = (contribution to overall weight)

The sum of all the contributions equals the observed atomic weight. For copper,

$$
\begin{array}{rcrcr}
0.69090 & \times & 62.9298 & = & 43.478 \\
0.30910 & \times & 64.9278 & = & 20.069 \\
& & & & \overline{63.547}
\end{array}
$$

This corresponds with the value of copper's atomic weight found on most periodic charts. Modern instrumentation allows the measurement of abundance and weight to five or six significant figures but normally only four significant figures need to be reported.

1.3. The Mole Concept

We encounter terms everyday that refer to a specific, understood quantity of things. A dozen donuts is understood to be 12 and a gross of pencils contains 144 pencils. When chemists refer to atoms, the numbers involved can be very large so we use the quantity mole to help keep track. *One mole of anything contains 6.0221×10^{23} units.* One mole of donuts would be 6.0221×10^{23} donuts (even the five billion people on Earth couldn't eat them all), and one mole of silicon contains 6.0221×10^{23} atoms of Si. The number of units in a mole, 6.0221×10^{23}, is called *Avogadro's number* and is sometimes abbreviated N_A.

$$N_A = 6.0221 \times 10^{23}$$

Amadeo Avogadro was the Italian scientist who in the early 1800's coined the word *molecule*, which means the specific combination of atoms given by a formula such as CO_2 or H_2O.

The atomic weight of an element is the weight of one mole of atoms of that element. One mole of silicon atoms weighs 28.09 grams. Stated another way, silicon has an atomic weight of 28.09 g/mole. By measuring a certain weight of an element we are also measuring a certain number of atoms.

Example 1.1. How many moles of phosphorus atoms are contained in 19.6 g of phosphorus?

One mole of P atoms weighs 30.97 g so 19.6 g of P must contain some fraction of a mole. The answer must have units of moles, which can be arrived at in several ways. Below is one method.

$$\frac{19.6 \text{ g P}}{31.0 \text{ g P/mol}} = 0.633 \text{ mol P}$$

Example 1.2. How many iron atoms are contained in a 1.0 lb piece of iron?

Your strategy should be to convert the weight in lbs to grams, then convert grams to moles of iron and finally to convert moles to the number of atoms using Avogadro's number. *Always include the units.* Every number used in a calculation will have units associated with

Chapter 1   Stoichiometry and the Basis of Atomic Theory

it, so write them down. The calculation considered here can be carried out as three separate steps, giving an intermediate answer from step 1 that is used in step 2, and an intermediate answer from step 2 that is used in step 3. From this comes the final answer. For example,

$$\frac{1.0 \text{ lb}}{454 \text{ g/ lb}^*} = 454 \text{ g},$$

$$\frac{454 \text{ gFe}}{55.85 \text{ gFe/ mol}} = 8.13 \text{ molFe}$$

$$8.13 \text{ mol Fe} \times 6.0221 \times 10^{23} \text{ atoms/mol} = 4.90 \times 10^{24} \text{ atoms of Fe}$$

Now let's turn our attention to substances other than elements. A substance denoted by a chemical formula that gives the specific ratio of atoms from which it is formed is usually called a molecule. Carbon dioxide (CO_2) describes a molecule composed of one carbon atom and two oxygen atoms. One mole of CO_2 molecules contains one mole of carbon atoms and two moles of oxygen atoms. The molecular weight of CO_2 is then given by the sum of the atomic weights of the elements multiplied by their appropriate subscripts.

Example 1.3. What is the molecular weight of CO_2?

1 mole C	=	12.01 g
2 mole O	=	2 (16.00) g
1 mole CO_2	=	44.01 g

Example 1.4. What is the molecular weight of octane, C_8H_{18}, a component of gasoline?

8 mole C	=	8 (12.01) g
18 mole H	=	18 (1.008) g
1 mole C_8H_{18}	=	114.224 g

A great many substances, however, do not exist as discrete molecules. Their formulas describe only the ratio of ions, not the intimate association of those elements. For example the formula for potassium fluoride, KF, indicates that one mole of potassium ions are required for each mole of fluoride ions, *but* there is not a KF molecule that exists under normal circumstances. Technically, KF does not have a molecular weight but rather it has a formula weight. Even so, chemists use the term molecular weight in this instance understanding that it describes the weight of non-molecular substances also.

---

* A list of some of the common conversions between English and SI units that you might need are: 1 pound = 453.6 grams, 1 inch = 2.54 cm, 1.06 quart = 1.00 liter.

Stoichiometry and the Basis of Atomic Theory   Chapter 1

Example 1.5. What is the formula weight or molecular weight of copper (II) sulfate pentahydrate, $CuSO_4 \cdot 5H_2O$?

$$
\begin{array}{rcl}
1 \text{ mole Cu} & = & 63.55 \text{ g} \\
1 \text{ mole S} & = & 32.06 \text{ g} \\
9 \text{ moles O} & = & 9\,(16.00) \text{ g} \\
10 \text{ moles H} & = & 10\,(1.008) \text{ g} \\
\hline
1 \text{ mole } CuSO_4 \cdot 5H_2O & = & 249.69 \text{ g}
\end{array}
$$

The $\cdot H_2O$ convention means that five moles of H_2O molecules are associated with each mole of Cu^{2+} and SO_4^{2-}.

Example 1.6. How many nitrogen atoms are contained in 0.052g of ammonium nitrate, NH_4NO_3?

First you must calculate the formula or molecular weight of NH_4NO_3.

$$1 \text{ mol } NH_4NO_3 = 80.05 \text{g}$$

Then convert grams of NH_4NO_3 to moles of NH_4NO_3, followed by the conversion to moles of N, then finally to atoms of N.

$$\frac{0.052 \text{ gNH}_4NO_3}{80.05 \text{ g/ mol}} = 6.5 \times 10^{-4} \text{ mol } NH_4NO_3$$

Since 1 mole of NH_4NO_3 contains 2 moles of N:

$$6.5 \times 10^{-4} \text{ mol } NH_4NO_3 \times 2 = 1.3 \times 10^{-3} \text{ mol N}$$

and since every mole of N contains an Avogadro's number of atoms:

$$1.3 \times 10^{-3} \text{ mol N} \times \frac{6.0221 \times 10^{23} \text{atoms}}{1 \text{ mol N}} = 7.8 \times 10^{20} \text{ atoms}$$

1.4. The Chemical Equation

While a balanced chemical equation tells us what reacts to form products and their ratios, it does not tell us *how* the reaction takes place, *how long* the reaction takes, or even if it goes to completion.

The left-hand side of an equation contains the reactants, the right-hand side the products. The numbers in front of each formula are the stoichiometric coefficients, indicating the relative number of moles required of each compound. A *simple stoichiometric equation* is the most descriptive type, giving the formulas for all compounds used in the reaction. In the case of ionic reactions however some species may be *spectators* because they are not changed during

Chapter 1 — Stoichiometry and the Basis of Atomic Theory

the course of the reaction. If spectator ions are removed we are left with a *net reaction equation*.

Usually the phase or condition of each product and reactant is given in parentheses following the formula. The choices for this notation are given below.

Phase	Symbol Used
solid	(s) or (ppt) or ↓ ; (ppt) is the abbreviation for precipitate
liquid	(l)
gas	(g) or ↑
dissolved in water	(aq) for aqueous

Example 1.7. Household baking soda is composed of sodium bicarbonate, $NaHCO_3$, which can be used as an antacid. Stomach acid is composed mainly of hydrochloric acid, HCl. What are the simple stoichiometric and the net reaction equations that describe this process?

simple stoichiometric equation: $NaHCO_{3(s)} + HCl_{(aq)} \rightarrow NaCl_{(aq)} + H_2O_{(l)} + CO_{2(aq)}$

net reaction equation $NaHCO_{3(s)} + H^+_{(aq)} \rightarrow Na^+_{(aq)} + H_2O_{(l)} + CO_{2(aq)}$

The Cl^- ion is a spectator ion.

Example 1.8. When ethane gas, C_2H_6, reacts with oxygen, O_2, water and carbon dioxide are produced. What is the balanced equation for this reaction?

unbalanced $C_2H_{6(g)} + O_{2(g)} \rightarrow H_2O_{(l)} + CO_{2(g)}$

balanced $2C_2H_{6(g)} + 7O_{2(g)} \rightarrow 6H_2O_{(l)} + 4CO_{2(g)}$

or

$C_2H_{6(g)} + 7/2\, O_{2(g)} \rightarrow 3H_2O_{(l)} + 2CO_{2(g)}$

You can't expect to know the normal phase of every compound, whether or not a salt precipitates or a gas is given off, in Chapter 1! But you should begin learning about the more common chemicals. Using a little common sense will help you. For instance, you already are aware that NaCl dissolves readily in water, as does $NaHCO_3$ (baking soda). In the previous example you may not have realized that ethane, C_2H_6, is a gas at room temperature but you should know that methane, CH_4, propane, C_3H_8, and butane, C_4H_{10}, are all gases. By "interpolation" of the data, ethane is expected to be a gas also. One of the more challenging tasks

Stoichiometry and the Basis of Atomic Theory   Chapter 1

ahead of you is learning to integrate your daily observations of the world with the theories presented in your chemistry class. It takes practice but if you are alert and inquisitive the task actually becomes fun.

1.5. Stoichiometric Relationships

The mole will now become the basic unit of our calculations. A chemical formula provides the mole ratios of all elements found in that compound, and a chemical equation provides the mole ratios of all compounds involved in the reaction. We now have an algebraic way to handle calculations.

Example 1.9. List some examples of the algebraic equivalences found in the following formula.

$$CH_3OH$$

1 mole of CH_3OH contains 1 mole of C atoms, 4 moles of H atoms and 1 mole of O atoms. In an equation, this is expressed as follows:

$$\text{moles } CH_3OH = \frac{\text{moles H}}{4}$$

$$\text{moles } CH_3OH = \text{moles O} = \text{moles C}$$

Example 1.10. List some examples of the algebraic equivalences found in the following balanced equation.

$$CH_3OH_{(l)} + 3/2\ O_{2(g)} \rightarrow CO_{2(g)} + 2H_2O_{(l)}$$

$$\text{moles } CH_3OH = \text{moles } \frac{O_2}{3/2}$$

$$\frac{\text{moles } H_2O}{2} = \text{moles } CO_2$$

$$\frac{\text{moles } O_2}{3/2} = \frac{\text{moles } H_2O}{2}$$

$$\frac{\text{moles } O_2}{\text{moles } H_2O} = \frac{3/2}{2}$$

Chapter 1 Stoichiometry and the Basis of Atomic Theory

1.6. Stoichimetric Calculations

Now let's see how we can do calculations relating weight, moles, atoms, and molecules. The method used here is essentially the same as used in the text where each step logically follows from the preceding one. Calculating each intermediate answer is not necessary once you understand the procedures. An alternate approach to solving stoichiometry problems using dimensional analysis can be found in Interchapter A, which immediately follows Chapter 1 in this study guide. You are encouraged to use whichever method you find comfortable, as the answers will be the same.

We shall begin by stating some definitions and facts needed for working stoichiometry problems. An *empirical formula* is one which uses the lowest whole number ratios for all elements in the compounds, while a *molecular formula* has the true ratios of atoms. Some examples of both are listed below.

Compound	Molecular Formula	Empirical Formula
ethene	C_2H_4	CH_2
glucose	$C_6H_{12}O_6$	CH_2O
formaldehyde	CH_2O	CH_2O
hydrogen peroxide	H_2O_2	HO

If a compound undergoes elemental analysis, the data determined is the empirical formula. A molecular formula can be determined from an empirical formula if the molecular weight is known.

Another useful fact to remember, although we'll learn to calculate it in Chapter 2, is the volume that a gas occupies at standard temperature and pressure (STP). At STP, which is 0°C and 1 atmosphere pressure, a gas will fill a 22.4 liter container. (For size comparison, a five gallon bottle of drinking water is 19 liters in volume.) If we can measure the volume occupied by an unknown gas at STP we can calculate how many moles of the gas we have, and if we know the weight of the sample we can find its molecular weight.

Now we are ready to work some problems. In Interchapter A, the same problems are shown using dimensional analysis.

Example 1.11. According to the following balanced equation, how many grams of KBr can be formed from 83.0 g of Br_2 in the presence of sufficient KOH?

$$6KOH_{(aq)} + 3Br_{2(l)} \rightarrow KBrO_{3(aq)} + 5KBr_{(aq)} + 3H_2O_{(l)}$$

The general sequence for arriving at the answer, in g of KBr is:

$$g\ Br_2 \rightarrow mol\ Br_2 \rightarrow mol\ KBr \rightarrow g\ KBr$$

Stoichiometry and the Basis of Atomic Theory   Chapter 1

In all cases we must calculate moles first because the balanced equation only gives us information in terms of moles.

$$\frac{83.0 \text{ g Br}_2}{159.8 \text{g Br}_{2/\text{mol}}} = 0.519 \text{ mol Br}_2$$

From the equation stoichiometry we see that:

$$\frac{\text{moles of Br}_2}{3} = \frac{\text{moles of KBr}}{5}$$

or

$$\text{moles of KBr} = \frac{5}{3} \text{moles of Br}_2$$

so that:

$$\frac{5}{3}(1.519 \text{ mol Br}_2) = 0.866 \text{ mol KBr}$$

and finally:

$$0.866 \text{ mol KBr} \times 119 \text{ g}\frac{\text{KBr}}{\text{mol}} = 103 \text{ g KBr}$$

Example 1.12. If 0.30 moles of PH_5 are reacted with 0.70 moles of O_2, what is the maximum amount of P_4H_{10} that can be produced according to the balanced equation below?

$$4 \text{ PH}_{5(g)} + 10 \text{ O}_{2(g)} \rightarrow \text{P}_4\text{O}_{10(s)} + 10 \text{ H}_2\text{O}_{(l)}$$

You should recognize this as a limiting reagent problem meaning one of the reactants will be completely consumed while there is an excess of another reactant. A good analogy is that of gloves: if you had 9 left-hand gloves and 6 right-hand gloves you could only make 6 complete pairs of gloves. Therefore the right-hand gloves are the limiting reagent and there is an excess of 3 left-hand gloves.

In this stoichiometry problem you will need to do two separate calculations, one for each reactant to find the maximum amount of products possible. The reactant that yields the *least* product is the limiting reagent and its yield is the correct answer.

$$\frac{\text{mol PH}_5}{4} = \frac{\text{mol P}_4\text{O}_{10}}{1}$$

$$\frac{0.30 \text{ mol PH}_5}{4} = 0.075 \text{ mol P}_4\text{O}_{10},$$

Chapter 1   Stoichiometry and the Basis of Atomic Theory

the maximum that can be formed from the PH_5

$$\frac{\text{mol } O_2}{10} = \frac{\text{mol } P_4O_{10}}{1}$$

$$\frac{0.70 \text{ mol } O_2}{10} = 0.070 \text{ mol } P_4O_{10},$$

the maximum that can be formed from the O_2.

The O_2 is the limiting reagent because it is consumed to make the 0.070 moles P_4O_{10} while the PH_5 remains in excess. Now the quantity of P_4O_{10} can be computed.

$$0.070 \text{ mol } P_4O_{10} \times \frac{284 \text{ g } P_4O_{10}}{1 \text{mol}} = 20 \text{ g } P_4O_{10}$$

Example 1.13. Limestone, $CaCO_3$, can be heated to form quicklime, CaO, and carbon dioxide gas. If 30.0 g of $CaCO_3$ is heated in an open container until half the original limestone has been decomposed, what is the weight of the remaining solid mixture?

First we must write and balance the chemical equation.

$$CaCO_{3(s)} \rightarrow CaO_{(s)} + CO_{2(g)}$$

At the half-way point we will have 15.0g $CaCO_3$ remaining, and 15.0g converted to products. The only solid product is CaO, so first we find out how much CaO can be made from 15.0g of $CaCO_3$.

$$\frac{15.0 \text{g} CaCO_3}{100.1 \text{ g } CaCO_3/ \text{ mol}} = 0.150 \text{ mol } CaCO_3$$

$$0.150 \text{ mol } CaCO_3 = 0.150 \text{ mol } CaO$$

$$0.150 \text{ mol } CaO \times \frac{56.1 \text{ g} CaO}{\text{mol}} = 8.42 \text{ g } CaO$$

total weights of solids = wt. $CaCO_3$ + wt. CaO = 15.0 g + 8.42 g = 23.4 g

---

Example 1.14. When 1.28 g of a compound with the formula $CaSO_4 \cdot xH_2O$ was heated to drive off all water of crystallization, 1.01 g of $CaSO_4$ remained. What is the value of x in the formula?

$$\text{total wt.} = \text{wt. } CaSO_4 + \text{wt. } H_2O$$

$$1.28 \text{ g} = 1.01 \text{ g} + \text{wt. } H_2O$$

$$0.27 \text{ g} = \text{wt. } H_2O$$

Stoichiometry and the Basis of Atomic Theory Chapter 1

$$\frac{0.27 \text{ g H}_2\text{O}}{18.0 \text{ g H}_2\text{O/mol}} = 0.015 \text{ mol H}_2\text{O}$$

We also need to know how many moles of $CaSO_4$ we have because we are essentially figuring out the molecular formula.

$$\frac{1.01 \text{ g CaSO}_4}{136.2 \text{ g CaSO}_4/\text{mol}} = 0.00742 \text{ mol CaSO}_4$$

$$\frac{\text{mol CaSO}_4}{\text{mol H}_2\text{O}} = \frac{0.00742}{0.015} = \frac{0.49}{1} = \frac{1}{2}$$

so for every mole of $CaSO_4$ there are 2 moles of H_2O and x = 2

Example 1.15. A sample, weighing 0.650 g, of a compound containing only nitrogen and hydrogen was burned completely in oxygen to yield 1.87g g of NO_2 and 0.731 g of H_2O. What is the empirical formula of the compound?

The general equation is:

$$N_xH_y + O_2 \rightarrow NO_2 + H_2O$$

Balanced, the equation becomes:

$$N_xH_y + (x+y/4)O_2 \rightarrow xNO_2 + y/2 H_2O$$

Now let's calculate what the values of x and y are based on the weights of NO_2 and H_2O obtained.

$$\frac{1.87 \text{ g NO}_2}{46.0 \text{ g NO}_2/\text{mol}} = 0.0407 \text{ mol NO}_2 = x$$

$$\frac{0.731 \text{ g H}_2\text{O}}{18.0 \text{ g H}_2\text{O/mol}} = 0.0407 \text{ mol H}_2\text{O} = y/2$$

So x = 0.0407 and y = 2 (0.0407), and $N_{0.0407}H_{0.0814}$ is the empirical formula, but we would prefer it in the lowest whole numbers possible: NH_2.

Example 1.16. A sample of aluminum metal weighing 16.4 g is added to excess hydrochloric acid, $HCl_{(aq)}$. What volume of $H_{2(g)}$ will be produced at STP?

As usual we must first write the balanced equation.

$$Al_{(s)} + 3HCl_{(aq)} \rightarrow 3/2 H_{2(g)} + AlCl_{3(aq)}$$

Now convert the starting quantity of Al to moles of Al, then to moles of H_2, and finally to

Chapter 1 Stoichiometry and the Basis of Atomic Theory

liters of H_2 using the STP relationship.

$$\frac{16.4 \text{ g Al}}{27.0 \text{ g Al/mol}} = 0.607 \text{ mol Al}$$

$$0.607 \text{ mol Al} \times \frac{3}{2} = \text{mol } H_2 = 0.911 \text{ mol } H_2$$

$$0.911 \text{ mol } H_2 \times \frac{22.4 \text{ liters @ STP}}{\text{mol of gas}} = 20.4 \text{ liters } H_2$$

Problems (answers can be found in Appendix C)

1. The earth currently is populated by 5 billion people. Express this number as moles of people.

2. Calculate the molecular or formula weight of each of the following:

    a) $(NH_4)_2SO_4$
    b) $Co(NO_3)_2 \cdot 6H_2O$

3. If an experiment calls for 0.200 moles of $Ca_3(PO_4)_2$, how many grams should be weighed out?

4. What percent by weight of acetic acid, CH_3COOH, is carbon? Oxygen? Hydrogen?

5. How many moles of sulfur are in 25.0 g of $Na_2S_2O_3$?

6. If the period at the end of a sentence has a mass of 1×10^{-6} grams and you assume it is carbon, how many carbon atoms are there in a period?

7. Now calculate the number of carbon atoms in the Hope Diamond. Assume the diamond is pure carbon and that it weighs 44.0 carats (1.0 carat = 0.20 g).

8. How many moles of sodium ions are contained in a 50 lb bag of water softener salt, NaCl?

9. What is the empirical formula of a compound that contains 30.5% nitrogen and 69.5% oxygen by weight?

10. Elemental analysis of a pure compound shows it to contain 7.52×10^{-4} moles of niobium for every 0.100g analyzed. The only other element present is oxygen. What is the empirical formula of the compound?

11. If 10.5g of a compound with the empirical formula CH_2 contains 0.150 moles of molecules, what is the correct molecular formula?

12. The empirical formula of a compound found by elemental analysis is CH_2O and its molecular weight is determined to be 150 g/mol. What is the molecular formula of the compound?

Stoichiometry and the Basis of Atomic Theory   Chapter 1

13. Suppose our scale of atomic weights was set to O = 100 g/mol. What would be the corresponding atomic weights of H, C and U?

14. What is the mass of an individual ^{1}H atom in grams? What is the mass of an individual ^{238}U atom in grams?

15. Balance the following equations, completing the product where indicated.

   a) $C_2H_2 + O_2 \rightarrow CO_2 + H_2O$

   b) $K_2CO_3 + H_2SO_4 \rightarrow K_2SO_4 + H_2O + CO_2$

   c) $NH_3 + O_2 \rightarrow N_2O_5 + H_2O$

   d) $C_2H_6O + O_2 \rightarrow CO_2 + ?$

   e) $Cr(OH)_2 \rightarrow H_2 + H_2O + Cr_2O_3$

   f) $NaH_2PO_4 \rightarrow Na_3P_3O_9 + H_2O$

16. Metallic iron is produced in a blast furnace by passing carbon monoxide gas over the iron ore, Fe_2O_3. Carbon dioxide is a product.

   a)   Write the balanced chemical equation for the above process.
   b)   How many moles of iron can be produced per mole of ore used?
   c)   If there is excess carbon monoxide, how many kg of Fe can be produced from 4 x 10^3 kg of iron ore?

17. If $HCl_{(aq)}$ and $MnO_{2(s)}$ are heated together, the products are water, chlorine gas, and $MnCl_2$.

   a)   Write the balanced chemical equation for this reaction.
   b)   How many grams of Cl_2 can be formed from 10.0 g MnO_2 with excess HCl?
   c)   How many liters of Cl_2 are formed in b) if measured at STP?
   d)   What is the maximum amount of $MnCl_2$ that can be made from 2.0 g HCl and 2.0 g MnO_2?

18. The element bromine occurs as two isotopes in nature; ^{79}Br with an atomic weight of 78.9183 and an abundance of 50.54%, as well as ^{81}Br with an atomic weight of 80.9163 and an abundance of 49.46%. What is the reported atomic weight for natural bromine?

19. The reported atomic weight for naturally occurring lithium is 6.941 g/mole. There are only two isotopes of significance; ^{6}Li (atomic weight = 6.01513) and ^{7}Li (atomic weight = 7.01601). What are the relative abundances of these two isotopes?

20. Naturally occuring germanium has the given isotopic components:

^{70}Ge (atomic weight = 69.9243), 20.52%

-21-

Chapter 1   Stoichiometry and the Basis of Atomic Theory

^{72}Ge (atomic weight = 71.9217), 27.43%

^{73}Ge (atomic weight = 72.9234), 7.76%

^{74}Ge (atomic weight = 73.9212), 36.54%

^{76}Ge (atomic weight = 75.9214), 7.76%

What is the atomic weight of natural germanium?

21. *A 4.000g sample of M_2S_3 was ignited in oxygen to yield 3.658 g of MO_2. What is element M?

22. *A gas mixture contains only ethane, C_2H_6, and propane, C_3H_8, in unknown amounts. The mixture is combusted completely with O_2. The products are 63.20% CO_2 and 36.80% H_2O by weight. What was the composition of the original mixture?

*Challenging problems that require several of the skills acquired in this chapter.

INTERCHAPTER A
DIMENSIONAL ANALYSIS

What follows is an alternative to the problem-solving method set forth in the text and in the main body of this study guide. Many students may find dimensional analysis a more direct technique for solving problems; indeed some chemistry courses teach only this method.

Dimensional analysis uses units to monitor the progress of a calculation and uses conversion factors and measured quantities to convert a given number and its unit to the desired unit. The strategy is similar to that used in Chapter 1, only the execution is different. Normally with this method intermediate answers are not calculated.

Here are some examples of how to use constants as conversion factors.

$$16 \text{ oz} = 1 \text{ lb so } \frac{16 \text{ oz}}{1 \text{ lb}} = 1 \text{ or } \frac{1 \text{ lb}}{16 \text{ oz}} = 1.$$

Since we can multiply any number by 1 without changing it, we can also multiply a number by 16oz/1 lb or 1 lb/16oz without changing it. Which way you use this conversion factor depends on the units desired.

Example 1. Convert 4.5 oz to pounds.

$$4.5 \text{ oz} \times \frac{1 \text{ lb}}{16 \text{ oz}} = 0.28 \text{ lb}$$

Notice we multiplied by 1 lb/16 oz so that the ounce units cancel, and pounds remain.

Example 2. Convert 15.3 lbs to ounces.

$$15.3 \text{ lb} \times \frac{16 \text{ oz}}{1 \text{ lb}} = 2.45 \times 10^2 \text{ oz}$$

Here we needed the answer in ounces so pound had to cancel, requiring the inverse of the previous conversion factor.

We will commonly use conversion factors that include moles. For example:

$$\frac{6.0221 \times 10^{23} \text{ atoms}}{1 \text{ mole}} \text{ or } \frac{1 \text{ mole}}{6.0221 \times 10^{23} \text{ atoms}}$$

Another possibility includes molecular or atomic weights.

Interchapter A   Dimensional Analysis

$$\frac{1 \text{ mole } H_2O}{18 \text{ g}} \quad \text{or} \quad \frac{18 \text{ g}}{1 \text{ mole } H_2O}$$

Stoichiometry of formulas and balanced equations also can lead to useful conversion factors.

$$\frac{2 \text{ moles } H}{1 \text{ mole } H_2O} \quad \text{or} \quad \frac{1 \text{ mole } H_2O}{2 \text{ moles } H}$$

Let's re-work many of the same problems used as examples in Chapter 1. This time we will solve them by dimensional analysis. Refer back to Chapter 1 if you wish to compare the methods used.

Example 3. How many iron atoms are contained in a 1.0 lb piece of iron?

Your strategy should be to convert pounds to grams, grams to moles, and moles to atoms. Each step requires use of a conversion factor.

$$1.0 \text{ lb} \times \frac{454 \text{ g}}{1.0 \text{ lb}} \times \frac{1 \text{ mol Fe}}{55.85 \text{ g}} \times \frac{6.0221 \times 10^{23} \text{ atoms}}{1 \text{ mol Fe}} = 4.9 \times 10^{24} \text{ atoms}$$

Example 4. How many nitrogen atoms are contained in 0.052g of ammonium nitrate, NH_4NO_3?

Your strategy should be to convert grams of NH_4NO_3 to moles of NH_4NO_3, then to moles of nitrogen, and finally to atoms of nitrogen.

$$0.052 \text{ g } NH_4NO_3 \times \frac{1 \text{ mol } NH_4NO_3}{80.05 \text{ g}} \times \frac{2 \text{ mol N}}{1 \text{ mol } NH_4NO_3} \times \frac{6.0221 \times 10^{23} \text{ atoms}}{1 \text{ mol N}} =$$

$$7.8 \times 10^{20} \text{ atoms of N}$$

Example 5. According to the following balanced equation, how many grams of KBr can be formed from 83.0 g of Br_2 in the presence of sufficient KOH?

$$6KOH_{(aq)} + 3Br_{2(l)} \rightarrow KBrO_{3(aq)} + 5KBr_{(aq)} + 3H_2O_{(l)}$$

We can solve this problem in one long conversion sequence where we change grams of Br_2 to moles of Br_2, then to moles of KBr, then to grams of KBr. The stoichiometry of the balanced equation gives the following conversion factors:

$$\frac{3 \text{ mol } Br_2}{5 \text{ mol KBr}} \quad \text{or} \quad \frac{5 \text{ mol KBr}}{3 \text{ mol } Br_2}$$

Dimensional Analysis   Interchapter A

because 3 moles of Br_2 produce 5 moles of KBr. This can be included as needed, in the solution.

$$83.0 \text{ g } Br_2 \times \frac{1 \text{ mol } Br_2}{159.8 \text{ g}} \times \frac{5 \text{ mol KBr}}{3 \text{ mol } Br_2} \times \frac{119 \text{ g}}{1 \text{ mol KBr}} = 103 \text{ g KBr}$$

You should recognize by now how valuable dimensional analysis can be. Most stoichiometry problems can be solved in one step through a series of appropriate conversion factors. You have a constant guide as to the next conversion factor needed at any point, as long as you carefully keep track of units.

Example 6. If 0.30 moles of PH_5 are reacted with 0.70 moles of O_2, what is the maximum amount of P_4H_{10} that can be produced according to the balanced equation below?

$$4PH_{5(g)} + 10\ O_{2(g)} \to P_4O_{10(s)} + 10H_2O_{(l)}$$

This is a limiting reagent problem so you must do two calculations and compare them.

$$0.30 \text{ mol } PH_5 \times \frac{1 \text{ mol } P_4O_{10}}{4 \text{ mol } PH_5} = 0.075 \text{ mol } P_4O_{10}$$

$$0.70 \text{ mol } O_2 \times \frac{1 \text{ mol } P_4O_{10}}{10 \text{ mol } O_2} = 0.070 \text{ mol } P_4O_{10}$$

The limiting reagent is O_2 so only 0.070 mol of P_4O_{10} can be produced with an excess of PH_5 being present.

Example 7. Limestone, $CaCO_3$, can be heated to form quicklime, CaO, and CO_2 gas. If 30.0 g of $CaCO_3$ is heated in an open container until half the original limestone has been decomposed, what is the weight of the remaining solid mixture?

$$CaCO_{3(s)} \to CaO_{(s)} + CO_{2(g)}$$

First calculate how many grams of CaO can be produced from the 15.0 grams of $CaCO_3$ reacted.

$$15.0 \text{ g } CaCO_3 \times \frac{1 \text{ mol } CaCO_3}{100.1 \text{ g}} \times \frac{1 \text{ mol CaO}}{1 \text{ mol } CaCO_3} \times \frac{56.1 \text{ g}}{1 \text{ mol CaO}} = 8.42 \text{ g CaO}$$

total weight of solids = 15.0 g $CaCO_3$ remaining + 8.42 g CaO = 23.4 g

---

Example 8. A sample of aluminum metal weighing 16.4g is added to excess hydrochloric acid, $HCl_{(aq)}$. What volume of $H_{2(g)}$ will be produced at STP?

Interchapter A   Dimensional Analysis

$$Al_{(s)} + 3HCl_{(aq)} \rightarrow 3/2\, H_{2(g)} + AlCl_{3(aq)}$$

$$16.4 \text{ g} \times \frac{1 \text{ mol Al}}{27.0 \text{ g}} \times \frac{3/2 \text{ mol } H_2}{1 \text{ mol Al}} \times \frac{22.4 \text{ L @ STP}}{1 \text{ mol } H_2} = 20.4 \text{ L of } H_2$$

    Once you have mastered dimensional analysis to the extent presented here, you will be able to solve problems with this method throughout the remaining chapters of the text. An additional interchapter will be used in this study guide to explore dimensional analysis as applied to solution stoichiometry.

CHAPTER 2
THE PROPERTIES OF GASES

2.1. The Gas Laws

Pressure is defined as force per unit area which for a gas measures the collisions of gas molecules with the walls of its container. There are many common units of pressure currently used by scientists. You will probably need to be familiar with most of them, shown below in their relationship to 1 atmosphere (atm) the unit most frequently used in chemistry.

$$1 \text{ atm} = 760 \text{ mm Hg} = 760 \text{ torr}$$

$$1 \text{ atm} = 1.013 \times 10^5 \text{ Pa} = 101.3 \text{ kPa} = 1.013 \text{ bar}$$

$$1 \text{ atm} = 14.7 \text{ psi (lb/in}^2\text{)}$$

$$1 \text{ atm} = 1.013 \times 10^6 \text{ dyne/cm}^2$$

Boyle's Law states that a gas sample at a constant temperature has a constant value of pressure times volume.

$$PV = \text{Constant}$$

so $P_1V_1 = P_2V_2$ for the same gas sample at the same temperature. This pressure-volume relationship should intuitively make sense to us since we know that if you put pressure on a gas it can be compressed into a smaller volume, as occurs in the down-stroke of a piston.

Example 2.1. If a soap bubble forms at an atmospheric pressure of 760 torr and has a volume of 10 milliliters, what will be the volume of the bubble when the atmospheric pressure is only 630 torr (Mexico City at an elevation of approximately 1 mile)? Assume no change in temperature.

$$P_1V_1 = P_2V_2$$

We can use Boyle's law, assuming there is no change in the temperature or quantity of gas inside the bubble. The units used for pressure and volume can be any that are convenient, providing they are consistent for both measurements.

$$(760 \text{ torr}) (10 \text{ mL}) = (630 \text{ torr}) V_2$$
$$12 \text{ mL} = V_2$$

The French scientist Jacques Charles, a balloon enthusiast, discovered that all gases expand by the same amount with a given rise in temperature. Conversely all gases shrink by the same amount with a given lowering of temperature. Charles, however, did not publish his findings and a few years later Gay-Lussac re-discovered the phenomenon. Thus *Charles' Law* is also known as *Gay-Lussac's Law* and states that for a given gas sample at a constant pressure, the volume has a linear dependence on temperature. If volume in any units is plotted against temperature, in °C, the line formed from measured data points will not pass through the origin. In other words, a gas at 0°C still has a measurable volume. But if a V vs T plot is extrapolated to the point where the volume of the gas equals zero, the corresponding

Chapter 2 — The Properties of Gases

temperature is predicted to be -273.15 °C. This temperature is called absolute zero or 0 Kelvin (K). *There is no temperature lower than 0 Kelvin.*

The mathematical expression of Charles' law uses the absolute temperature scale.

$$\frac{V}{T} = \text{constant}$$

Volume may be expressed in any convenient units but T must be in Kelvin where:

$$T\,(K) = t\,(°C) + 273.15.$$

The useful form of Charles' law is:

$$\frac{V_1}{T_1} = \frac{V_2}{T_2}$$

for a gas sample as constant pressure.

Example 2.2. A balloon is filled with 2.61 liters of air on a very warm day when the temperature is 40.3 °C. What would be the volume of the same balloon on a very cold day when the temperature is -5.3 °C?

$$\frac{V_1}{T_1} = \frac{V_2}{T_2}$$

$$\frac{2.61\,L}{(273.15 + 40.3)K} = \frac{V_2}{(273.15 - 5.3)K}$$

$$2.23\,L = V_2$$

The gas laws discovered by Boyle, Charles, and Gay-Lussac can be combined to show how pressure, volume, and temperature all interrelate for a specified gas sample. The result is the *combined gas law*:

$$\frac{P_1 V_1}{T_1} = \frac{P_2 V_2}{T_2}$$

But what if the gas sample changes? How can we incorporate moles into this relationship? The text shows the simple derivation of the *ideal gas equation* from the combined gas law, giving us what is probably the best known equation in chemistry,

$$PV = nRT$$

where n = gas moles, and R = universal gas constant whose numerical value depends on the units used for P, V, and T. As we saw in Chapter 1, one mole of an ideal gas at STP occupies a volume of 22.4 liters. Using this observation, R can be calculated as:

$$R = 0.08206\ \text{L-atm/mol-K}$$

or

$$R = 8.3144\ \text{m}^3\text{-Pa/mol-K}$$

The Properties of Gases   Chapter 2

depending on whether SI units or the more common units of pressure and volume are used.

Remember that the ideal gas equation assumes "ideal" gas behavior, the definition of which we will see later in this chapter. While many real gases under real conditions behave as though they are ideal, we should not be surprised to learn than there are also many real gases and real conditions that do not behave so perfectly.

Example 2.3. What volume will 2.1 moles of an ideal gas occupy at 27 °C and 0.957 atmospheres of pressure?

$$PV = nRT$$

Be sure that your units for R agree with the measured units. The temperature must be in Kelvins.

$$(0.957 \text{ atm}) V = (2.1 \text{ mol}) \left(0.08206 \frac{\text{L-atm}}{\text{mol-K}}\right)(273 + 27 \text{ K})$$

$$V = 54 \text{ L}$$

Example 2.4. Assuming ideal gas behavior, what volume will 1.0 g of CH_4 occupy at STP?

$$PV = nRT$$

The ideal gas equation still applies but we must remember that moles is equal to grams/molecular weight.

$$n = \frac{1.0 \text{g } CH_4}{16.0 \text{g/ mol}} = 6.25 \times 10^{-2} \text{ mol}$$

$$(1.0 \text{ atm}) V = (6.25 \times 10^{-2} \text{ mol}) \left(0.08206 \frac{\text{L-atm}}{\text{mol-K}}\right)(273 \text{K})$$

$$V = 1.4 \text{ L}$$

An ideal gas consists of molecules that act independent of one another and of any other gas molecules present. This means that the total pressure of a gas mixture is just the sum of all the individual pressures of the gas components. This is *Dalton's Law of Partial Pressure* and is mathematically expressed below.

$$P_T = P_A + P_B + P_C + \ldots$$

This means that the total pressure of our atmosphere is given by the sum of component pressures.

$$P_{air} = P_{N_2} + P_{O_2} + P_{Ar} + P_{CO_2} + P_{H_2O} + \ldots$$

Chapter 2 The Properties of Gases

For each gas pressure we can substitute from the ideal gas equation:

$$P_A = \frac{n_A RT}{V}$$

When this substitution is made for each gas and the ratio of P_A, the partial pressure, to P_T, the total pressure, is taken, we find that it is equal to the ratio of moles of A to the total gas moles. This ratio is called

$$\frac{P_A}{P_T} = \frac{n_A}{n_T} = \chi_A$$

the mole fraction, χ_A. Expressed in another way the equation becomes:

$$P_A = \chi_A P_T$$

Example 2.5. A gas mixture is made that contains 0.20 mol He, 0.10 mol Ar, and 0.35 mol Xe. At a temperature of 300K this mixture has a pressure, in some container, of 1130 torr. What is the mole fraction of each gas and its partial pressure? What is the volume of the container?

$$\chi_{He} = \frac{n_{He}}{n_T} = \frac{0.20}{(0.10+0.20+0.35)} = 0.31$$

$$\chi_{Ar} = \frac{n_{Ar}}{n_T} = \frac{0.10}{0.65} = 0.15$$

$$\chi_{Xe} = \frac{n_{xe}}{n_T} = \frac{0.35}{0.65} = 0.54$$

$$\sum_i \chi_i = 1$$

by definition so you should double check your calculation to be sure it has been worked properly.

$$P_{He} = \chi_{He} P_T = 0.31\ (1130\ \text{torr}) = 350\ \text{torr}$$

$$P_{Ar} = \chi_{Ar} P_T = 0.15\ (1130\ \text{torr}) = 170\ \text{torr}$$

$$P_{Xe} = \chi_{Xe} P_T = 0.54\ (1130\ \text{torr}) = 610\ \text{torr}$$

Again, you can check your work by being sure the sum of partial pressures equals 1130 torr. Last, calculate the volume of the container using the ideal gas equation and the P_T and n_T values.

$$PV = nRT$$

$$\left(\frac{1130\ \text{torr}}{760\ \text{torr/atm}}\right) V = (0.65\ \text{mol}) \left(0.08206 \frac{\text{L}-\text{atm}}{\text{mol}-\text{K}}\right) (300\ \text{K})$$

The Properties of Gases   Chapter 2

$$V = 11 \text{ L}$$

Example 2.6. The density of a gas is generally expressed as grams/liter. What is the density of CO at STP?

$$PV = nRT = \left(\frac{g}{mw}\right) RT$$

Rearranging to solve for density:

$$P\frac{mw}{RT} = \frac{g}{V} = \text{gas density}$$

$$\frac{(1 \text{atm})\, 28 \text{g/mol}}{\left(0.08206 \frac{\text{L-atm}}{\text{mol-K}}\right) 273 \text{ K}} = 1.25 \frac{g}{L}$$

Example 2.7. What is the molecular weight of an ideal gas, 40.0 grams of which occupy 12.0 liters at 25 °C and 740 torr?

$$PV = nRT = \frac{g}{MW} RT$$

$$\left(\frac{740 \text{ torr}}{760 \text{ torr/atm}}\right) 12.0 \text{L} = \left(\frac{40.0 \text{ g}}{mw}\right)\left(0.08206 \frac{\text{L-atm}}{\text{mol-K}}\right)(298 \text{ K})$$

$$\frac{83.7 \text{ g}}{\text{mol}} = mw$$

2.2. The Kinetic Theory of Gases

The kinetic theory of gases leads to the theoretical derivation of Boyle's law based on the assumptions that gas molecules are of negligible size, and that they move randomly and independently of one another except during collisions. Boyle's law is easily derived, and along with it several other mathematical relationships of interest emerge.

$$\overline{E}_{KE} = \frac{3}{2} RT, \quad \overline{E}_{KE} = \text{average kinetic energy per mole of gas}$$

$$\sqrt{\overline{c^2}} = \text{root-mean-square speed}$$

$$\sqrt{\overline{c^2}} = \frac{\sqrt{3RT}}{M}$$

Chapter 2 The Properties of Gases

$$M = \text{molar mass in } \frac{\text{kg}}{\text{mol}}$$

These equations show how temperature and speed, or temperature and kinetic energy are related. We intuitively expect that raising the temperature will speed up the molecules (and raise their kinetic energy) but the square root dependence on temperature seems less obvious. Similarly the qualitative way in which speed varies with the mass of a molecule (light molecules move faster than heavy molecules at the same temperature) is predictable though probably not the quantitative dependence.

Both *effusion* (the rate with which gas particles pass through a hole) and *diffusion* (the rate with which gas particles spread out through available space) experimentally take advantage of the difference in speed for two gases of different molecular weights. The ratio of effusion rates or diffusion rates is the same as the ratio of rms speed, and if the temperature is held constant, the ratio reduces to the equation below.

$$\frac{\overline{c}_1}{\overline{c}_2} = \frac{\text{effusion rate}_1}{\text{effusion rate}_2} = \frac{\text{diffusion rate}_1}{\text{diffusion rate}_2} = \frac{\sqrt{mw_2}}{\sqrt{mw_1}}$$

Example 2.8. How much faster can an H_2 molecule diffuse than an O_2 molecule at the same temperature?

$$\frac{\text{rate } H_2}{\text{rate } O_2} = \frac{\sqrt{mw\ O_2}}{\sqrt{mw\ H_2}} = \frac{\sqrt{32}}{\sqrt{2}} = \sqrt{16} = 4$$

So an H_2 molecule, on the average, moves four times faster than an O_2 molecule.

Example 2.9. What is the rms speed of an O_2 molecule at $0\,°C$?

$$\text{rms speed} = \sqrt{\overline{c^2}} = \frac{\sqrt{3\,RT}}{M} = \sqrt{\frac{3\left(8.314\,\frac{\text{kg-m}^2}{\text{s}^2\text{-mol-K}}\right)273\,\text{K}}{32 \times 10^{-3}\,\frac{\text{kg}}{\text{mol}}}}$$

$$\text{rms speed} = 461\,\frac{\text{m}}{\text{s}}\ (\text{thats about 1000 miles per hour})$$

Measuring the temperature of a known gas sample allows us to calculate its root-mean-square speed but this is only an average, of a sort, of all the speeds. Many molecules will be moving faster than the rms speed and even more will be moving slower. There is a statistical distribution of speeds that is *not* symmetric about the center axis as a Gaussian distribution would be. The molecular speeds follow the Maxwell-Boltzmann distribution seen in figures 2.12 and 2.13 in the main text. Raising the temperature of a gas sample changes the distribution of speeds by increasing the number of speeds available to the molecules. This changes the shape of the distribution curve rather than just shifting it.

When chemists deal with molecules rather than moles, the Maxwell-Boltzmann distribution plays a very important role in predicting chemical behavior.

The temperature of a substance is raised by the addition of heat. The measure of how much is needed per mole of substance to raise its temperature 1°C (or 1K) is called the heat capacity. For gases, heat capacities are predictable and depend on the number of atoms per molecule. We shall consider only monoatomic gases (like He, Ne, Ar etc.) and diatomic gases (like O_2, H_2, Cl_2 etc.).

These heat capacities are of two types, one measured at constant volume and the other measured at constant pressure, C_v and C_p respectively. For monoatomic gases:

$$C_v = \frac{3}{2} R \text{ per mole}, \quad C_p = \frac{5}{2} R \text{ per mole}$$

For diatomic gases:

$$C_v = \frac{5}{2} R \text{ per mole}, \quad C_p = \frac{7}{2} R \text{ per mole}$$

Diatomic molecules must necessarily have higher heat capacities because there are more ways to distribute added heat. A diatomic molecule can move through the surrounding volume (translation) or it can rotate in distinct ways. It can also vibrate, though this motion is usually negligible in its contribution to gas heat capacities. It is convenient, as well as indicative of ideal behavior, that most gases have C_v and C_p values closely approximated by the formulas above.

2.3 Nonideal Gases

We learned that several assumptions were made concerning ideal gases during the development of kinetic molecular theory: molecules occupy negligible volume and molecules have no interactions except during collisions. While these assumptions were helpful, even necessary, to derive the mathematical equations we have seen thus far, many gases under normal circumstances behave non-ideally. However by adjusting the ideal gas equation to take into account the fact that molecules *do* have interactions, we can arrive at a more accurate relationship, usually called the *van der Waals equation*.

$$(P + \frac{a}{\overline{V}^2})(\overline{V} - b) = RT$$

where $\overline{V}$ = volume per mole of gas
a and b are constants, varying with the gas used.

The terms a and b are simply correction factors for real gases. Though the volume of a gas molecule is very small it frequently isn't negligible. In a gas sample under high pressure the molecular density may be quite large, so that the molecular volume is actually a significant portion of the container volume. Obviously molecular volume is a less important factor at low pressures where the molecules may occupy little of the available volume. But in any case a gas molecule has a finite diameter and, if it is treated as a hard sphere, volume. This volume is not available to be occupied by any other molecule, it is *excluded volume*. The excluded volume, b, can be calculated geometrically from the diameter and as would be expected, small gas molecules like He and H_2 have small b values (23.8 mL and 26.6 mL per mole respectively).

Chapter 2   The Properties of Gases

While CH_4 and CO_2 have somewhat larger values (43.1 mL and 42.9 mL per mole respectively). This excluded volume is simply subtracted from the total volume (volume of the container) to yield the available volume in which the molecules may move. Thus the (V-b) term in the van der Waals equation. Table 2.5 in the text lists other common gases and their b values.

Now let's tackle the second false assumption about gases, that the molecules don't interact. Pressure exerted by a gas can be understood, on the molecular scale, as a measure of the gas molecules colliding with the walls of the container. If gas molecules exert attractive forces on one another, then as a gas molecule nears the wall it will be "held back" by the other gas molecules around it, which means there will be fewer collisions with the walls and the measured pressure will be *less* than if there were no inter-molecular forces. Such is the case. So the P in the ideal gas equation must be corrected for these weak, but numerous interactions. The correction factor depends on the density of the molecules, thus the volume of the container is important. The strength of the attractions is symbolized by *a*, and the term represents the adjusted pressure.

$$\left(P + \frac{a}{\overline{V}^2}\right)$$

The effects of molecular volume are smallest at low pressures and the effects of intermolecular attractions are smallest when the temperature is high. At high temperatures the molecules move so fast and have so much kinetic energy that the weak attractive forces between them become negligible. This means that even nonideal gases behave ideally at sufficiently low pressures and high temperatures.

Example 2.10. Using the data from Table 2.5 in the text, what fraction of the total volume is excluded for CO molecules at STP?

An ideal gas at STP has a volume of 22.4 liters. A mole of CO molecules excludes 0.0395 liters according to its b value.

$$\frac{0.0395 L}{22.4 L} \times 100 = 0.176 \text{ \% volume excluded at STP}$$

Example 2.11. Compare the pressure calculated for 15.0 g of N_2 in a 0.500 L container at 150 °C using the ideal gas equation with that from van der Waals equation.

ideal treatment: PV = nRT

$$P(0.500 L) = \left(\frac{15.0 g}{28.0 g/mol}\right)\left(0.08206 \frac{L-atm}{mol-K}\right)(423 K)$$

$$P = 37.2 \text{ atm}$$

nonideal treatment:

$$\left(P + \frac{a}{\overline{V}^2}\right)\left(\overline{V} - b\right) = nRT$$

The Properties of Gases

where $\overline{V}$ = volume per mole,

$$\overline{V} = \frac{0.500\text{L}}{(15.0\text{g}/\ 28.0\text{g}/\text{mol})} = 0.933\text{L}\frac{\text{L}}{\text{mol}}$$

$$\left(P + \frac{1.35\text{L}^2 - \text{atm}/\text{mol}^2}{(0.933\text{L}/\text{mol})^2}\right)\left(0.933\frac{\text{L}}{\text{mol}} - 0.0386\frac{\text{L}}{\text{mol}}\right) = \left((0.08206\frac{\text{L-atm}}{\text{mol-K}}\right)(423\text{K})$$

$$P = 37.4 \text{ atm}$$

Even the van der Waals equation is not general enough to be applied to all gases under all conditions. This is because molecules can only be approximated by the hard-sphere model and also because at close distances molecules repel each other while they attract each other when farther apart. These additional complications to our treatment of gases lead to a still more general form of the gas equation called the *virial equation of state*:

$$\frac{P\overline{V}}{RT} = 1 + \frac{B(T)}{V} + \frac{C(T)}{V^2} + \frac{D(T)}{V^3} + \ldots$$

where the virial coefficients $B(T)$, $C(T)$, $D(T)$, etc., are functions of temperature and the particular gas.

The attractive and repulsive forces between neutral gas molecules require a bit more explanation. Two gas molecules when far apart have no interaction with each other. As they get closer together there is a net attraction between them due mainly to the positively charged nuclei attracting not only their own nearby electrons but also those electrons of the other molecule, albeit more weakly. This attraction leads to a lowering of the overall potential energy as seen in figure 2.1.

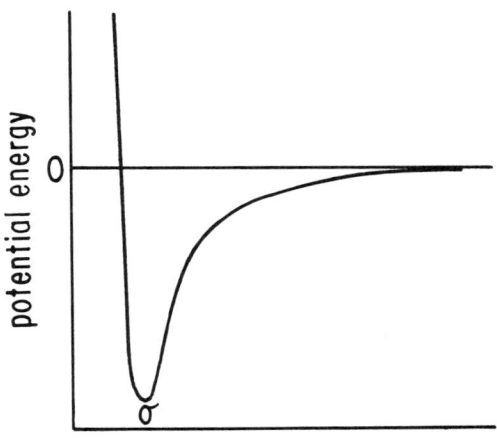

Figure 2.1.

When the molecules get very close together the repulsion of the electron clouds and of the nuclei becomes much stronger and the potential energy rises abruptly. The mathematical equation that describes this curve shows that the attractive forces depend on inter-molecular

Chapter 2 — The Properties of Gases

distance, r, to the sixth power while the repulsions depend on the inter-molecular distances to the twelfth power. This *Lennard-Jones potential function* also depends on the closest approach of the two molecules, σ, and on the energy minimum, ϵ.

$$\phi = 4\epsilon\left(\left(\frac{\sigma}{r}\right)^{12} - \left(\frac{\sigma}{r}\right)^{6}\right)$$

If many gas molecules attract one another sufficiently, either because they are close together (high pressure) or their kinetic energy is low enough that ϕ becomes important (low temperature) the gas will become a liquid. Although gases and liquids are both fluid, their molecular organization is quite different.

There are certain conditions under which it is impossible to condense a gas to a liquid. The temperature above which the liquid phase cannot be formed no matter how much pressure is applied is called the critical temperature, T_c. The critical pressure, P_c, and critical volume, V_c, are the corresponding pressure and volume at T/dc/u for the coexistence of gas and liquid. These critical parameters can be used to calculate the van der Waals factors a and b for gases (see problem 2.23 in the text).

Since gas molecules under reasonable conditions of pressure and temperature exclude so little of the available volume (see example 2.10) yet move so fast (see example 2.9) it is interesting to ask how often do gas molecules collide and how far can a gas molecule travel between collisions? There is so much empty space between the molecules that you may be surprised to learn how frequent collisions are.

$$\text{collisions per second for a molecule} = \sqrt{2}\pi\, \rho^2\, \overline{C}\, n^*$$

where p = hard-sphere molecular diameter
$\overline{c}_*$ = average speed
n^* = molecules per volume

At 25 °C and 1.0 atm, most gas molecules collide about 10^{10} times! This means that in every cubic centimeter of volume there are approximately 10^{29} collisions per second. In order for molecules to collide that often, the distance between collisions, the mean free path, must be very small.

The equation for mean free path calculations is:

$$\lambda = \text{mean free path} = \frac{10^3\, RT}{\sqrt{2}\pi\, \rho^2\, N_A\, P}$$

where N_A = Avogardro's number

Example 2.12. What is the mean free path for an N_2 molecule, $\rho \simeq \sigma = 3.74 \times 10^{-8}$ cm, at 25 °C in a laboratory vacuum line kept at 10^{-2} atm?

$$\lambda = \frac{10^3\, RT}{\sqrt{2}\pi\, \rho^2\, N_A\, P}$$

The Properties of Gases   Chapter 2

$$\lambda = \frac{10^3 \frac{cm^3}{L}\left(0.08206 \frac{L-atm}{mol-K}\right)298}{\sqrt{2}\pi\,(3.74 \times 10^{-8}\,cm)^2 (6.02 \times 10^{23}\frac{molecules}{mol})(10^{-4}\,atm)}$$

$$\lambda = 6.5 \times 10^{-2}\,cm$$

We generally associate the property of viscosity with liquids. Molasses is viscous while water is not very viscous. Viscosity is the resistance to flow freely due to the molecular friction. But gases also have viscosity. Adjacent layers of gas molecules move against each other with varying degrees of difficulty, depending on the gas. The viscosity, as measured by the coefficient of viscosity η is:

$$\eta = \frac{5}{16}\sqrt{\frac{K}{\pi}}\left(\frac{\sqrt{mT}}{\rho^2}\right)$$

where k = Boltzmann constant

m = molecular mass

Although one might predict that as a gas is compressed (that is, pressure is increased) it should become more viscous, we can see that η does not depend on pressure and indeed this is experimentally proven to be true at pressures of 1 atm or lower. This prediction, followed by laboratory verifications, showed the power of Maxwell's kinetic theory of gases to scientists of the later 1800's.

Once again, however, we find that treating molecules as hard spheres that do not interact is a simplification. Allowing for both these complications by including the Lennard-Jones potential of interactions, a much better fit of the experimental data can be accomplished.

Problems

1. The newest sphygmomanometers measure blood pressure in kilopascals instead of the traditional torr. If a person has a blood pressure of 120/80 (which is 120 torr/80 torr) what will the readings be in kPa?

2. If one mole of a gas at 20°C occupies 205 mL of volume, what is the pressure of the gas? What must the pressure be if the same gas sample occupies 650 mL at 144°C?

3. An ideal gas has a volume of 1.50 L at 127°C and 900 torr. What is the gas volume at 327°C and 300 torr?

4. A 2.15 L steel vessel contains 17.5g of neon at a pressure of 7.30 atm. What is the temperature of the gas?

5. When 4.00 moles of an ideal gas were placed in a container at -84°C, the pressure was found to be 0.576 atm. How many moles of gas must be added or removed to produce a pressure of 0.423 atm at 298K?

Chapter 2 The Properties of Gases

6. What is the molecular weight of a gas if 0.600g in a 1.20L container at 55.0°C exerts a pressure of 0.110 atm?

7. Consider two gas samples of equal volume, one is O_2, the other is N_2 at the same pressure and temperature. Assuming both gases behave ideally, compare the following for the samples:

   a) the density of the gases
   b) the number of grams
   c) the moles
   d) the number of molecules
   e) the average molecular velocity
   f) the average kinetic energy

8. What is the molecular formula of a gas with the empirical formula CH and a density of 1.72 g/L at 30°C and 0.658 atm?

9. When ethane is combusted according to the equation below, 23.5 L of C_2H_6 is reacted with oxygen at 57°C and 756 torr. How many liters of CO_2 are generated at this same pressure and temperature?

$$C_2H_{6(g)} + 7/2\, O_{2(g)} \rightarrow 2CO_{2(g)} + 3H_2O_{(l)}$$

10. How many liters of H_2 are needed to react with 185g of I_2 at STP, according to the following equation?

$$H_{2(g)} + I_{2(s)} \rightarrow 2HI_{(g)}$$

11. A gas mixture contains 4.5 mol N_2, 6.2 mol O_2, and 1.6 mol CO. What is the mole fraction of each gas? What is the partial pressure of each gas if the total pressure is 943 kPa?

12. Consider one mole of helium gas at 3.0 atm and 310 K.

    a) What is the average translational kinetic energy of the mole of gas?
    b) What is the average translational kinetic energy of a helium atom?
    c) What is the root-mean-square speed of the helium atoms?

13. Two sealed flasks are connected to each other by a stopcock. In flask 1, at 25°C, the volume is 2.50 L and there is a pressure of 4.00 atm of N_2. Flask 2 contains 1.10 L of H_2 at 25°C, and a pressure of 1.90 atm. Then the stopcock is opened and the gases are allowed to mix.

    a) Assuming no change in temperature, what is the final pressure in the flasks?
    b) What is the mole fraction of H_2? N_2?

The Properties of Gases   Chapter 2

14. CH_4 molecules at 300K have a rms speed of 630 m/sec. What is the molecular weight of a gas that has a rms speed of 376 m/sec at the same temperature?

15. At what temperature will O_2 molecules have the same rms speed as H_2 molecules at 273°C?

16. Hydrogen and nitrogen react according to the reaction below. If 5.0 atm H_2 and 3.0 atm N_2 are mixed together, what is the total pressure in the container *after* the reaction is complete? (assume no volume or temperature change)

$$3H_{2(g)} + N_{2(g)} \rightarrow 2NH_{3(g)}$$

17. Four volumes of a compound containing only nitrogen and hydrogen is burned in excess O_2. Eight volumes of NO and eight volumes of H_2O gas are formed, all measured at the same pressure and temperature. What is the formula of the compound?

18. A solid sample containing a mixture of $Al(OH)_3$ and Al_2O_3 weighs 3.98 grams. Upon heating to 1000K, the $Al(OH)_3$ decomposes according to the equation shown below. The water vapor that was collected at 1000 K had a pressure 2900 torr in a 550 mL container. What was the percentage of $Al(OH)_3$ in the original sample?

$$2Al(OH)_{3(s)} \rightarrow Al_2O_{3(s)} + 3H_2O_{(g)}$$

19. Calculate the rms speed for a nitrogen molecule at 1.0 atm and 300K. How does this compare with the speed of sound at 300K? v_{sound} @300K = 345 m/sec

20. What is the mean free path of an argon atom at STP? At 1000K and 1 atm? At 273K and 100 atm? Use the Lennard-Jones σ value for Ar.

21. Three gases, X, Y and Z, have the following van der Waals constants. Which gas behaves closest to ideal at STP?

	$a \left(\dfrac{L^2-atm}{mol^2} \right)$	$b \left(\dfrac{L}{mol} \right)$
X	1.1	0.0280
Y	13.0	0.0390
Z	3.5	0.0450

CHAPTER 3
LIQUIDS AND SOLUTIONS

3.1. Kinetic Theory of Liquids

In many respects liquids are similar to the other condensed phase, solids. Both liquids and solids, because their molecules are already close together, cannot be further compressed to any extent. Certainly liquid molecules have more freedom than solid molecules because a liquid is fluid and takes the shape of its container, but liquid molecules lack the complete freedom of a gas where each molecule is essentially an independent entity. The molecules in the condensed phases are governed by the forces exerted by their neighbors, mainly attractive in nature.

The greatest similarity between liquid and gas molecules comes from the astute observation by Robert Brown that small particles in a liquid are in constant random motion; *Brownian motion* was brought about by collisions with surrounding particles. This macroscopic property is indicative of the molecular properties, but because liquid molecules are more densely packed, they must experience a shorter mean-free-path than gas molecules. Both however, have a kinetic energy equal to 3/2 kT per molecule or 3/2 RT per mole.

3.2. Phase Equilibria

Many liquids and solids when placed in a container will evaporate until a certain partial pressure of the gas phase compound is reached. This partial pressure of the gas phase above the condensed phase is a constant value for each compound at any given temperature, and is called the *equilibrium vapor pressure*. At *equilibrium*, there is no net change in the partial pressure of the gas yet the situation described is dynamic, with some gas molecules condensing while some liquid or solid molecules are evaporating. Both processes just occur at the same rate at equilibrium. An equilibrium is indicated in a chemical equation by the use of double arrows ($\rightleftarrows$), showing that the reaction is going in both directions simultaneously, as illustrated below.

$$A_{(l)} \rightleftarrows A_{(g)}$$
$$\text{or}$$
$$A_{(g)} \rightleftarrows A_{(l)}$$

The vapor pressure of a substance depends on only the nature of the substance and on temperature, but does *not* depend on the external air pressure or volume of the container. For example, on a warm day when the equilibrium temperature equals 30°C, the vapor pressure of H_2O is 31.82 torr whether we confine a small amount of water to a beaker with a watch glass over it or we talk about the air over Lake Erie. In both instances equilibrium will not be reached until the partial pressure of the water vapor equals 31.82 torr. This is the basis for the relative humidity readings given by meteorologists. If the relative humidity is 90%, then the partial pressure of the water vapor is only 90% of the equilibrium vapor pressure. A 90% relative humidity reading at 30°C means that H_2O = 28.64 torr. The variation of the vapor pressure with temperature is easily calculated, (we will return to that particular subject shortly), but it is important to note that many substances, even in the solid phase, have appreciable vapor pressures. Even ice *sublimes* (evaporates directly from the solid phase) though its vapor pressure is low, about 4.6 torr at 0°C.

Generally substances are classified according to their vapor pressures; *volatile* substances, like water, have easily measured, significant vapor pressures, while *non-volatile* substances have small, sometimes negligible vapor pressures.

Liquids and Solutions — Chapter 3

Examples of volatile substances: ether, gasoline, ethanol, chloroform, napthalene (an ingredient in moth balls), all gases.

Examples of non-volatile substances: paraffin, sugar, mercury, glycerin, most salts.

Example 3.1. Zinc metal dissolves in hydrochloric acid to produce hydrogen gas. If an apparatus is constructed to collect the H_2 as it bubbles through water, and a total pressure of 778 torr in a 500 mL container is collected at 26 °C, how many moles of H_2O at 26 °C is 25.21 torr.

First we must realize that the total pressure of gas collected is a sum of the partial pressures of H_2O and H_2.

$$P_T = P_{H_2O} + P_{H_2}$$

$$778 \text{ torr} = 25.21 \text{ torr} + P_{H_2}$$

$$753 \text{ torr} = P_{H_2}$$

then the moles of H_2 can be calculated from the ideal gas equation.

$$PV = nRT$$

$$\left(\frac{753 \text{torr}}{760 \text{torr}/ 1\text{atm}}\right)(0.500\text{L}) = n\left(0.08206 \frac{\text{L}-\text{atm}}{\text{mol}-\text{K}}\right)(273 + 26 \text{ K})$$

$$2.02 \times 10^{-2} = n\ H_2$$

Chemists like to be as specific as possible when describing how a reaction is carried out. We saw in Chapter 2 that the heat capacity of a substance depends on whether we are dealing with a constant volume or a constant pressure process. This means the *system* must be distinguished from the *surroundings*. The system is the specific reaction container, chemical, or environment we are studying while everything else in the universe constitutes the surroundings. It sounds rather vague but will become clearer as we use the two terms.

When phase changes are carried out, only the substance itself is considered to be the system. When water is boiled we know we must put heat into the system. That heat will come from the surroundings. For one mole of water, 44 kJ of heat is required for vaporization under normal cirumstances and this heat can be included in a balanced chemical reaction that describes the process. This heat can be treated

$$H_2O_{(l)} + 44 \text{ kJ} \rightarrow H_2O_{(g)}$$

stoichiometrically so that if *two* moles of water are vaporized the heat required must be 88 kJ.

$$2\ H_2O_{(l)} + 88 \text{ kJ} \rightarrow 2\ H_2O_{(g)}$$

We can also view the reaction as happening in the reverse direction. When one mole of water vapor condenses to liquid, 44 kJ of heat is released.

$$H_2O_{(g)} \rightarrow H_2O_{(l)} + 44 \text{ kJ}.$$

Chapter 3 — Liquids and Solutions

When a reaction or system releases heat we say it is *exothermic*. When a reaction or system absorbs heat we say it is *endothermic*. If the heat involved in a process is measured under constant pressure conditions we call that heat *enthalpy*, H. The enthalpy or heat content of a substance depends on the external conditions (temperature, pressure, quantity) as well as internal conditions (the nature of the substance itself). The enthalpy of a substance usually is given in units of kJ/mol. Absolute enthalpies are difficult to measure but changes in enthalpy are easily determined, and must be equal to the product enthalpies minus the reactant enthalpies.

$$\Delta H = H \text{ (products)} - H \text{ (reactants)}$$

Usually standard conditions are employed, so all gases have a pressure of 1 atm. This standard condition is denoted by a superscript zero.

$$\Delta H^0 = H^0 \text{ (products)} - H^0 \text{ (reactants)}$$

A ΔH^0 value that is negative indicates an exothermic process where the system gives off heat. A positive ΔH^0 value indicates heat is added to the system. Enthalpy values can be treated the same as any other product or reactant when equations are added or subtracted.

Example 3.2. Calculate the enthalpy of sublimation for Xe using the data given below. Give the answer in kJ/mol as well as J/g.

$$\Delta H^0 \text{ vap} = 12.6 \text{ kJ/mol}$$

$$\Delta H^0 \text{ melt} = 2.30 \text{ kJ/mol}$$

We shall begin by writing the chemical equation for sublimation and recognizing that it is simply the sum of the equations for melting and vaporization.

$$Xe_{(s)} \rightarrow Xe_{(l)} \quad \Delta H^0 = 2.30 \text{ kJ/mol}$$

$$Xe_{(l)} \rightarrow Xe_{(g)} \quad \Delta H^0 = 12.6 \text{ kJ/mol}$$

$$Xe_{(s)} \rightarrow Xe_{(g)} \quad \Delta H^0 = 14.9 \text{ kJ/mol}$$

The equations and the ΔH^0 values are simply added.

$$\Delta H^0 = \frac{14.9 \text{ kJ/mol}}{131.3 \text{ g/mol}} = 0.113 \text{ kJ/g or } 113 \text{ J/g}$$

Qualitatively it makes sense that ΔH^0 vap is greater than ΔH^0 melt. Only some of the molecular attractions must be broken to turn a highly ordered solid into a slightly less ordered liquid, but virtually all molecular forces must be overcome in order to release the liquid molecules to the gas phase.

We also expect that a system will tend to move toward lower energy, like a ball rolling down a hill, and that when a system reaches equilibrium it will be in an energy well. Although energy is not exactly the same thing as enthalpy, the two functions are closely related so we might expect that systems tend toward lower enthalpy. If that were the case all exothermic reactions would be spontaneous and all gases would automatically condense to liquids, and all liquids would solidify. Obviously a lowering in energy or enthalpy is not the only driving force behind a reaction. Many endothermic reactions are spontaneous at 25°C, like the melting of

Liquids and Solutions — Chapter 3

ice. The enthalpy of $H_2O_{(l)}$ is considerably greater than that of $H_2O_{(s)}$. So what drives the reaction?

There is a second driving force that must be considered. Systems tend to increase in disorder. This increase in disorder is evident on both the macroscopic scale (your dorm room or apartment spontaneously becomes untidy after it is cleaned; the reverse never happens) and on the microscopic scale (liquid water molecules are more disorganized than ice molecules). The term that describes the amount of disorder or randomness is *entropy*. Some reactions are driven by an increase in entropy while others are driven by a decrease in enthalpy. These two properties of a system together determine spontaneity. Entropy is treated numerically in Chapter 8.

As was mentioned previously, the vapor pressure of a pure substance depends on the temperature. As the temperature of a liquid is raised the molecules increase in average kinetic energy ($E = 3/2\ RT$) and more of the molecules will have sufficient energy to escape to the gas phase. The process is endothermic but there is an increase in entropy. The temperature where the vapor pressure equals atmospheric pressure, which is normally 1 atm, is the normal boiling point. Certainly volatile liquids should be easier to boil than non-volatile liquids and thus have a lower boiling point. This is experimentally true for most chemicals. The relationship between vapor pressure and temperature is logarithmic. The ln of vapor pressure varies in a straight line with the inverse of temperature. The slope of this line gives $\Delta H^0_{vap}/R$. In equation form this is:

$$\ln\frac{(VP)_{T_2}}{(VP)_{T_1}} = \frac{\Delta H^\circ_{vap}}{R}\left\{\frac{1}{T_1} - \frac{1}{T_2}\right\}$$

If we assume ΔH^0_{vap} doesn't vary with temperature, and that it is always a positive value, then as T increases, so must the VP.

Example 3.3. At an altitude of 5000 ft., approximately that of Denver, Colorado, the air pressure is 630 torr. What is the approximate boiling point of water at this altitude? Assume that ΔH^0_{vap} for water is constant over the temperature range used, and is 44.0 kJ/mol.

This problem is easily solved if you remember that water boils at 373 K (100 °C) when its vapor pressure equals the 1.00 atm external pressure. In Denver, water will boil when its vapor pressure is 0.829 atm (630 torr/760 torr atm^{-1}).

$$\ln\frac{(VP)_{T_2}}{(VP)_{T_1}} = \frac{\Delta H^\circ_{vap}}{R}\left\{\frac{1}{T_1} - \frac{1}{T_2}\right\}$$

$$\ln\left\{\frac{0.829}{1}\right\} = \frac{44.0\ kJ/mol}{8.314 \times 10^{-3}\ kJ/mol\text{-}K}\left\{\frac{1}{373K} - \frac{1}{T_2}\right\}$$

$$368\ K = T_2 = 95\ °C$$

Chapter 3 — Liquids and Solutions

Example 3.4. What is ΔH^0_{vap} for a liquid that has a vapor pressure of 0.083 atm at 25 °C, and 0.360 atm at 90 °C?

$$\ln\frac{(VP)_{T_2}}{(VP)_{T_1}} = \frac{\Delta H^o{}_{vap}}{R}\left\{\frac{1}{T_1} - \frac{1}{T_2}\right\}$$

$$\ln\left\{\frac{0.360}{0.083}\right\} = \frac{\Delta H^o{}_{vap}}{8.314 \times 10^{-3} \text{ kJ/mol-K}}\left\{\frac{1}{298K} - \frac{1}{363K}\right\}$$

$$4.43 \text{ kJ/mol} = \Delta H^0{}_{vap}$$

$\Delta H^0{}_{vap}$ values should always be positive. Chemists would rarely use only two data points to do a calculation such as this. Rather, the vapor pressure would be followed with temperature and ln VP *vs* 1/T data would be graphed. The $\Delta H^0{}_{vap}$ could then be obtained from the slope of the line.

What if the line, ln VP *vs* 1/T, were not straight? Then the $\Delta H^0{}_{vap}$ is not constant over the temperature range chosen. We have thus far assumed $\Delta H^0{}_{vap}$ to be the same for a given substance regardless of the temperature at which the phase change occurs. In fact, ΔH values may change somewhat with temperature. We shall see how to take this into account as well as why it is so, in Chapter 8.

Thus far we have not discussed the phase change from solid to liquid or the reverse. We have mentioned that many solids also have measurable vapor pressures even at low temperatures. The definition of *freezing point* or *melting point*, depending on which way you approach the problem, is the temperature where the vapor pressure of the solid phase equals the vapor pressure of the liquid phase. In other words both the solid and liquid phase are in equilibrium with the gas phase, thus they are in equilibrium with each other. As expected, the vapor pressure of a solid rises as the temperature rises, but the curve for this temperature dependence is different than the vapor pressure *vs* temperature curve of the liquid. Where these two curves intersect, as can be seen in figure 3.3 in the text, there is melting or freezing. As you can tell, the *triple point*, the temperature where only the three phases are in equilibrium, must be close to the freezing point, which is measured under normal atmospheric pressure conditions.

The freezing point of a substance frequently shows some pressure dependence. Most substances are more dense in the solid phase than the liquid phase so when pressure is placed on a liquid/solid system, the liquid freezes in order to decrease its volume. The result is that the melting or freezing point is raised for substance as pressure increases. Water, however, is a major exception. Ice is less dense than liquid water. As the pressure on an ice/water system is raised the ice melts in order to decrease its volume. The freezing point for H_2O is lowered as pressure increases.

When all three curves, vapor pressure of solid vs. temperature, vapor pressure of liquid vs. temperature, and solid/liquid equilibrium pressure vs. temperature are combined, we have a *phase diagram*. The phase diagram of water can be seen in figure 3.5 in the text.

Liquids and Solutions   Chapter 3

Example 3.5. What physical changes take place in a piece of ice originally at -5.0 °C and 0.5 atm when the temperature is raised but the external pressure is held constant? Refer to figure 3.5 in the text.

The ice remains solid as the temperature is raised until the solid/liquid interface line is reached. The temperature where this intersection occurs is slightly less than the triple point (actually it is 0.005 °C). After all the ice has melted the temperature increase is continued until the liquid/gas intersection is reached. Boiling occurs because the vapor pressure of the water equals the external pressure of 0.5 atm. The corresponding temperature for this phase change is 82 °C. These changes can be followed by drawing a horizontal line straight across the phase diagram at 0.5 atm.

3.3. Types of Solutions

Solutions are mixtures that are homogeneous even at the molecular scale. A *solute* is dissolved in a *solvent*. The solvent is generally the portion of the solution that is the greater in terms of weight or volume.

In sea water for example, the solvent is H_2O and the solutes are NaCl, NaBr, NaI, $CaCl_2$, etc. Sometimes it is hard to decide which component is the solute and which is the solvent. This is true of car antifreeze which is 50% water by weight and 50% ethylene glycol. In this case we usually call the more volatile component, water, the solvent.

Solutions can be of several varieties that include all possible phases as solvents and solutes. Examples are shown below.

Solution Type	Name	Main Components
solid in liquid	maple syrup	sugars in water
liquid in liquid	gin	ethanol in water
gas in liquid	soda water	CO_2 in water
gas in gas	air	O_2 in N_2
gas in solid	-	H_2 in Pd or Pt
solid in solid	brass	zinc in copper
liquid in solid	dental amalgam	mercury in tin or silver

Solutions are *ideal* if $\Delta^0_{soln.}$ is zero, which means that the solvent molecules interact with solute molecules exactly the same way as they interact with other solvent molecules. For ideal gases these interactions are zero so all gas/gas solutions should be ideal or nearly so. But many other solutions warm up or cool down when formed because the heat of solution may be exothermic or endothermic depending on the molecular interactions in the solvent, solute, and solution. The rule of thumb used by chemists is "like dissolves like" which means that substances with similar molecular interactions will dissolve in each other.

| Chapter 3 | Liquids and Solutions |

Chemists classify these interactions as polar or nonpolar. A *polar* molecule is one in which there is some partial charge separation. Though the molecules remain neutral, one side of the molecule can develop a partial negative charge and then there must be a compensating positive charge developed elsewhere on the molecule. An extreme polar molecule would actually exhibit complete charge separation, or formation of ions, as in salts. *Nonpolar* molecules show no appreciable charge separation, like the hydrocarbons in gasoline and oil. Water is a polar solvent which will not dissolve most nonpolar solutes. So the saying "like dissolves like" can be rephrased "polar dissolves polar, nonpolar dissolves nonpolar".

Because water is polar it dissolves most substances composed of ions, like NH_4Cl or $CaSO_4$. It also solubilizes most acids and bases because these substances generate H^+ or OH^- ions in aqueous systems. When an aqueous solution contains a great many free ions, as in salt water, then the solution strongly conducts electricity and we say that the NaCl is a *strong electrolyte*. An aqueous system that contains few free ions, as in vinegar, will not be a good electrical conductor and the solute CH_3COOH is a *weak electrolyte*. Because sugar water doesn't conduct electricity at all, we say sugar is a *non-electrolyte*.

How much solute is present in the solvent is called the *concentration*. Concentration can be expressed in many different units. Examples of the most common units are given below.

Mole Fraction indicates what fraction of the total moles belongs to an individual component.

$$\text{mol fraction A} = X_A = \frac{\text{mol of A}}{\text{total mol}}$$

The total mole fractions of all components must sum to 1.

Example 3.6. A solution is made by dissolving 10.0 g of NaBr in 150 g of H_2O. What are the mole fractions of each component?

$$\frac{10.0 \text{ g NaBr}}{103 \text{ g/mol}} = 0.971 \text{ mol NaBr}$$

$$\frac{150 \text{ g } H_2O}{18.0 \text{ g/mol}} = 8.333 \text{ mol } H_2O$$

$$X_{NaBr} = \frac{\text{mol NaBr}}{\text{mol NaBr} + \text{mol } H_2O} = \frac{0.971}{(0.971 + 8.333)} = 0.104$$

$$X_{H_2O} = 1 - X_{NaBr} = 1 - 0.104 = 0.896$$

So the solution is 10.4% NaBr, by moles, and 89.6% H_2O by moles.

Molality indicates how many moles of solute are present per kilogram of solvent.

$$\text{molality} = m = \frac{\text{mol solute}}{\text{kg solvent}}$$

Liquids and Solutions   Chapter 3

Example 3.7. What is the molality of a solution made by dissolving 63.0g of glucose, $C_6H_{12}O_6$, in 1.0 liter of water? The density of water can be assumed to be 1.0 g/mL.

1.0 L of H_2O weighs 1.0 kg since the density is 1.0 g/mL (prove this to yourself).

So the molality is:

$$\frac{\text{mol glucose}}{1 \text{ kg water}} = \frac{(63.0 \text{ g glucose})/(180 \text{ g mol}^{-1})}{1.0 \text{ kg } H_2O} = 0.35 \text{ m}$$

When molality is used the solute is added to a certain amount of solvent. This is quite different from the more useful unit *molarity*.

Molarity indicates how many moles of solute are present per 1 liter of *total solution*.

$$\text{molarity} = M = \frac{\text{mol solute}}{\text{liter solution}}$$

Example 3.8. 19.6 g of KF are dissolved in water. The total volume of the solution is adjusted by adding enough water to get exactly 500 mL of solution. What is the molarity of the KF?

$$\frac{\text{mol KF}}{\text{L soln}} = \frac{(19.6 \text{ g KF})/(58.1 \text{ g mol}^{-1})}{0.500 \text{ L}} = 0.675 \text{ M}$$

Formality indicates how many formula weights of a solute are present per liter of total solution. Remember that a salt does not really consist of molecules so there isn't a true molecular weight, but rather a formula weight. Thus a salt concentration should be given in formality instead of molarity. Since we have chosen to use molecular weights to apply to salts, though not strictly correct, we will also use molarity for salt solutions as in example 3.8.

Normality indicates how many equivalent weights of a solute are present per liter of total solution. This unit can be somewhat ambiguous since it depends on *how* the solute is to be used. The equivalent weight of HNO_3 depends on whether the H^+ is involved in the pertinent reaction or the NO_3^- is involved. For example, the two equations below use HNO_3 for different purposes, in equation (1) the H^+ is transferred to an OH^-.

(1) $HNO_3 + NaOH \rightarrow H_2O + NaNO_3$

In this case 1 mole of HNO_3 produces 1 mole of H_2O so there is 1 equivalent of H^+/mol HNO_3. Now look at equation (2) where there are electrons transferred to nitrogen.

Chapter 3   Liquids and Solutions

$$(2) \quad 4\,H^+ + NO_3^- + 3\,Ag \rightarrow 3\,Ag^+ + 2\,H_2O + NO$$

In this equation, 1 mole of HNO_3 produces 3 moles of Ag^+ so there are 3 equivalents of electrons/mol HNO_3. That is why normality should not be used to express concentration unless the specific use of the solution is clearly stated.

Percentage units also can be ambiguous unless it is known what percentage is used; options are by weight, by volume, or even by moles (see mole fraction).

Example 3.9. Alcohol is generally rated by proof, which equals 2 x (percent by volume) at 60°F. What is the percent by volume of alcohol in bourbon that is 86 proof?

$$\% \text{ volume alcohol} = \frac{\text{proof}}{2} = \frac{86}{2} = 43\%$$

So 43% of the total volume is ethanol. Note that this is different from adding 43 mL of ethanol to 57 mL of water. The sum of the solute volume and solvent volume does not necessarily equal the total volume. You should not assume volumes to be additive unless told so.

One concentration unit frequently can be changed to another provided you know molecular weights and densities.

Example 3.10. When 20.0 mL of ethanol are added to 40.0 mL of water, all at 30°C, the final solution has a volume of 58.3 mL. The density of ethanol is 0.789 g/mL and the density of water is 1.00 g/mL. What is the percent by volume alcohol, percent by weight alcohol, and the density of the final solution?

$$\% \text{ volume} = \frac{20.0 \text{ mL ethanol}}{58.3 \text{ mL soln}} \times 100 = 34.3\% \text{ ethanol}$$

$$\% \text{ weight} = \frac{\text{g ethanol}}{\text{g soln.}} \times 100$$

$$20.0 \text{ mL ethanol} \times \frac{0.789 \text{ g}}{\text{mL}} = 15.8 \text{ g ethanol}$$

$$10.0 \text{ mL H}_2\text{O} \times \frac{1.00 \text{ g}}{\text{mL}} = 40.0 \text{ g H}_2\text{O}$$

$$\% \text{ weight} = \frac{15.8 \text{ g ethanol}}{(15.8 + 40.0) \text{ g soln}} \times 100 = 28.3\% \text{ ethanol}$$

$$\text{solution density} = \frac{\text{g soln}}{\text{mL soln}} = \frac{(15.8 + 40.0) \text{ g}}{58.3 \text{ mL}}$$

Liquids and Solutions	Chapter 3

$$d = 0.957 \frac{g}{mL}$$

3.4. Solution Stoichiometry

Now that we know how to express the quantity of solute or solvent in a solution we can incorporate solutions into our stoichiometry calculations.

Example 3.11. How many moles of sulfate ions are contained in 510 mL of a 0.64 M $Fe_2(SO_4)_3$ solution?

$$\text{molarity} = \frac{\text{mol } Fe_2(SO_4)_3}{\text{L soln.}}$$

$$\text{mol } Fe_2(SO_4)_3 = 0.510 \text{ L} \times \frac{0.64 \text{ mol}}{L} = 0.33$$

Since there are 3 times as many sulfate ions as iron (III) sulfate formula units:

$$\text{mol } SO_4^{2-} = 3 \times 0.33 = 0.99 \text{ mol}$$

Example 3.12. What is the molarity of Na^+ in a solution made by mixing 433.2 mL of 1.255 M Na_3PO_4 and 1.525 L of 0.825 M NaCl? Assume volumes are additive.

First you must calculate how many moles of Na^+ are contributed by each solution.

Soln 1: $1.255 \text{ M } Na_3PO_4 = \dfrac{\text{mol } Na_3PO_4}{0.4332 \text{ L}}$

$$\text{mol } Na_3PO_4 = 0.544$$

$$\text{mol } Na^+ = 3 \times 0.544 = 1.63$$

Soln 2: $0.825 \text{ M NaCl} = \dfrac{\text{mol NaCl}}{1.525 \text{ L}}$

$$\text{mol NaCl} = 1.26$$

$$\text{mol } Na^+ = 1.26$$

combination: $Na^+ \text{ molarity} = \dfrac{\text{mol } Na^+ \text{ total}}{\text{L total}} =$

Chapter 3   Liquids and Solutions

$$\frac{(1.63 + 1.26) \text{ mol Na}^+}{(0.4332 + 1.525) \text{ L}} = 1.48 \text{ M}$$

Example 3.13. How would you dilute 5.0 mL of 6.0 M HCl to make 0.10 M HCl?

Since the dilution will involve only the addition of water, the number of moles of HCl remains constant, only the solution volume changes.

$$6.0 \text{ M HCl} = \frac{\text{mol HCl}}{0.0050 \text{ L}}$$

mol HCl = 3.0×10^{-2} which can also be expressed as 30 millimol

The final solution still contains the same 3.0×10^{-2} mol HCl, only the final volume is unknown.

$$0.10 \text{ M HCl} = \frac{3.0 \times 10^{-2} \text{ mol HCl}}{\text{L}}$$

$$\text{L} = 0.30 \text{ or } 300 \text{ mL}$$

Be careful here! The 300 mL is the total volume of the final solution. In order to make it, 295 mL of water is added to 5.0 mL of 6.0 M HCl.

Example 3.14. What volume of 0.238 M H_3PO_4 is required to neutralize (completely react the H^+ and OH^-) 0.492 g of Ca$(OH)_2$?

Your first step should be to write the balanced reaction so that you can clearly see the stoichiometric relationship between the reactants. Remember that the reaction of an acid with a base forms water and a salt.

$$2 \text{ H}_3\text{PO}_4 + 3 \text{ Ca(OH)}_2 \rightarrow 6 \text{ H}_2\text{O} + \text{Ca}_3(\text{PO}_4)_2$$

$$\frac{0.492 \text{ g Ca(OH)}_2}{74.1 \text{ g/mol}} = 6.64 \times 10^{-3} \text{ mol Ca(OH)}_2$$

$$\text{mol H}_3\text{PO}_4 = \frac{2}{3} \text{ molCa(OH)}_2 = 4.43 \times 10^{-3} \text{ mol}$$

$$0.238 \text{ M H}_3\text{PO}_4 = \frac{4.43 \times 10^{-3} \text{ mol H}_3\text{PO}_4}{\text{X}}$$

$$\text{X} = 1.86 \times 10^2 \text{ L or } 18.6 \text{ mL of the solution}$$

Liquids and Solutions   Chapter 3

3.5. Henry's Law & Raoult's Law

The vapor pressure of pure liquids and solids was discussed earlier in this chapter and we saw how vapor pressure varied with temperature. Now we will see how vapor pressure of a solution can be determined from the vapor pressures of the solute and solvent, if we know their concentrations.

It can be easily shown that the amount of gas that will dissolve in a liquid varies with the pressure of the gas. Carbonated water can be made by using a high pressure of CO_2 over water. Each gas has a unique variation with pressure that is measured as K_H, Henry's law constant.

$$K_H = \frac{P_{gas}}{X_{gas\ in\ solution}}$$

K_H, not surprisingly, changes with temperature.

If, instead of considering the solute we consider the solvent, another relationship can be found. When comparing a pure solvent with a solution that contains a non-volatile solute we expect the vapor pressure of the pure solvent to be greater since proportionally more of the molecules are volatile. As the solute concentration increases, the vapor pressure of the solution should decrease. The mathematical relationship is given in Raoult's law.

$$P_{solvent} = X_{solvent}\ P^0_{solvent}$$

where $P^0_{solvent}$ is the vapor pressure of the pure solvent.

Example 3.15. Maple syrup can be regarded as an aqueous sucrose solution that contains 943 g of sucrose, $C_{12}H_{22}O_{11}$, per liter of water. If pure water has a vapor pressure of 23.8 torr at 25 °C, what is the vapor pressure of maple syrup?

$$X_{H_2O} = \frac{mol\ H_2O}{mol\ H_2O\ +\ mol\ sucrose}$$

If we assume the density of H_2O = 1.0 g/mL then 1.0 L of water weighs 1000 g at 25 °C.

$$X_{H_2O} = \frac{1000\ g/\ 18\ g\ mol^{-1}}{(1000\ g/\ 18\ g\ mol^{-1}) + (943\ g/\ 342\ g\ mol^{-1})} = 0.953$$

$$P_{H_2O} = X_{H_2O} P^0_{H_2O}$$

$$P_{H_2O} = 0.953\ (23.8\ torr) = 22.7\ torr$$

3.6. Ideal Solution Theory

As mentioned earlier, ideal solutions have no ΔH^0_{soln} because the solute/solute interactions are similar to the solvent/solvent and solute/solvent interactions. There are several properties of ideal solutions that depend only on the quantity of solute dissolved. These are called

Chapter 3 — Liquids and Solutions

colligative properties. Very dilute nonideal solutions also can be treated ideally.

The colligative properties are vapor pressure lowering, boiling point elevation, freezing point depression, and osmotic pressure. In the previous section we calculated how the vapor pressure of a solution changes with mole fraction. The *vapor pressure lowering* is merely the difference $P^0_{solvent} - P_{solvent}$, or $\Delta P_{solvent}$.

$$\Delta P_{solvent} = X_{solute} P^0_{solvent}$$

measuring the vapor pressure lowering caused by an unknown solute is one method of determining molecular weight.

Because the vapor pressure changes with solute concentration we should also expect the boiling point and freezing point of the solution to be different from the pure solvent. Both these temperatures depend on the equilibrium of vapor pressures of two phases.

The boiling point of a solution is greater than that of the pure solvent, as a direct result of the vapor pressure lowering. It necessarily requires a higher temperature in order for the solution vapor pressure to equal the 1 atm external pressure. This boiling point elevation is usually calculated from molality, which is closely related to mole fraction.

$$\Delta T_b = K_b m$$

where ΔT = the change in boiling point
K_b = boiling point constant, unique for each solvent
m = molality of solute particles

Example 3.16. An antifreeze solution for car radiators is usually about 50% by weight ethylene glycol ($C_2H_6O_2$) and 50% water. This same mixture prevents boil-over in the radiator also. Calculate the boiling point of a solution made by mixing 2.0 kg H_2O with 2.0 kg ethylene glycol. (Is this boiling point the same as if 1.0 kg of each component is mixed)?

$$K_b \text{ of } H_2O = 0.51 \,°C \text{ (or K)}/m$$

$$m = \frac{\text{mol solute}}{\text{kg solvent}} = \frac{(2.0 \times 10^3 \text{ g glycol}/ 62 \text{ g mol}^{-1})}{2.0 \text{ kg water}} = 16$$

$$\Delta T_b = K_b m$$

$$\Delta T_b = \left(\frac{0.51°C}{m}\right)(16m) = 8.2°C$$

Therefore the boiling point of the solution is elevated 8.2 °C.

$$T_b = 108.2 \,°C \text{ or } 381.2 \text{ K}$$

The freezing point of a solution is less than that of the pure solvent, again as a direct result of the vapor pressure lowering. The freezing point is where the solid vapor pressure is the same

Liquids and Solutions   Chapter 3

as the vapor pressure of the liquid phase, which corresponds on a VP vs T plot to the intersection of the solid VP curve and the liquid VP curve. Since the solution vapor pressure curve is uniformly lower than the pure solvent curve, the intersection with the solid curve is at a lower temperature (see figure 3.8 in the text). The vapor pressure of the solid doesn't change because only the solvent freezes, leaving the solute in the remaining liquid solution. This is why icebergs are fresh water, floating in sea water.

The calculation of the freezing point depression is similar to that for the change in boiling point.

$$\Delta T_f = K_f m$$

The freezing point constant, K_f, is unique for the solvent and has units of °C (or K)/m.

Example 3.17. What is the freezing point of the same antifreeze solution used in example 3.16?

$$K_f \text{ of } H_2O = 1.86 \text{ °C}/m$$

$$\Delta T_f = K_f m = 1.86°C/m(16\ m)$$

$$\Delta T_f = 29.8 \text{ °C}$$

This means the freezing point is lowered by 29.8 °C.

$$T_f = 29.8 \text{ °C or } 243.2 \text{ K}$$

The examples we have seen so far involved a non-volatile molecule that dissolves in water without any ionization. What is the effect of dissolving NaCl in water? Each mole of NaCl that dissolves yields 2 moles of solute particles Na^+ and Cl^-; each ion or particle affects the colligative properties equally; a 1 molal NaCl solution is 2 molal in particles, and the colligative properties are changed accordingly.

Example 3.18. In cold-weather climates, rock salt (NaCl) is dispersed over ice to melt it. At 0°C, 35.7g of NaCl can dissolve in 100g of H_2O. At what temperature will this solution begin to freeze?

$$\text{molality of NaCl} = \frac{35.7\ g/\ 58.5\ g\ mol^{-1}}{0.1\ kg\ H_2O} = 6.10$$

$$\text{molality of particles} = 2\ (6.10) = 12.20$$

$$\Delta T_f = K_f m = 1.86(12.20)$$

$$\Delta T_f = -22.7$$

-53-

Chapter 3 — Liquids and Solutions

$$T_f = -22.7\,°C \text{ or } 250.3\,K$$

Example 3.19. Sulfurous acid is a weak acid that only dissociates 12% in water when its concentration is 1 molal. What is the freezing point of such a solution?

$$H_2SO_3 \rightarrow H^+ + HSO_3^-$$

if 1 m H_2SO_3 is only 12% dissociated then the concentration of H^+ and of HSO_3^- will each be 0.12 m, while the H_2SO_3 remaining is 0.88 m.

$$\text{molality of particles} = m_{H_2SO_3} + m_{H^+} + m_{HSO_3^-}$$

$$m = 0.88 + 0.12 + 0.12$$

$$m = 1.12$$

$$\Delta T_f = K_f\, m = 1.86(1.12) = 2.08$$

$$T_f = -2.08\,°C$$

There is one more colligative property to be explored here. To introduce *osmotic pressure*, let's set up the following experiment: one container half-filled with pure water and one container half-filled with salt water, placed together under a sealed glass dome or bell jar. After some time has passed you will notice the volume of pure water decreasing and the volume of salt water increasing. After sufficient time, there will be only one solution, diluted salt water. Why? What drives this process? The answer is vapor pressure. The pure water and the salt water both have different vapor pressures. The two liquids can never both attain their equilibrium vapor pressures since there can't be two different P_{H_2O} values at the same time. So as the pure liquid water evaporates, the vapor pressure exceeds that of the salt water and the water vapor condenses. Eventually only one solution remains, though its volume has increased.

If the experiment is carried out where the water, instead of evaporating and condensing to pass from one solution to the other, can move through a membrane that allows the water but not the solute to flow, the water will move from the higher concentration to the lower concentration, pushing against gravity (see figure 3.11 in the text). The osmotic pressure can be calculated from the density of the solution and the height it is raised. Expressed another way, osmotic pressure is how much pressure must be exerted on a solution in order to stop the flow of water (osmosis) through the membrane. Osmotic pressure is given the symbol π and can be found by the equation below.

$$\pi = cRT$$

where c = concentration of particles in molarity

$$R = \text{ideal gas constant, } 0.082\,\frac{L-atm}{mol-K}$$

T = temperature in K

Liquids and Solutions   Chapter 3

Drinking water can be prepared from solutions unsuitable for drinking by using *reverse osmosis*. By applying more than the osmotic pressure, pure solvent can be made to flow out of a solution.

Example 3.20. A sea water sample was found to be a 0.69 M salt solution, mostly due to NaCl. Estimate π at 20°C, assuming all the salts to yield 2 particles in solution.

$$\pi = cRT = \left(2 \times 0.69 \frac{\text{mol}}{\text{L}}\right)\left(0.08206 \frac{\text{L-atm}}{\text{mol-K}}\right)(293K)$$

$$\pi = 33 \text{ atm}$$

That means that > 33 atm would be needed to reverse the direction of osmosis and get pure water from the sea water!

If an ideal solution is composed of two or more volatile components, both contribute to the vapor pressure. Dalton's law says the total vapor pressure is the sum of the partial vapor pressure.

$$P_T = P_1 + P_2 + \ldots$$

Combining this equation with Raoult's law, we have:

$$P_T = X_1 P_1^0 + X_2 P_2^0 + \ldots$$

Example 3.21. What is the vapor pressure at 50°C for a solution of 1.0 mole of carbon tetrachloride, CCl_4, and 0.60 mole of benzene, C_6H_6. Assume ideal behavior.

$$P_{C_6H_6} = 270 \text{ torr @ } 50°C$$
$$P_{CCl_4} = 305 \text{ torr @ } 50°C$$

$$P_T = P_{C_6H_6} + P_{CCl_4} = X_{C_6H_6} P^0_{C_6H_6} + X_{CCl_4} P^0_{CCl_4}$$

$$P_T = \left(\frac{0.6}{1.6}\right) 270 \text{ torr} + \left(\frac{1}{1.6}\right) 305 \text{ torr} = 292 \text{ torr}$$

3.7. Nonideal Solutions

Although the mathematical treatment of nonideal solutions is beyond the scope of this course we should keep in mind that most solutions do not behave ideally. Thus calculations of colligative properties should be handled cautiously. It can be very instructive to investigate how a solution deviates from the ideal situation. The kind of interaction that exists between solute and solvent molecules can frequently be inferred from the magnitude of negative or positive deviations from Raoult's law.

Chapter 3   Liquids and Solutions

3.8. Solubility

If a salt is *soluble* in water it will dissolve to a concentration of 0.01 M. Any salt that cannot dissolve to that extent is *insoluble*. In either instance when the maximum amount of salt is in the water the solution is said to be *saturated*. Additional salt would be present only as a *precipitate* or undissolved.

The solubility rules presented in the text may not seem to be predictable or orderly but learning them now will make your chemistry studies easier and more rewarding. Take some time and learn the seven empirical rules.

The solubility of most salts depends on temperature, but predictably so. Let's take an example of a salt, AX, that dissolves endothermically ($\Delta H^\circ_{soln} > 0$). We can include heat as a reactant in this case.

$$\text{heat} + AX_{(s)} + H_2O \rightarrow A^+_{(aq)} + X^-_{(aq)}$$

In order to produce more of the products, we need only to add more reactants. Since heat is a reactant, addition of heat, easily accomplished by raising the temperature, will result in more product. This is true for any salt with a $\Delta H^0_{soln} > 0$, raising the temperature increases its solubility.

The reverse is true of a salt that dissolves exothermically, where heat is a product.

$$BY_{(s)} + H_2O \rightarrow B^+_{(aq)} + Y^-_{(aq)} + \text{heat}$$

Now raising the temperature will not produce more product but inhibit the reaction, making the salt less soluble.

Problems

1. Benzene, C_6H_6, has a vapor pressure of 75.6 torr at 20°C. What is the normal boiling point of benzene? ($\Delta H^0_{vap} = 30.8$ kJ/mol).

2. Chlorobenzene, C_6H_5Cl, has a vapor pressure of 100 torr at 71°C and 400 torr at 110°C. What is the heat of vaporization for chlorobenzene per mole? What is the normal boiling point?

3. At what temperature will water boil on a mountain where the atmospheric pressure is only 0.692 atm? ($\Delta H^0_{vap} = 44.0$ kJ/mol)

4. What is the highest temperature at which the vapor pressure of a substance can be measured?

5. What are the molality, and mole fractions of a solution containing 31.0g of methanol, CH_3OH, in 426 mL of water?

6. An average bottle of wine contains 12% ethanol, C_2H_5OH, by volume. The density of ethanol at 20°C is 0.789 g/mL and the density of wine is 0.983 g/mL. What is the ethanol concentration expressed in molarity, molality, and mole fraction?

7. How many milliliters of 6.0 M HCl contain 275 millimoles of H^+?

Liquids and Solutions — Chapter 3

8. How much water would you add to 0.50 L of 0.42 M Na_2SO_4 to get a 0.30M solution of Na^+? Assume volumes are additive.

9. How many liters of 6.0M H_2SO_4 can be made from a 2 L bottle of concentrated H_2SO_4, 18M? Assume volumes are additive.

10. Using the following balanced net ionic equation, calculate the volume of 0.10 M $CuSO_4$ needed to completely dissolve a 27.0g piece of Zn.

$$Cu^{2+}_{(aq)} + Zn_{(s)} \rightarrow Cu_{(s)} + Zn^{2+}_{(aq)}$$

11. If 200 mL of a 1.4M solution of KI is diluted with water to 850mL, what is the molarity of KI?

12. Copper dissolves in nitric acid to produce both NO_2 and NO gases. What total volume of gas at STP can be produced from a 5.0g piece of copper in 100mL of 6.0 M HNO_3?

$$4\,Cu_{(s)} + 12HNO_{3(aq)} \rightarrow 2NO_{2(g)} + 2NO_{(g)} + 6H_2O_{(l)} + 4\,Cu(NO_3)_{2(aq)}$$

13. What is the vapor pressure at 298K of a solution containing 0.390 mole of a sugar dissolved in 500 g of H_2O? (V.P. H_2O @ 298 K = 23.8 torr)

14. What is the vapor pressure of an aqueous solution at 25°C that is 25% by weight glucose, $C_6H_{12}O_6$? (V.P. H_2O @ 25°C = 23.8 torr)

15. A solution that is 50% by weight hexane, C_6H_{14}, and 50% benzene, C_6H_6, is made. At 50°C the vapor pressure of pure hexane is 100 torr, and that of pure benzene is 260 torr. What is the mole fraction of each component? What is the total vapor pressure of the solution at 50°C? Assume ideal behavior.

16. The vapor pressure of a solution that contains 5.26 g of a non-volatile solute dissolved in 118 g of H_2O is 23.56 torr at 25°C. What is the molecular weight of the solute?

17. What is the freezing point of a solution that is 10% by weight LiBr in water?

18. Rank the following aqueous solutions from highest to lowest freezing point.

10 g CH_3OH
10 g $C_6H_{12}O_6$ (glucose)
10 g LiCl
10 g $BaBr_2$

19. A solution that contains an unknown amount of napthalene, $C_{10}H_8$, in benzene boils at 76.7°C. The normal boiling point of benzene is 80.2°C. How many grams of napthalene are contained per 100 g of benzene.

20. Sulfuric acid, H_2SO_4, completely dissociates into H^+ and HSO_4^-. The bisulfate ion, HSO_4^- dissociates an additional 10% to H^+ and SO_4^{2-}. What is the boiling point of a 1.0 m aqueous solution of sulfuric acid?

21. Human blood cells at 37°C have an average osmotic pressure of 7.7 atm. Intravenous solutions must closely match this osmotic pressure, and are called isotonic. What concentrations of glucose, $C_6H_{12}O_6$, will be isotonic with blood? What concentrations of NaCl

Chapter 3   Liquids and Solutions

will be isotonic with blood?

22. What is the minimum pressure that must be applied at 25°C to obtain pure water by reverse osmosis from an aqueous solution that is 0.26 M in NaCl and 0.063 M in $MgSO_4$?

23. The molecular weight of large molecules is frequently calculated from an osmotic pressure experiment. If 1.7g of a protein, dissolved in 100 mL of solution, has an osmotic pressure of 11.5 torr at 25°C, what is the molecular weight of the protein?

24. Can you explain, in terms of Henry's law, why champagne bubbles vigorously when opened?

25. Predict the products and indicate whether or not each product is soluble in water. All solutions are mixed in equimolar amounts.

    a) $Ba(OH)_2 + H_2SO_4$

    b) $NaCl + AgNO_3$

    c) $Pb(OAc)_2 + H_2S$

    d) $NaOH + Fe(NO_3)_3$

26. The solubility of $CaCl_2$ in water decreases as the temperature is raised. Is ΔH^0_{soln} positive or negative? $CaCl_2 + H_2O$ is used in first-aid applications. Is it contained in hot packs or cold packs?

INTERCHAPTER B
DIMENSIONAL ANALYSIS REVISITED

The latter portion of Chapter 3 deals with solution stoichiometry. Calculations of this sort lend themselves to solving by dimensional analysis. To illustrate how this is done, several of the appropriate examples from the preceding chapter will be re-worked using this technique.

Example 1. How many moles of sulfate ions are contained in 510 mL of a 0.64 M $Fe_2(SO_4)_3$ solution?

$$\frac{0.64 \text{ mol } Fe_2(SO_4)_3}{1 \text{ L}} \times 0.510 \text{ L} \times \frac{3 \text{ mol } SO_4^{2-}}{1 \text{ mol } Fe_2(SO_4)_3} = 0.99 \text{ mol } SO_4^{2-}$$

Example 2. What is the molarity of Na^+ in a solution made by mixing 433.2 mL of 1.255 M Na_3PO_4 and 1.525 L of 0.825 M NaCl? Assume volumes are additive.

$$\frac{1.255 \text{ mol } Na_3PO_4}{1 \text{ L}} \times 0.4332 \text{ L} \times \frac{3 \text{ mol } Na^+}{1 \text{ mol } Na_3PO_4} = 1.631 \text{ mol } Na^+$$

$$\frac{0.825 \text{ mol NaCl}}{1 \text{ L}} \times 1.525 \text{ L} \times \frac{1 \text{ mol } Na^+}{1 \text{ mol NaCl}} = 1.26 \text{ mol } Na^+$$

$$\frac{2.89 \text{ mol } Na^+}{(1.525 + 0.4332) \text{ L}} = 1.48 M$$

Example 3. How would you dilute 5.0 mL of 6.0 M HCl to make 0.10 M HCl?

$$\frac{6.0 \text{ mol HCl}}{1 \text{ L}} \times 5.0 \times 10^{-3} \text{ L} \times \frac{1 \text{ L}}{0.10 \text{ mol HCl}} = 0.30 \text{ L of soln.}$$

This means 0.295L or 295 mL of water was added to the 5.0 mL.

Example 4. What volume of 0.238M H_3PO_4 is required to neutralize (completely react the H^+ and OH^-) 0.492 g of $Ca(OH)_2$?

$$2H_3PO_4 + 3Ca(OH)_2 \rightarrow 6H_2O + Ca_3(PO_4)_2$$

$$0.492 \text{g } Ca(OH)_2 \times \frac{1 \text{ mol } Ca(OH)_2}{74.1 \text{ g}} \times \frac{2 \text{ mol } H_3PO_4}{3 \text{ mol } Ca(OH)_2} \times \frac{1 \text{ L}}{0.238 \text{ mol } H_3PO_4}$$

$$= 0.0186 \text{ L or } 18.6 \text{ mL.}$$

CHAPTER 4
CHEMICAL EQUILIBRIUM

4.1. The Nature of Chemical Equilibrium

Suppose we put some pure dinitrogen tetroxide gas, N_2O_4, in a sealed glass container attached to a manometer. Initially the gas sample is colorless and clear, but in a very short time we will notice that the pressure in the vessel increases and that the sample becomes red-brown in color. Obviously a reaction is taking place, in this case $N_2O_{4(g)} \rightarrow 2\ NO_{2(g)}$. As the N_2O_4 disappears and the colored NO_2 appears, the increase in gas moles raises the pressure. Eventually the reaction will appear to halt, there will be no more change in color or pressure.

What if the reverse reaction is carried out? If we could put pure nitrogen dioxide in a vessel we would find that the color fades and the pressure decreases with time. This is due to the reaction $2\ NO_{2(g)} \rightarrow N_2O_{4(g)}$. Eventually no more macroscopic changes will take place in this system either. But what we have done is approach the same reaction from two different directions. We can combine the two chemical equations with uni-directional arrows into one equation.

$$N_2O_4 \rightleftarrows 2\ NO_2$$
or
$$2\ NO_2 \rightleftarrows N_2O_4$$

In either case, when there is no further *net* change we say the system has reached *equilibrium*. At equilibrium the concentrations of the products and reactants no longer change because the forward reaction proceeds at exactly the same rate as the reverse reaction. This dynamic condition means that molecules are still converted from product to reactant and vice versa with no observable change. We are unable to predict at this point how long it takes for a system to reach equilibrium. The NO_2/N_2O_4 system described needs only a few seconds at room temperature to arrive at equilibrium but other reactions may take days, years or only microseconds to do the same.

The equilibrium condition is the most stable situation, thus *the spontaneous direction of a reaction is toward equilibrium, regardless of the starting point.* As an analogy (a crude one at best) picture two hills with a valley between them. A ball rolled down hill #1 travels part of the distance to hill #2 but finally will come to rest in the valley. A ball rolled from hill #2 will travel part of the distance to hill #1 but will eventually come to rest in the same valley. Similarly, starting a reaction with pure reactants, or with pure products, will eventually lead to the same equilibrium condition because it is the most stable situation no matter how it is approached.

4.2 The Equilibrium Constant

Upon reaching equilibrium there exists a particular relationship between the quantities of reactants and products, called the *equilibrium constant*, K. The equilibrium constant is all the product concentrations at equilibrium multiplied together divided by all the reactant concentrations multiplied together. Let's use $N_2O_4 \rightleftarrows 2\ NO_2$ as our first example.

Chemical Equilibrium — Chapter 4

$$K_c = \frac{[NO_2][NO_2]}{[N_2O_4]} = \frac{[NO_2]^2}{[N_2O_4]}$$

where K_c = equilibrium constant based on concentrations

[ ] = concentration in molarity

Since there were two NO_2 moles produced as products the $[NO_2]$ was included twice in the equilibrium constant. This is the same as raising the $[NO_2]$ to a power equal to its stoichiometric coefficient. For reactions of gases, K can also be expressed in pressure units. The values of K_p and K_c are *not* equal for most reactions but they are related.

$$K_p = \frac{(P_{NO_2})^2}{P_{N_2O_4}}$$

$$K_p = K_c (RT)^{\Delta n}$$

or

$$K_c = \frac{K_p}{(RT)^{\Delta n}}$$

where Δn = the change in gas moles, n (products) - n (reactants) You will note that if $\Delta n = 0$ (that is, there are the same number of gas moles on the product and reactant sides of the equation) then $K_c = K_p$.

Pure solids and liquids that are involved in a reaction have a constant concentration in themselves. For instance the concentration of $H_2O_{(l)}$ is 55.5 mol/L (calculate it for yourself). Because this concentration doesn't change, or changes very little, it is also treated as part of the numerical value of K, and not as a variable.

Example 4.1. What is the form of K for the following balanced reaction?

$$H_2 + CO_2 \rightleftarrows CO + H_2O$$

$$K_c = \frac{[CO][H_2O]}{[H_2][CO_2]} \quad \text{or} \quad K_p = \frac{(P_{CO})(P_{H_2O})}{(P_{H_2})(P_{CO_2})}$$

In this case the H_2O is a gas, not a solvent, and so is included in K.

Chapter 4 Chemical Equilibrium

Example 4.2. What is the form of K for the following balanced reaction?

$$H^+ + OH^- \rightleftharpoons H_2O$$

$$K_c = \frac{1}{[H^+][OH^-]}$$

where K_c already includes the $[H_2O]$.

Example 4.3. What is the form of K for the following balanced reaction, and how does it compare with that in Example 4.2?

$$H_2O \rightleftharpoons H^+ + OH^-$$

$$K = [H^+][OH^-]$$

This K is the inverse of that in example 4.2, because the equation is reversed.

Example 4.4. How does K change if a reaction is multiplied by 2?

(1) $1/2\ A + B \rightleftharpoons 2C$

$$K_{c_1} = \frac{[C]^2}{[A][B]}$$

(2) $A + 2B \rightleftharpoons 4C$

$$K_{c_2} = \frac{[C]^4}{[A][B]^2}$$

The relationship between K_{c1} and K_{c2} is:

$$(K_{c_1})^2 = K_{c_2} \text{ or } K_{c_1} = \sqrt{K_{c_2}}$$

Example 4.5. Find K for $3\ O_{(g)} \rightleftharpoons O_{3(g)}$ in terms of K_1 and K_2 below.

(1) $2O_{(g)} \rightleftharpoons O_2$ $K_1 = \dfrac{P_{O_2}}{(P_o)^2}$

(2) $3O_{2(g)} \rightleftharpoons 2O_{3(g)}$ $K_2 = \dfrac{(P_{O_3})^2}{(P_{O_2})^3}$

Chemical Equilibrium Chapter 4

We must algebraically manipulate the chemical equations so that we can add them up to get the desired reaction.

$$3/2 \ (2\ O \rightleftarrows O_2) \ (K_1)^{\frac{3}{2}}$$

$$1/2 \ (3\ O_2 \rightleftarrows 2\ O_3) \ (K_2)^{\frac{1}{2}}$$

$$3\ O + 3/2\ O_2 \rightleftarrows 3/2\ O_2 + O_3$$

$$3\ O \rightleftarrows O_3 \quad K = (K_1)^{\frac{3}{2}} \times (K_2)^{\frac{1}{2}}$$

When two reactions are added, the equilibrium constants are multiplied. Let's double-check to be sure this is the case.

$$K = \frac{P_{O_3}}{(P_O)^3} = \left\{\frac{P_{O_2}}{(P_O)^2}\right\}^{\frac{3}{2}} \times \left\{\frac{(P_{O_3})^2}{(P_{O_2})^3}\right\}^{\frac{1}{2}}$$

$$K = (K_1)^{\frac{3}{2}} \times (K_2)^{\frac{1}{2}}$$

Indeed, this gives the correct result.

Example 4.6. A steel reaction vessel contains the following equilibrium partial pressures:

$P_{H_2} = 0.050$ atm, $P_{N_2} = 0.120$ atm, $P_{NH_3} = 3.20$ atm, at 298K. What are the values of K_p and K_c for $N_{2(g)} + 3\ H_{2(g)} \rightleftarrows 2\ NH_{3(g)}$ at this temperature?

$$K_p = \frac{(P_{NH_3})^2}{(P_{N_2})(P_{H_2})^3} = \frac{(3.20)^2}{(0.120)(0.050)^3} = 6.8 \times 10^5$$

The value of K_c can be found in two ways. Each partial pressure in atm can be converted to concentration in mol/L using the ideal gas law so we can use the relationship:

$$K_c = \frac{K_p}{(RT)^{\Delta n}}$$

$$K_c = \frac{6.8 \times 10^5}{(0.0821 \times 298)^{-2}} = 4.1 \times 10^8$$

Neither K value has units associated with it because each partial pressure or concentration used is really a unitless activity (see Ch. 3).

Chapter 4   Chemical Equilibrium

The value of K for a reaction is a constant that depends only on the temperature, but what does it mean if a reaction has a very large or a very small value of K? The significance of the equilibrium constant value is easy to discern in terms of products and reactants. For example, the following reaction has a large K.

$$N_{2(g)} + 3H_{2(g)} \rightarrow 2NH_{3(g)} \quad K_p = 6.8 \times 10^5$$

This means that, at equilibrium, there is a greater quantity of products than reactants. The NH_3 is favored at equilibrium over N_2 and H_2.

Just the opposite is the case for the dissolving of AgCl in water.

$$AgCl_{(s)} \rightarrow Ag^+_{(aq)} + Cl^-_{(aq)} \quad K_c = 1.7 \times 10^{-10}$$

In this instance, K is very small indicating there are only small quantities of the products at equilibrium. Most of the AgCl is not dissolved.

We can generalize that when $K \gg 1$ the products are favored at equilibrium, and little of the reactants remain.

K $\ll$ 1   there is little product formation, and the reactants are favored at equilibrium

K $\simeq$ 1   approximately comparable amounts of products and reactants exist at equilibrium

---

Example 4.7. Which salt dissolves to the greater extent in water at 25 °C, $CaSO_4$ or $BaSO_4$?

$$BaSO_{4(s)} \rightleftarrows Ba^{2+}_{(aq)} + SO_4^{2-}_{(aq)} \quad K = 1.5 \times 10^{-9}$$

$$CaSO_{4(s)} \rightleftarrows Ca^{2+}_{(aq)} + SO_4^{2-}_{(aq)} \quad K = 2.4 \times 10^{-5}$$

$$K_{BaSO4} = [Ba^{2+}][SO_4^{2-}]$$

$$K_{CaSO4} = [Ca^{2+}][SO_4^{2-}]$$

Since K for $BaSO_4$ is less than K for $CaSO_4$ we know that there must be less $BaSO_4$ dissolved at equilibrium than $CaSO_4$. This is an easy comparison since both reactions involve the same numbers of moles of products and reactants but the comparison gets a bit more complicated if there are squared or cubed terms in K, as we shall see.

Remember that equilibrium is a condition that can be approached from the forward or the reverse direction. K is a true constant because the same relative amounts of products and reactants exist at equilibrium regardless of the starting point. But how can we predict the direction of a reaction from any given starting point? There are two extreme cases where this prediction is easy, starting with only pure reactants or starting with only pure products.

Let's illustrate this with the reaction

$$N_2O_{4(g)} \rightleftarrows 2NO_{2(g)}$$

$$K_p = \frac{(P_{NO_2})^2}{P_{N_2O_4}} = 80 \text{ at } 100°C$$

If we start with only a quantity of N_2O_4, and no NO_2, then the reaction *must* go forward to reach equilibrium. However if we started only with NO_2, and no N_2O_4, the reaction must proceed in the reverse direction in order to satisfy the equilibrium constant. But what about a starting condition of $P_{N_2O_4} = P_{NO_2} = 1.0$ atm? Logically we insert these non-equilibrium values into K_p to see if they satisfy the constant. If non-equilibrium values are used, we call the reaction quotient Q.

$$Q = \frac{(P_{NO_2})^2}{P_{N_2O_4}}$$

Q may contain any set of values for NO_2 and N_2O_4 pressures, while K may only contain equilibrium values. If $P_{NO_2} = P_{N_2O_4} = 1.0$ atm, then Q = 1. Comparing Q and K we see that we need to increase the P_{NO_2} relative to the $P_{N_2O_4}$ in order to move toward equilibrium. Therefore the reaction must proceed in the forward direction in order to reach equilibrium. The relationship between Q and K can be summarized and generalized:

if Q < K the reaction must proceed forward in order to reach equilibrium (left to right)

if Q > K the products are in excess the reaction must proceed in reverse to reach equilibrium (right to left)

if Q=K the reaction is at equilibrium

Example 4.8. A certain reaction $A + 2B \rightleftarrows 3C$ has an equilibrium constant, K_c, of 5×10^{-2} at 25°C. If the starting concentrations are [A] = 0 M and [B] = 0.10 M will the reaction proceed spontaneously?

$$Q = \frac{[C]^3}{[A][B]^2} = \frac{(0.08)^3}{(0)(0.10)^2} \rightleftarrows \text{ because of 0 in the denominator}$$

Q > K so the reaction must proceed from right to left before it reaches equilibrium.

4.3 External Effects on Equilibrium

Equilibrium is a dynamic, reversible state. If a change or stress is made to a system at equilibrium, the system will react so as to relieve the stress and re-establish equilibrium. This is called *Le Chatelier's principle*. A non-scientific analogy might be a spring. When not connected to anything nor stressed in any way, the coils of a spring are in an equilibrium position. Stress can be applied by compressing the spring, which offsets this stress by closing the gaps

Chapter 4 — Chemical Equilibrium

between the coils - a new equilibrium is established. Removing the stress results in the previous equilibrium state.

Stress can be applied to a chemical system by changing the volume, partial pressures, or concentrations of reactants. Any of these changes may lead to a shift in the direction of reaction but will result in equilibrium being re-established. The value of K remains the same in spite of these changes.

A temperature change will also shift the direction of a reaction, however K depends on temperature and the new equilibrium constant will be different. Let's follow the logic behind how all these changes can affect a reaction at equilibrium.

The sample system we shall investigate is:

$$O_{2(g)} + 2NO_{(g)} \rightleftharpoons 2NO_{2(g)} + \text{heat}$$

$$K_p = \frac{(P_{NO_2})^2}{(P_{O_2})(P_{NO})^2}$$

We shall consider the system to be at equilibrium ($Q = K$). Now, what will happen if we add more O_2 thereby raising the P_{O_2}? Q becomes ss than K because P_{O_2} is in the denominator. The reaction must shift from left to right to compensate for this change and get back to equilibrium. Adding quantities of any compound on the left-hand side of the equation will shift the reaction from left to right provided the compound is not a pure solid or liquid. (Solids and liquids don't change in concentration and don't appear in the reaction quotient. Therefore the addition or removal of solids or liquids does not stress the system, except for H_2O as seen in a later example.)

What if the total volume of the equilibrium system above is increased? Providing the temperature remains constant, increasing the volume leads to a corresponding decrease in each partial pressure. For instance, if the volume is doubled each partial pressure is cut in half. In this particular example it means that both the denominator and numerator get smaller. Q is now greater than K because the denominator is affected more due to its greater dependence on partial pressures. Therefore the reaction must shift from right to left to re-establish equilibrium. A more simplistic way of looking at this volume effect is to recognize that if a greater volume is available to a gas system, the side of the equation with more gas moles will be favored. Those systems that don't include gases and don't have equal moles of gases on both sides of the equation show no volume effects.

What about temperature changes? Look back at the balanced equation we are dealing with; heat appears as a product since the reaction is exothermic. By analogy with the other changes we have investigated, we can see that by raising the temperature of the system (ie. adding heat) we essentially are adding extra product. This results in a shift in the direction of reaction from right to left. This time K actually decreases in value, so when the new equilibrium is established it satisfies a different equilibrium constant value. All exothermic reactions have the same response: K decreases as T increases or K increases as T decreases. All endothermic reactions show the opposite trends: K increases as T increases or K decreases as T decreases. Table 4.1 gives a summary of the effects due to changes on our sample system.

Chemical Equilibrium Chapter 4

Table 4.1. $O_{2(g)} + 2NO_{(g)} \rightleftarrows 2NO_{2(g)}$ + heat

Change Made	Direction the Reaction Must Proceed to Return to Equilibrium	Effect on K
add O_2 or NO	→	none
add NO_2	←	none
remove O_2 or NO	←	none
remove NO_2	→	none
raise the temperature	←	decreases
lower the temperature	→	increases
increase the volume	←	none
decrease the volume	→	none
increase pressure by adding inert gas	no shift	none

One change that can be made but hasn't been mentioned yet is the addition of a catalyst. A catalyst speeds up both the forward and reverse reactions so that equilibrium is established more quickly. *A catalyst does not change the position of equilibrium nor K.*

Example 4.9. Predict how the following equilibrated system will change in order to reduce the following stresses.

$$PbCl_{2(s)} \rightleftarrows Pb^{2+}_{(aq)} + 2\, Cl^-_{(aq)}$$

$$\Delta H° = K = [Pb^{2+}][Cl^-]^2$$

a) adding more $PbCl_{2(s)}$
b) warming the system
c) diluting the system with more water

Because the solid $PbCl_2$ doesn't appear in Q or K, changing the amount of the salt has no effect, provided that some is present. The answer for a) is no effect.

Since the reaction is endothermic, the addition of heat should raise the value of K. Indeed, it is easily shown that more $PbCl_2$ dissolves in warm water than in cold water (part b).

Part c) is a bit tricky. Since water is neither a product nor a reactant and doesn't appear in Q or K you might say that adding water has no effect. But remember that the ion concentrations, in mol/L, will decrease as water is added. So Q < K and the reaction shifts →.

Surely this makes qualitative sense since more water should be able to dissolve more salt.

4.4 Free Energy and Equilibria in Nonideal Solutions

We saw in an earlier chapter that the spontaneity of a reaction depends on two thermodynamic factors, enthalpy and entropy. These two factors may reinforce each other or vie for control of a reaction. A decrease in enthalpy is favorable and an increase in entropy is favorable. The combination of these two terms gives us a criterion for spontaneity called *free energy*. At constant temperature and pressure (the most common conditions) the free energy is represented as G and the change in free energy as ΔG. A system moves toward a lowering of free energy so $\Delta G = 0$ describes a spontaneous reaction while $\Delta G = -$ applies to non-spontaneous reactions. Eventually a reaction will reach the free energy minimum which means it is at equilibrium. The relationship between K and ΔG^0 (standard state free energy change) is logarithmic.

$$\Delta G^0 = RT \ln K$$

or

$$K = e^{-\Delta G^0 / RT}$$

This means that when K>1, $\Delta G^0 = 0$ and when K<1, $\Delta G^0 = +$. Occasionally K = 1 and $\Delta G^0 = 0$.

Free energy will be discussed in much greater detail in Chapter 8, but as with ΔH^0 values, ΔG^0 for a reaction can be calculated from tabulated values at 25 °C.

$$\Delta G^0_{rxn} = \sum \Delta G^0_f \text{ (products)} - \sum \Delta G^0_f \text{(reactants)}$$

This is a convenient way to find equilibrium constants.

Try not to lose sight of the fact that solutions behave ideally only at very low concentrations, usually less than 0.1 M. At higher concentrations ionic solutions can only be approximated by our treatment of equilibria. Consideration of all the possible errors and the corresponding correction factors for nonideal behavior is beyond the scope of this text.

4.5 Calculations Using the Equilibrium Constant

The remainder of this chapter will, through examples, show how to work with K values and partial pressures mathematically. Frequently we shall be interested in knowing what is the extent of reaction, f, which is also called the fraction of reaction. We will also use reaction stoichiometry extensively. Problems involving solutions will be treated quantitatively in the next chapter while gas phase reactions will be covered below.

There are several different methods, all correct, for solving equilibrium problems. If a student fully understands the material, all methods are equally easy to use. The beginning student, however usually needs an organized approach to problem solving. For this reason we will make use of the tabular method of arranging data. You may or may not feel the need to solve problems this way; the option is yours. It will nonetheless be instructive to follow the procedure through several examples to acquaint yourself with it.

Example 4.10. The reaction $Cl_{2(g)} \rightleftarrows 2Cl_{(g)}$ has a $K_p = 2.48 \times 10^{-5}$ at 1200 K. If 1.10 atm of

Chemical Equilibrium   Chapter 4

Cl_2 is introduced into an empty reaction vessel at 1200 K, what are the partial pressures of Cl_2 and Cl at equilibrium? What fraction of Cl_2 has dissociated?

The table of data we construct will necessarily involve some unknowns. Beneath the balanced equation, we will tabulate how much of each substance we have at the start, how much change and in which direction, and finally the equilibrium quantities, which are simply the sum of the starting quantity and the change.

	$Cl_{2(g)}$	$\rightleftarrows$	2 $Cl_{(g)}$
start	1.10 atm		0 atm
change	-x		+2x
at equilibrim	(1.10-x) atm		2x atm

We don't know how much Cl_2 dissociates, (thus x is used) but we do know that for every Cl_2 that reacts, 2 Cl will be formed, or 2x. We know the reaction must proceed left to right in order to reach equilibrium because $Q_{start} > K$. We can use the equilibrium values to substitute in K, whose value we are given.

$$K = \frac{(P_{Cl})^2}{P_{Cl_2}} \qquad 2.48 \times 10^{-5} = \frac{(2x)^2}{(1.10 - x)}$$

Now we have a quadratic equation which can be algebraically expanded and x can be found using the quadratic formula. We can save ourselves some time and reduce our chance of making errors by recognizing that since $K \ll 1$, very little Cl_2 must be dissociated at equilibrium. Thus x is small and 1.10-x is nearly 1.10.

$$\text{since } K \ll 1, \quad x \ll 1.10$$
$$\text{so } 1.10 - x \simeq 1.10$$

Now the equation becomes easier to solve:

$$2.48 \times 10^{-5} = \frac{(2x)^2}{1.10} = \frac{4x^2}{1.10}$$

$$2.61 \times 10^{-3} = x$$

We can find the partial pressures at equilibrium:

$$P_{Cl_2} = 1.10 - x \simeq 1.10 \text{ atm}$$

$$P_{Cl} = 2x = 5.22 \times 10^{-3} \text{ atm}$$

You can see that our simplification was acceptable because 1.10- x does equal 1.10 to the allowed number of significant figures. You should also double check that these partial pressures give the measured K value of 2.48×10^{-5}. We still have one last calculation to do. The fraction of Cl_2 that dissociated will be very small, but it can be found.

$$f = \frac{\text{amount dissociated}}{\text{starting amount}} = \frac{x}{1.10} = 0.00237$$

Chapter 4											Chemical Equilibrium

Example 4.11. A mixture of 0.350 atm of NO and 0.300 atm of Cl_2 is prepared at 500 K. The reaction proceeds according to the balanced equation below. At equilibrium the total pressure is 0.520 atm. What is K_p at 500 K for this reaction?

	$2NO_{(g)}$	$\rightleftarrows$	$Cl_{2(g)}$	+	$2NOCl_{(g)}$
start	0.350		0.300		0
change	$-2x$		$-x$		$+2x$
at equilibrium	$(0.350-2x)$		$(0.300-x)$		$2x$

$$P_T = 0.520 \text{ atm} = (0.350-2x) + (0.300-x) + 2x$$

$$0.130 = x$$

$$P_{NO} = (0.350 - 2x) = 0.090 \text{ atm}$$

$$P_{Cl_2} = (0.300 - x) = 0.170 \text{ atm}$$

$$P_{NOCl} = 2x = 0.260 \text{ atm}$$

$$K_p = \frac{(P_{NOCl})^2}{(P_{NO})^2 (P_{Cl_2})} = \frac{(0.260)^2}{(0.090)^2 (0.170)} = 49.1$$

Example 4.12. If 1.50 atm of $PCl_{5(g)}$ is introduced into a reaction flask at 573 K, 89.3% of the PCl_5 dissociates to $PCl_{3(g)}$ and $Cl_{2(g)}$. What is K_p for the reaction at this temperature?

$$PCl_{5(g)} \rightleftarrows PCl_{3(g)} + Cl_{2(g)}$$

	$PCl_{5(g)}$	$\rightleftarrows$	$PCl_{3(g)}$	+	$Cl_{2(g)}$
start	1.50		0		0
change	$-0.893(1.50)$		$+(0.893)1.50$		$+0.893(1.50)$
at equilibrium	0.16		1.34		1.34

$$K_p = \frac{(P_{Cl_2})(P_{PCl_3})}{(P_{PCl_5})} = \frac{(1.34)^2}{(0.16)} = 11.2$$

Note: Don't sacrifice common sense in order to fit a problem to your solving technique. This can be more difficult than it sounds. Let's see what difficulties might arise and how common sense will help extricate you.

Example 4.13. At some temperature $K_p = 6.25 \times 10^9$ for the reaction below. If at this

Chemical Equilibrium — Chapter 4

temperature, 0.30 atm of Cl_2 and 0.81 atm of F_2 are mixed together, what are all the partial pressures at equilibrium?

$$Cl_{2(g)} + 3F_{2(g)} \rightleftarrows 2ClF_{3(g)}$$

	$Cl_{2(g)}$	$+3F_{2(g)} \rightleftarrows$	$2\,Cl\,F_{3(g)}$
initial	0.30	0.81	0
change	−x	−3x	+2x
at equilibrium	(0.30−x)	(0.81−3x)	2x

$$K_p = \frac{(P_{ClF_3})^2}{(P_{Cl_2})(P_{F_2})^3}$$

$$6.25\times 10^9 = \frac{(2x)^2}{(0.30 - x)(0.81 - 3x)^3}$$

Now we have a problem. Since K >> 1 we cannot assume that x << 0.30 or 0.80. Therefore we will have an equation of the form $ax^4+bx^3+cx^2+dx+e=0$ to solve. It doesn't sound very inviting does it? But since K_p is so very large, maybe we shouldn't use this approach at all. A large equilibrium constant means that there will be mainly products at equilibrium, and little of any reactants left. In other words, the reaction goes virtually to completion. The easier approach to this problem is to treat the reaction as going to completion, then reversing just a little in order to satisfy K_p.

$$Cl_2 + 3\,F_2 \rightleftarrows 2\,Cl\,F_3$$

	Cl_2	$+3F_2 \rightleftarrows$	$2ClF_3$
start	0.30	0.81	0
change	−0.27	−0.81	+0.54
at completion	0.03	0	0.54

Since F_2 is the limiting reagent, only 0.27 atm of the Cl_2 can be used. We have no F_2, which means that Q > K, and the reaction must proceed from right to left.

$$Cl_2 + 3\,F_2 \rightleftarrows 2\,ClF_3$$

Chapter 4 　　　　　　　　　　　　　　　　　　　　　　　　　　　　Chemical Equilibrium

	Cl_2	$+3F_2$	$\rightleftharpoons$ $2ClF_3$
new start	0.03	0	0.54
change	+x	+3x	-2x
at equilibrium	(0.03+x)	3x	(0.54-2x)

$$6.25 \times 10^9 = \frac{(0.54-2x)^2}{(0.03+x)(3x)^3}$$

The equation still seems messy but now since we know the reaction can only satisfy K_p if it proceeds slightly in the reverse direction, we can assume 2x << 0.54 and x << 0.03.

$$6.25 \times 10^9 = \frac{(0.54)^2}{(0.03)(3x)^3}$$

$$5.77 \times 10^{-11} = x^3$$

$$3.86 \times 10^{-4} = x$$

$$P_{Cl_2} = 0.03 + x \simeq 0.03 \text{ atm}$$

$$P_{F_2} = 3x = 1.16 \times 10^{-3} \text{ atm}$$

$$P_{ClF_3} = 0.54 - 2x \simeq 0.54 \text{ atm}$$

Problems

1. For each of the following balanced equations write the form of the equilibrium constant.

    a) $2N_2O_{(g)} + 3O_{2(g)} \rightleftharpoons 4NO_{2(g)}$

    b) $2Cl^-_{(aq)} + 4H^+_{(aq)} + MnO_{2(s)} \rightleftharpoons Cl_{2(g)} + 2H_2O_{(l)} + Mn^{2+}_{(aq)}$

    c) $2SO_{2(g)} + O_{2(g)} \rightleftharpoons 2SO_{3(g)}$

    d) $SO_{2(g)} + 1/2\, O_{2(g)} \rightleftharpoons SO_{3(g)}$

    e) $SO_{3(g)} \rightleftharpoons SO_{2(g)} + 1/2\, O_{2(g)}$

    f) $4PCl_{3(g)} \rightleftharpoons P_{4(s)} + 6Cl_{2(g)}$

    g) $2Al(OH)_{3(s)} \rightleftharpoons Al_2O_{3(s)} + 3H_2O_{(g)}$

    h) $Br_{2(l)} \rightleftharpoons Br_{2(g)}$

2. $K_p = 50.1$ for the following reaction. What is K_c?

$$2NO_{(g)} + Cl_{2(g)} \rightleftharpoons 2NOCl_{(g)}$$

-72-

Chemical Equilibrium   Chapter 4

3. If $K_p = 3.8 \times 10^{-2}$ for $2NH_{3(g)} \rightleftarrows N_{2(g)} + 3H_{2(g)}$ what is K_p, at the same temperature for $N_{2(g)} + 3/2 H_{2(g)} \rightleftarrows NH_{3(g)}$?

4. Nitrogen dioxide can be made from any of the three equations below at 25°C. If all three processes are assumed to come to equilibrium fast when started with 1.0 atm of all reactants, which reaction will produce the most NO_2?

    a) $N_2O_{4(g)} \rightleftarrows 2NO_{2(g)}$ $K_p = 0.20$

    b) $2NO_{(g)} + O_{2(g)} \rightleftarrows 2NO_{2(g)}$ $K_p = 4.2 \times 10^{12}$

    c) $N_{2(g)} + 2O_{2(g)} \rightleftarrows 2NO_{2(g)}$ $K_p = 1.3 \times 10^{-18}$

5. For the three equations in problem #4 above, what are the values of K_c at 25°C?

6. The equilibrium constant for the reaction $H_{2(g)} + I_{2(g)} \rightleftarrows 2HI_{(g)}$ was measured at two different temperatures. At T_1, $K = 46$ and at T_2, $K = 617$. If $T_2 < T_1$, what is the sign of ΔH^0 for this reaction?

7. Refer to the balanced equation below and predict, using Le Chatelier's principle what effect, if any, the following changes will make on the system at equilibrium.

    $$2Cl^-_{(aq)} + 4H^+_{(aq)} + MnO_{2(s)} \rightleftarrows Cl_{2(g)} + 2H_2O_{(l)} + Mn^{2+}_{(aq)} + \text{heat}$$

    a) increase $[Cl^-]$
    b) add MnO_2
    c) remove Cl_2
    d) add water
    e) lower the temperature

8. Calculate K_p for $C_{(s)} + CO_{2(g)} \rightleftarrows 2CO_{(g)}$ from the following data.

    $C_{(s)} + 2H_2O_{(g)} \rightleftarrows CO_{2(g)} + 2H_{2(g)}$ $K_1 = 3.8$

    $H_2O_{(g)} + CO_{(g)} \rightleftarrows H_{2(g)} + CO_{2(g)}$ $K_2 = 1.4$.

    Assume no temperature change.

9. What are K_c and K_p for the reaction $CaCO_{3(s)} \rightleftarrows CaO_{(s)} + CO_{2(g)}$ if at equilibrium 0.20 moles of $CaCO_3$, 0.10 moles of CaO and 0.012 moles of CO_2 are present in a 10L flask at 1000 K?

10. For the reaction $I_{2(g)} \rightleftarrows 2I_{(g)}$, $K_p = 3.08 \times 10^{-3}$ at 1000 K. Suppose you inject 0.50 mole of I_2 into a 2.0 liter box at 1000 K. What will be the final equilibrium pressures of I_2 and I?

11. 2.00 atm of NH_3 is introduced into a reaction flask and allowed to undergo partial decomposition at high temperature to N_2 and H_2. At equilibrium, 1.00 atm of NH_3 remains. What is K_p?

Chapter 4   Chemical Equilibrium

$$2NH_{3(g)} \rightleftarrows N_{2(g)} + 3H_{2(g)}$$

12. At a particular temperature a flask was filled with 1.75 atm of pure N_2O_4. After equilibration it was found that 32% of the N_2O_4 had dissociated. What is K_p for the reaction?

$$N_2O_{4(g)} \rightleftarrows 2NO_{2(g)}$$

13. The reaction $I_{2(g)} + Br_{2(g)} \rightleftarrows 2IBr_{(g)}$ has a K = 4.0 at 100°C. If an equal number of moles of I_2 and Br_2 are placed in a flask so that the total pressure is 60 torr, what will be the equilibrium vapor pressures of all substances?

14. Calcium bicarbonate decomposes when heated, according to the following equation.

$$Ca(HCO_3)_{2(s)} \rightleftarrows CaO_{(s)} + H_2O_{(g)} + 2CO_{2(g)}$$

At some high temperature the equilibrium vapor pressure of the water is 0.15 torr. What is the partial pressure of CO_2? What is K_p? If at this same temperature the P_{CO_2} is held constant at 250 torr, what would be the P_{H_2O} value?

15. Consider the following reaction and equilibrium constant at 723 K.

$$3H_{2(g)} + N_{2(g)} \rightleftarrows 2NH_{3(g)}$$

$K_p = 4.28 \times 10^{-5}$

If the following mixture is made up, what will happen?

$$P_{H_2} = 2.00 \text{ atm}, P_{N_2} = 5.00 \text{ atm}, P_{NH_3} = 0.100 \text{ atm}.$$

Is there a reaction and if so, in which direction?

16. The air we breathe is approximately 79% N_2 and 21% O_2. A primary air pollutant is NO, produced mainly in car exhaust.

$$N_{2(g)} + O_{2(g)} \rightleftarrows 2NO_{(g)}$$

At 25°C, $K_p = 3 \times 10^{-31}$ and at 2000°C (temperature attained in an internal combustion engine) $K_p = 4 \times 10^{-4}$.

a) What is the equilibrium partial pressure of NO at 25°C?
b) What is the equilibrium partial pressure of NO at 2000°C?
c) Is the reaction endothermic or exothermic?

17. The water-gas shift reaction below is carried out at a temperature where all the components are gases. An equilibrium mixture has $P_{CO_2} = 0.83$ atm, $P_{H_2} = 1.10$ atm, $P_{CO} = 0.27$ atm, $P_{H_2O} = 0.10$ atm. An additional 1.0 atm of CO is injected into the reaction flask. What are the new equilibrium partial pressures?

$$CO_{(g)} + H_2O_{(g)} \rightleftarrows CO_{2(g)} + H_{2(g)}$$

CHAPTER 5
IONIC EQUILIBRIA IN AQUEOUS SOLUTIONS

5.1. Sparingly Soluble Salts

Many salts, like $NaCl$ or KNO_3 dissolve readily in water and their solubility is given in g/L. There are also a great many salts that do not dissolve appreciably in water (see the solubility rules of Chapter 3) like $AgCl$ or $PbSO_4$. Because these sparingly soluble salts have such low solubilities in g/L, chemists usually refer to the corresponding equilibrium constant of solubility, K_{sp} or as it is commonly called, the *solubility product*. The form of K_{sp} is written in the same manner as any other equilibrium constant.

$$Ca_3(PO_4)_{2(s)} \rightleftarrows 3Ca^{2+}_{(aq)} + 2PO^{3-}_{4(aq)}$$

$$K_{sp} = [Ca^{2+}]^3 [PO_4^{3-}]^2$$

Because the sparingly soluble salt is a solid with a constant concentration, it is already included in the K_{sp} value. As you might expect, sparingly soluble salts have K_{sp} values $<<1$, indicating only small amounts dissolve.

Word problems involving solubility products are worked similarly to the other equilibrium problems we have already dealt with, and hopefully mastered. The problems are, however, likely to be worded differently and the value of the solubility of the salt in g/L or mol/L is commonly used.

Example 5.1. 100 mL of water will dissolve 3.0 mg of $PbSO_4$ at 25 °C. What is the K_{sp} of $PbSO_4$ at this temperature?

$$PbSO_{4(s)} \rightleftarrows Pb^{2+}_{(aq)} + SO^{2-}_{4(aq)}$$

$$K_{sp} = [Pb^{2+}][SO_4^{2-}]$$

From the stoichiometry we can see that every mole of $PbSO_4$ that dissolves yields 1 mole of Pb^{2+} and 1 mole of SO_4^{2-}.

$$\frac{3.0 \text{ mg } PbSO_4 \times 1 \text{ g}/ 10^3 \text{mg}}{303 \text{ g } PbSO_4/ \text{mol}} = 9.9 \times 10^{-6} \text{ mol } PbSO_4$$

9.9×10^{-6} mol $PbSO_4$ dissolved → 9.9×10^{-6} mol Pb^{2+} and
9.9×10^{-6} mol SO_4^{2-}

Chapter 5 — Ionic Equilibria in Aqueous Solutions

$$[Pb^{2+}] = [SO_4^{2-}] = \frac{9.9 \times 10^{-6} \text{mol}}{0.100 \text{ L}} = 9.9 \times 10^{-5} \text{ M}$$

$$K_{sp} = [Pb^{2+}][SO_4^{2-}] = (9.9 \times 10^{-5})^2 = 9.8 \times 10^{-9}$$

Example 5.2. The solubility of $Ca_3(PO_4)_2$ in pure water is 9.9×10^{-5} M at 25 °C. What is the K_{sp}?

This problem can be worked with an equilibrium table if you desire, with the initial condition being just the salt and water, prior to any dissolving.

$$Ca_3(PO_4)_{2(s)} \rightleftarrows 3Ca^{2+}_{(aq)} + 2PO^{3-}_{4(aq)}$$

start	constant	0	0
change	-9.9×10^{-5} M	$+3(9.9 \times 10^{-5}$ M$)$	$+2(9.9 \times 10^{-5}$ M$)$
equilibrium	constant	2.97×10^{-4} M	1.98×10^{-4} M

$$K_{sp} = [Ca^{2+}]^3[PO_4^{3-}]^2 = (2.97 \times 10^{-4})^3 (1.98 \times 10^{-4})^2$$

$$K_{sp} = 1.0 \times 10^{-18}$$

Since such small quantities of these salts dissolve in pure water, even minute amounts of impurities can disturb the equilibrium. For instance, tap water in many parts of the country contains added F^- to help prevent dental cavities. The solubility of CaF_2 and MgF_2 in this tap water is certainly less, according to Le Chatelier's principle, than in pure water. Since hard water contains some Ca^{2+} and Mg^{2+} ions, the F^- levels must be monitored in order to keep the fluoride salts from precipitating. This is called the *common ion effect*, where one or more ions generated by the sparingly soluble salt is also present in the solution from another source.

Example 5.3. Calculate the molar solubility of AgCl at 25 °C in a 0.040 M solution of NaCl. K_{sp} of AgCl = 1.6×10^{-10}.

Since NaCl is not a sparingly soluble salt it does not have a K_{sp}. A 0.040 M solution of NaCl completely dissociates to give 0.040 M Na^+ and 0.040 M Cl^- solutions. The Cl^- is the common ion which inhibits the dissolving of AgCl.

$$AgCl_{(s)} \rightleftarrows Ag^+_{(aq)} + Cl^-_{(aq)}$$

Ionic Equilibria in Aqueous Solutions   Chapter 5

Let s = solubility of AgCl in mol/L.

$$s = [Ag^+] \text{ at equilibrium}$$

$$0.040 + s = [Cl^-] \text{ at equilibrium}$$

$$K_{sp} = 1.6 \times 10^{-10} = [Ag^+][Cl^-] = s(0.04 + s)$$

$s \ll 0.040$ because K_{sp} is so small, so $0.040 + s \simeq 0.040$

$$1.6 \times 10^{-10} = s(0.040)$$

$$4.0 \times 10^{-9} \text{ M} = s = \text{solubility of AgCl in a 0.040 M } Cl^- \text{ solution}$$

Checking the assumption, s is much less than 0.040.

What if the common ion isn't like Cl^- which is common to one soluble state (NaCl) and one insoluble salt (AgCl)? What if the common ion can form two or more sparingly soluble salts? If this is the case we must work with all the appropriate simultaneous equilibria.

Example 5.4. A solution is 0.10 M Ag^+ and 0.10 M Ba^{2+}. If SO_4^{2-} is added to the solution, which will precipitate first, Ag_2SO_4 or $BaSO_4$? (Use the K_{sp} values given.) What concentrations of SO_4^{2-} can cause only one salt to precipitate?

$$K_{sp} \quad BaSO_4 = 1.1 \times 10^{-10}$$

$$K_{sp} \quad Ag_2SO_4 = 1.2 \times 10^{-5}$$

Let's first solve for the critical $[SO_4^{2-}]$ for each salt, that is, at what concentration of sulfate will precipitation start?

$$K_{sp} = 1.1 \times 10^{-10} = [Ba^{2+}][SO_4^{2-}] = (0.10)[SO_4^{2-}]$$

$$1.1 \times 10^{-9} = [SO_4^{2-}], \text{ which means}$$

Chapter 5   Ionic Equilibria in Aqueous Solutions

that $BaSO_4$ will precipitate only when

$$[SO_4^{2-}] \geqslant 1.1 \times 10^{-9} \text{ M} \quad .$$

$$K_{sp} = 1.2 \times 10^{-5} = [Ag^+][SO_4^{2-}] = (0.10)^2 [SO_4^{2-}]$$

$$1.2 \times 10^{-3} = [SO_4^{2-}], \text{ which means}$$

that Ag_2SO_4 will precipitate only when

$$[SO_4^{2-}] \geqslant 1.2 \times 10^{-3} \text{ M}$$

From these calculations we see that $BaSO_4$ will precipitate before Ag_2SO_4, and will be the only precipitate as long as the $[SO_4^{2-}]$ is $\geqslant 1.1 \times 10^{-9}$ M and $< 1.2 \times 10^{-3}$ M. This provides a *selective precipitation* of the $BaSO_4$.

When the K_{sp} values are closer together, the calculation becomes a bit more complicated.

5.2. Acids and Bases

In aqueous systems, the most useful definition of an *acid* is a substance that can be a proton donor. A *base* is a proton acceptor. This is the *Brønsted-Lowry definition*, and while it is not general enough for use in all solvents, it is sufficient for aqueous solutions. The ability to donate or accept a proton, which is just an H^+, is measured against H_2O. So when we say that HNO_3 is a strong acid, we mean that is donates virtually all available H^+ to H_2O, as shown in the equation below.

$$HNO_{3(aq)} + H_2O_{(l)} \rightleftarrows H_3O^+_{(aq)} + NO_{3(aq)}^-$$

Although a double arrow is included in the equation, the products of this reaction are favored to such an extent that essentially no $HNO_{3(aq)}$ remains. In other words, $K \gg 1$. It is so large as to be unmeasurable in water.

A Brønsted-Lowry base accepts a proton, as the H_2O did in the above equation. A common base is sodium bicarbonate, the main ingredient in baking soda, which reacts with water as shown:

$$NaHCO_{3(aq)} + H_2O \rightleftarrows Na^+_{(aq)} + H_2CO_{3(aq)} + OH^-$$

The bicarbonate ion, HCO_3^-, accepts the H^+ from the water. The equilibrium constant for this reaction is measurable because HCO_3^- is only a moderate strength base. We would look at the form of K by using only the *net ionic reaction*.

$$HCO_{3(aq)}^- + H_2O_{(l)} \rightleftarrows H_2CO_{3(aq)} + OH^-_{(aq)}$$

Ionic Equilibria in Aqueous Solutions

As with all other K values, the pure solid and liquid concentrations are part of the numerical constant. Since this K describes a reaction of a base with water it is called a K_b.

$$K_b = \frac{[H_3CO_3][OH^-]}{[HCO_3^-]}$$

A similar K for acids, a K_a is used for moderate and weak acids, such as HF.

$$HF_{(aq)} + H_2O_{(l)} \rightleftarrows H_3O^+_{(aq)} + F^-_{(aq)}$$

$$K_a = \frac{[H_3O^+][F^-]}{[HF]}$$

Traditionally the H^+ transferred in the reaction is shown unattached to any water molecules, though in fact, it is highly solvated. For this reason we will use H_3O^+ and H^+ interchangeably.

$$K_a = \frac{[H_3O^+][F^-]}{[HF]} = \frac{[H^+][F^-]}{[HF]}$$

The relative strength of an acid or base is easily discerned from its K_a or K_b value. There are only a few very strong acids and bases and their equilibrium constants are so high that we can assume 100% dissociation. It is recommended that you remember these acids and bases, which are listed in Table 5.1.

Table 5.1. Strong Acids and Bases in Water

Common Strong Acids ($K_a \gg 1$)	Common Strong Bases ($K_b \gg 1$)
HCl	LiOH
HBr	NaOH
HI	KOH
HNO_3	RbOH
H_2SO_4	CsOH
$HClO_4$	$Ca(OH)_2$
$HClO_3$	$Sr(OH)_2$
	$Ba(OH)_2$

Chapter 5 — Ionic Equilibria in Aqueous Solutions

Because an acid donates a proton to water or to some other species, every acid reaction must also be a base reaction. So if H_2SO_4 is the acid in the equation below, H_2O is the base. In the reverse reaction, H_3O^+ is the acid and HSO_3^- is the base.

$$\underset{\text{acid}}{H_2SO_{4(aq)}} + \underset{\text{base}}{H_2O_{(l)}} \rightleftarrows \underset{\text{acid}}{H_3O^+_{(aq)}} + \underset{\text{base}}{HSO_{3(aq)}^-}$$

It is said that H_2SO_4 and HSO_3^- are a *conjugate acid/base pair*. Likewise, H_3O^+ and H_2O are a conjugate acid/base pair. Conjugate acid/base pairs differ only by a H^+; thus, H_2O and OH^- or HCO_3^- and CO_3^{2-} are also conjugates.

Example 5.5. What is the conjugate acid of NH_3?

It is NH_4^+.

Both a strong acid like HCl and a weak acid like CH_3COOH are *monoprotic*, meaning there is one ionizable or replaceable H^+ per molecule. These acids vary drastically in their strength, however HCl donates all its protons to water while CH_3COOH donates very few. Consequently the K_a for CH_3COOH is much less than the K_a for HCl. Another way to indicate the acidity or basicity of a solution is to state its concentration of H^+ (or H_3O^+). Even in strong acids the $[H^+]$ may be less than 1 M, so a logarithmic scale of $[H^+]$ was developed called the *pH scale*.

$$pH = -\log[H^+] = -\log[H_3O^+]$$

So a solution with $[H^+] = 10^{-2}$ M will have a pH = 2. The normal limits of the pH scale are 0-14, with 0-7 defining acid solutions and 7-14 defining basic solutions. Sometimes negative pH values will be encountered or values greater than 14. Taking the negative of the logarithm of a number is common practice in quantitative chemistry, so we also have pOH, pK_a and pK_b.

$$pOH = -\log[OH^-]$$

$$pK_a = -\log K_a$$

$$pK_b = -\log K_b$$

The pH and pOH of a solution are closely related since both the H^+ and OH^- affect the equilibrium of water.

$$2H_2O \rightleftarrows H_3O^+ + OH^-$$

or

Ionic Equilibria in Aqueous Solutions   Chapter 5

$$H_2O \rightleftarrows H^+ + OH^-$$

The equilibrium constant for this self-ionization is called K_w.

$$K_w = [H^+][OH^-] = [H_3O^+][OH^-]$$

$$K_w = 1 \times 10^{-14} \quad \text{at} \quad 25°C$$

As the $[H^+]$ increases, the $[OH^-]$ must decrease, or vica versa. In pure water there are equal amounts of hydronium and hydroxide ions, both at 10^{-7} M. So, in pure water, pH = pOH = 7. At all times the K_w must be satisfied, and pK_w = pH + pOH = 14.

If an acid is added to water, the $[H^+]$ concentration increases from its previous value of 10^{-7} M. Consequently the $[OH^-]$ must decrease, according to Le Chatelier's principle. This is another form of the common ion effect. Addition of an acid (or a base) to water suppresses its self-ionization.

Example 5.6. What is the $[H^+]$ and the pH of a 0.043 M solution of HCl?

Since HCl is a strong acid, the $[H^+]$ = 0.043 M. The pH = $-\log[H^+]$ = 1.37.

Example 5.7. What is the $[OH^-]$ and the pH of a 0.011 M solution of $Ca(OH)_2$?

We can easily calculate the $[OH^-]$ because $Ca(OH)_2$ is a strong base.

$$[OH^-] = 0.022 \text{ M}$$

$$pOH = 1.66$$

$$pH = 14 - pOH = 12.34$$

A short discussion of pH and significant figures is in order here. Because pH is a logarithmic term, only the digits after the decimal place are significant. The digits before the decimal only place the power of 10. Thus a pH = 1.37 or a pOH = 12.34 both have only 2 significant figures. To make sure this is clear, let's work a problem backwards.

Chapter 5 Ionic Equilibria in Aqueous Solutions

Example 5.8. The pH of a solution is measured to be 6.265. What is [H$^+$] for this solution?

$$pH = -\log[H^+]$$

$$6.265 = -\log[H^+]$$

so $\quad 10^{-6.265} = [H^+]$

or $\quad$ antilog $(-6.265) = [H^+]$

or $\quad$ INVLOG $(-6.265) = [H^+]$

$$5.43 \times 10^{-7} \text{ M} = [H^+]$$

The pH value contained 3 significant figures (0.265) while the 6 indicated the power of 10. Thus the [H$^+$] also should contain 3 significant figures.

Just as [H$^+$] and [OH$^-$] show an inverse relationship to one another, so do the K_a and K_b of a conjugate acid/base pair. Let's use a weak acid, acetic acid, to show this. CH$_3$COOH is the acid and CH$_3$COO$^-$ is the conjugate base. Since CH$_3$COOH does not donate protons to a great extent to water, the conjugate base CH$_3$COO$^-$ will easily accept a proton from water.

(1) $\quad$ CH$_3$COOH + H$_2$O $\rightleftarrows$ CH$_3$COO$^-$ + H$_3$O$^+$

(2) $\quad$ CH$_3$COO$^-$ + H$_2$O $\rightleftarrows$ CH$_3$COOH + OH$^-$

for equation 1 $\quad K_a = \dfrac{[\text{CH}_3\text{COO}^-][\text{H}_3\text{O}^+]}{[\text{CH}_3\text{COOH}]}$

for equation 2 $\quad K_b = \dfrac{[\text{CH}_3\text{COOH}][\text{OH}^-]}{[\text{CH}_3\text{COO}^-]}$

If we multiply $K_a \times K_b$ we get K_w.

Ionic Equilibria in Aqueous Solutions   Chapter 5

$$K_a \times K_b = \frac{[CH_3COO^-][H_3O^+]}{[CH_3COOH]} \times \frac{[CH_3COOH][OH^-]}{[CH_3COO^-]} = [H_3O^+][OH^-]$$

$K_a \times K_b = K_w$ for any conjugate acid/base pair.

Example 5.9. K_a for acetic acid is 1.8×10^{-5}. What is K_b for the acetate ion?

$$K_a \times K_b = K_w$$

$$K_b = \frac{1 \times 10^{-14}}{1.8 \times 10^{-5}} = 5.6 \times 10^{-10}$$

5.3 Numerical Problems

Calculations involving only a strong acid or a strong base in water are straightforward (see examples 5.6 and 5.7). If the acid or base is weak, the equilibrium constant must be used. We can use the same methods that were appropriate for gas-phase equilibria and, as before, we frequently can make simplifying assumptions and use some common sense.

Example 5.10. Calculate the pH of a 0.10 M solution of HCN in water. $K_a = 6.2 \times 10^{-10}$.

Obviously, we need to know the $[H^+]$ in order to find the pH. Since HCN is an acid, albeit weak, the pH of this solution should be <7. Let's use an equilibrium table, taking 0.10 M to be the starting concentration of the undissociated acid, HCN.

$$HCN + H_2O \rightleftharpoons H_3O^+ + CN^-$$

	HCN	+ H$_2$O	$\rightleftharpoons$	H$_3$O	+ CN$^-$
start	0.10 M	constant		10^{-7}	0
change	-x	-		+x	+x
equilibrium	(0.10-x)	constant		(10^{-7}+x)	x

The starting $[H_3O^+]$ is that of pure water, 10^{-7} M.

$$K_a = 6.2 \times 10^{-10} = \frac{[H_3O^+][CN^-]}{[HCN]} = \frac{(10^{-7}+x)(x)}{(0.10-x)}$$

Chapter 5 — Ionic Equilibria in Aqueous Solutions

Expanding the algebraic equation yields a quadratic equation which can be solved using:

$$x = \frac{-b \pm \sqrt{b^2 - 4ac}}{2a}$$

when the equation is in the form:

$$ax^2 + bx + c = 0$$

But here is where common sense comes in. A $K_a \simeq 10^{-10}$ is very small so we can assume $x \ll 0.10$ and $(0.10 - x) \simeq 0.10$. The next assumption concerns the term $(10^{-7} + x)$. Is either term negligible compared to the other? We already said x is small but is it smaller than 10^{-7}? No, it isn't, otherwise HCN wouldn't be an acid. The $[H_3O^+]$ in an acid will be $> 10^{-7}$ M, so $x > 10^{-7}$. But maybe $10^{-7} \ll x$ and $(10^{-7} + x) \simeq x$. This is an assumption worth testing.

$$\text{Assume } x \ll 0.10 \quad \text{so} \quad (0.10 - x) \simeq 0.10$$

$$10^{-7} \ll x \quad \text{so} \quad (10^{-7} + x) \simeq x$$

$$6.2 \times 10^{-10} = \frac{(x)(x)}{0.10}$$

$$7.87 \times 10^{-6} \text{ M} = x = [H_3O^+] = [CN^-]$$

Now double check the assumptions. Indeed $x \ll 0.10$, but 10^{-7} is not $\ll x$. That means either the quadratic formula must be used or the method of successive approximations outlined in the text. In any case the value of x will not change much.

$$6.2 \times 10^{-10} = \frac{(10^{-7} + x)(x)}{0.10}$$

expands to: $\quad x^2 + 10^{-7}x - 6.2 \times 10^{-11} = 0$

Solving for x we see that $x = 7.82 \times 10^{-6}$, which is the same, to two significant figures, so the answer obtained above using the faulty assumption.

$$pH = -\log [H_3O^+] = -\log (7.8 \times 10^{-6})$$

$$pH = 5.11$$

Ionic Equilibria in Aqueous Solutions

Example 5.11. Calculate the pH of a 0.10 M solution of NH_3. $K_b = 1.8 \times 10^{-5}$.

	NH_3	+	H_2O	$\rightleftarrows$	NH_4^+	+	OH^-
start	0.10		constant		0		10^{-7}
change	-x		-		+x		+x
equilibrium	(0.10-x)		constant		x		(10+x)

$$K_b = 1.8 \times 10^{-5} = \frac{[NH_4^+][OH^-]}{[NH_3]} = \frac{(x)(10^{-7}+x)}{(0.10-x)}$$

Common sense should dictate the following assumptions (remember if the assumptions are valid you save a lot of time but if they are not valid you can always go back and solve the equation in a more formal manner).

$$\text{since} \quad K_b \ll 1, \quad x \ll 0.10$$

$$\text{so} \quad (0.10-x) \simeq 0.10$$

$$\text{and} \quad 10^{-7} \ll x \quad \text{so} \quad (10^{-7}+x) \simeq x$$

$$1.8 \times 10^{-5} = \frac{x^2}{0.10}$$

$$1.3 \times 10^{-3} \, M = x = [NH_4^+] = [OH^-]$$

Double checking the assumptions shows they are valid.

$$[OH^-] = 1.3 \times 10^{-3} \, M$$

$$[H^+][OH^-] = K_w = 1 \times 10^{-14}$$

$$[H^+] = \frac{1 \times 10^{-14}}{1.3 \times 10^{-3}} = 7.7 \times 10^{-12} \, M$$

Chapter 5 Ionic Equilibria in Aqueous Solutions

$$pH = -\log[H^+] = -\log(7.7 \times 10^{-12}) = 11.11$$

This answer makes sense. A base should have a pH>7.

Calculations involving only a weak acid in water or a weak base in water can be handled as in the two examples above. But what about substances which we don't normally consider to be acidic or basic salts? If an acid is strong like HCl or HNO_3, the conjugate base is very weak like Cl^- or NO_3^-. However, if the acid is weak like HCN or CH_3COOH, the conjugate bases CN^- or CH_3COO^- will be moderately strong. Thus a solution of NaCN or $NaCH_3COO$ will affect the pH through *hydrolysis*. We generally consider hydrolysis to be the reaction of one or more of the salt ions with water.

Example 5.12. What is the pH of a 0.10 M solution of sodium acetate, $NaCH_3COO$?

We will use NaAc to mean the salt sodium acetate. First we should recognize that a 0.10 M solution of NaAc yields 0.10 M Na^+ and 0.10 M Ac^- in water. Each ion must be considered for its effects on the pH. Na^+ is the very weak conjugate acid of the strong base NaOH. That means Na^+ will be merely a spectator ion, because it has little tendency to abstract an OH^- from water. The Ac^- is another story. HAc is a weak acid so Ac^- is a moderately strong base, which means it can accept a proton from any acid present. The strongest acid in the solution is H_2O.

The hydrolysis reaction is:

$$Ac^- + H_2O \rightleftarrows HAc\ OH^-$$

The reaction produces hydroxide ions, so the NaAc solution should have a pH>7. We will need the K_b for Ac^- which we can find from the K_a for HAc.

$$K_a \times K_b = K_w$$

$$(1.8 \times 10^{-5}) \times K_b = 1 \times 10^{-14}$$

$$K_b = 5.6 \times 10^{-10}$$

Now the problem can be solved in the usual way.

	Ac^-	+ H_2O	$\rightleftarrows$ HAc	+ OH^-
start	0.10	constant	0	10^{-7}
change	-x	-	+x	+x
equilibrium	(0.10-x)	constant	x	$(10^{-7}+x)$

Ionic Equilibria in Aqueous Solutions   Chapter 5

$$K_b = \frac{[HAc][OH^-]}{[Ac^-]} = \frac{x(10^{-7}+x)}{(0.10-x)}$$

since K_b is so small, $x \ll 0.10$

so $(0.10-x) \simeq 0.10$. Also assume that

$10^{-7} \ll x$, so $(10^{-7} + x) \simeq x$.

$$5.6 \times 10^{-10} = \frac{x^2}{0.10}$$

$$7.5 \times 10^{-6} \text{ M} = x = [HAc] = [OH^-]$$

The first assumption was certainly valid. The second assumption seems to be faulty since 10^{-7} is not $\ll 7.5 \times 10^{-6}$. As in example 5.10, a more exact treatment is needed to find x. Returning to the original K_b expression, if the quadratic formula is used, $x = [OH^-] = [HAc] = 7.4 \times 10^{-6}$ M.

$$[H^+] = \frac{K_w}{[OH^-]} = \frac{1 \times 10^{-14}}{7.4 \times 10^{-6}} = 1.4 \times 10^{-9}$$

$$pH = -\log[H^+] = 8.87$$

As expected the NaAc solution is basic.

Sample problems have been shown for pure acids, bases, or salts in water. Many solutions however are mixtures of two of these components. If a solution contains a weak acid plus a salt of its conjugate base, it is called a *buffer*. Likewise, a solution of a weak base plus the salt of its conjugate acid is also a buffer. Examples of buffer solutions would be HCN + NaCN, or NH_3 + NH_4Cl. The word buffer is appropriate for such solutions because small amounts of added acid or base will not change the pH by much. The solution is resistant to pH change because there are comparable quantities of a conjugate acid/base pair, so any additional H^+ or OH^- simply shifts the reaction (Le Chatelier again) to accommodate the extra amount.

The ability to buffer pH is a property of weak acids and bases. The equilibrium position of strong acids and bases lies so far to one side of the equation that it cannot easily shift the direction of the reaction. Many physiological systems are buffered because of reactions that can only take place in a narrow pH range. Human blood, for instance, maintains a pH between 7.0 and 7.9.

Chapter 5   Ionic Equilibria in Aqueous Solutions

A rather dramatic, but simple, exhibition of buffering capacity can be seen by comparing two solutions, 1.0 M HCl/1.0 M NaCl with 1.0 M HAc/1.0 M NaAc.

          1.0 M HCl/1.0 M NaCl   1.0 M HAc/1.0 M NaAc

$[H^+] = 1.0$ M $K_a = 1.8 \times 10^{-5} = \dfrac{[H_3O^+][Ac^-]}{[HAc]}$

pH = 0 $1.8 \times 10^{-5} \equiv \dfrac{[H_3O^+](1.0)}{(1.0)}$

1.8×10^{-5} M $= [H_3O^+]$

pH = 4.74

Now let's dilute these two solutions to 0.10 M.

          0.10 M HCl/0.10 M NaCl   0.10 M HAc/ 0.10 M NaAc

$[H^+] = 0.10$ M $K_a = \dfrac{[H_3O^+][Ac^-]}{[HAc]}$

pH = 1.00 $1.8 \times 10^{-5} = \dfrac{[H_3O^+](0.10)}{(0.10)}$

$1.8 \times 10^{-5} = [H_3O^+]$

4.74 = pH

The strong acid system changes pH with concentration while in the weak system. Only the ratio of $[Ac^-]/[HAc]$ (salt/acid) controls the pH.

Example 5.13. A buffer with a pH = 4.00 is desired, using benzoic acid and sodium benzoate. The concentration of benzoic acid, $C_{16}H_5COOH$, that is to be used is 0.0500 M. How should 1.0 L of this buffer be prepared?

$$K_a \text{ for } C_6H_5COOH = 6.46 \times 10^{-5}$$

$$K_a = \dfrac{[H_3O^+][C_6H_5COO^-]}{[C_6H_5COOH]} = \dfrac{[H_3O^+][\text{salt}]}{[\text{acid}]}$$

if the pH = 4.00, $[H_3O^+] = 10^{-4}$ M

$$6.46 \times 10^{-5} = \frac{10^{-4}[\text{salt}]}{(0.0500)}$$

$$[\text{salt}] \equiv 0.0323 \text{ M}$$

To make 1.0 L of solution that is 0.0500 M in C_6H_5COOH and also 0.0323 M in NaC_6H_5COO,

$$4.65 \text{ g } NaC_6H_5COO + 6.10 \text{ g } C_6H_5COOH$$

dissolved in 1.0 L of total solution

How do you choose a good buffer system? The greatest buffering capacity is found when [salt] = [acid] or [salt] = [base] in a conjugate acid/base system. This means, for the best acid buffer:

$$\frac{[\text{salt}]}{[\text{acid}]} = 1 \quad , \quad pH = pK_a$$

or for the best base buffer:

$$\frac{[\text{salt}]}{[\text{base}]} = 1 \quad , \quad pOH = pK_b$$

If you refer to Table 5.2 you could choose an appropriate buffer at pH = 2.00 (oxalic acid/oxalate salt) or at pH = 6.00 (H_2CO_3/ HCO_3^-). And a solution that is 1.0 M in HAc and in Ac^- would have a pH = 4.74.

Table 5.2.

K_a	pK_a	Acid
5.9×10^{-2}	1.23	oxalic
1.4×10^{-4}	3.85	lactic
1.8×10^{-5}	4.74	acetic
4.3×10^{-7}	6.37	carbonic
6.2×10^{-10}	9.21	hydrocyanic

Chapter 5 — Ionic Equilibria in Aqueous Solutions

The following problem is a very good exercise to help you understand how and why a buffer works.

Example 5.14. Let's consider 1.0 L of a buffer that is 1.0 M in lactic acid and 1.0 M in sodium lactate.

- a) What is the pH?
- b) What is the pH after addition of 100 mL of 1.0 M HCl?
- c) What would be the pH if 100 mL of NaOH were added instead of HCl?

    a)   because [salt] = [acid], the pH = pK_a

$$pH = 3.85$$

    b)   when acid is added the reaction shifts to the left ($\leftarrow$) in order to adjust

$$H_2O + HLac \rightleftarrows H_3O^+ + Lac^-$$

If 0.100 L of 1.0 M HCl is added, then 0.10 mol of H_3O^+ is added. When the reaction shifts left this will decrease the moles of Lac^- by 0.10 mol, and increase HLac by 0.10 mol. This means that all the added H_3O^+ reacts with Lac^- for HLac. Remember that the volume is changing, as well as moles.

$$[HLac] = \frac{\text{initial mol} + \text{mol formed}}{\text{total volume}}$$

$$= \frac{1.0 \text{ mol} + 0.10 \text{ mol}}{1.1 \text{ L}} = 1.0 \text{ M}$$

$$[Lac^-] = \frac{\text{initial mol} - \text{mol reacted}}{\text{total volume}}$$

$$= \frac{1.0 \text{ mol} - 0.10 \text{ mol}}{1.1 \text{ L}} = 0.82 \text{ M}$$

$$K_a = 1.4 \times 10^{-4} = \frac{[H_3O^+][Lac^-]}{[HLac]} = \frac{[H_3O^+](0.82)}{(1.0)}$$

$$1.71 \times 10^{-4} \text{ M} = [H_3O^+]$$

$$pH = 3.77$$

c) if adding acid lowered the pH slightly, then adding base should raise the pH slightly. Adding 100 mL of 1.0 M NaOH essentially adds 0.10 mol OH^-. This has the same effect as if 0.10 mol H_3O^+ was removed; the reaction shifts to the right (→) in order to compensate.

$$HLac + H_2O \rightleftarrows H_3O^+ + Lac^-$$

The OH^- that is added reacts with 0.10 mol of H_3O^+ to form water. The reaction must shift right in order to replace the H_3O^+ used. Therefore the $[Lac^-]$ increases and the [HLac] decreases.

$$[HLac] = \frac{\text{initial mol} - \text{mol reacted}}{\text{total volume}}$$

$$= \frac{1.0 \text{ mol} - 0.10 \text{ mol}}{1.1 \text{ L}} = 0.82 \text{ M}$$

$$[Lac^-] = \frac{\text{initial mol} + \text{mol formed}}{\text{total volume}}$$

$$= \frac{1.0 \text{ mol} + 0.10 \text{ mol}}{1.1 \text{ L}} = 1.0 \text{ M}$$

$$K_a = \frac{[H_3O^+][Lac^-]}{[HLac]}$$

$$1.4 \times 10^{-4} = \frac{[H_3O^+](1.0)}{0.82}$$

$$1.15 \times 10^{-4} \text{ M} = [H_3O^+]$$

$$pH = 3.94$$

Certainly a buffer has limits to the amount of acid or base to which it can adjust. If 1.0 mol of H_3O^+ or OH^- had been added, all the salt or acid would have been used and the buffer capacity would be destroyed.

Chapter 5
Ionic Equilibria in Aqueous Solutions

5.4. Summary of Net Reaction Equations

The generalized formulas for calculating pH or pOH for weak acids, weak bases, salts and buffers are presented in compact form in the main text. Be sure that you understand how each was derived.

5.5. Exact Treatment of Ionic Equilibria

In many instances it will be impossible to make simplifying assumptions in order to solve an ionic equilibrium problem. We have already worked examples in which the ionization of water cannot be ignored. In general, problems such as this can be handled by writing three equations with three unknowns, or even five equations with five unknowns. The equilibrium constant expression is always one of these equations. The others are *mass balance* and *charge balance* equations. Charge balance equations are based on the fact that any solution is electrically neutral. The anion concentrations must equal the cation concentrations. Mass balance equations are based on stoichiometry.

Example 5.15. Write the mass balance and charge balance equations for a buffer solution of HF and NaF.

Charge balance:

$$\sum [\text{anions}] = \sum [\text{cations}]$$

$$[F^-] + [OH^-] = [H_3O^+] + [Na^+]$$

Mass balance:

At equilibrium, the $[Na^+]$ must be the same as the starting $[F^-]$ because the Na^+ comes from only one source.

$$[Na^+] = [F^-] \quad .$$

Since F^- comes only from the HF and the salt, F^-, the sum of those concentrations must always be equal.

$$[F^-]_0 + [HF]_0 = [HF] + [F^-]$$

These equations, along with the K_a expression for HF and K_w, will suffice to solve exactly for all concentrations under any circumstances.

5.6. Special Topics in Acid-Base Equilibria

As you look at several varieties of acid-base equilibria you should note that in all instances you need a firm *qualitative* understanding of equilibrium in order to solve the problem *quantitatively*. You must know when simplifications are appropriate. Be sure that you comprehend each sample problem before proceeding to the next one.

Ionic Equilibria in Aqueous Solutions Chapter 5

An acid-base *titration* is simply the progressive addition of an acid to a base or vice versa. At the *equivalence point*, stoichiometric amounts of the acid and base have reacted. At any point in the titration other than the equivalence point, there is an excess of either acid or base, which will determine the pH. At the equivalence point there is no excess acid or base, only the salt formed, which may in turn affect the pH through hydrolysis.

There are four general types of acid-base titrations:

1. Strong acid plus strong base. The salt produced in a reaction like this is neutral; the pH at the equivalence point is 7.

   Example

   $$HCl + NaOH \rightarrow NaCl + H_2O$$

2. Strong acid plus weak base. The salt produced here will be acidic in nature; the pH at the equivalence point will be less than 7.

   Example

   $$HCl + NH_3 \rightarrow NH_4Cl$$

3. Weak acid plus strong base. The salt produced is basic; the pH at the equivalence point will be greater than 7.

   Example

   $$HAc + NaOH \rightarrow NaAc + H_2O$$

4. Weak acid plus weak base. The pH of the equivalence point will depend on the relative strength of the acid and base used.

   Example

   $$HAc + NH_3 \rightarrow NH_4Ac$$

In order to calculate the pH at any point during a titration we must be concerned with the balanced chemical reaction or the acid + base → salt + H_2O. Then we will calculate which reagent, if any, is in excess. The pH will be controlled by that reagent. Don't forget that as the two solutions are added during the titration, the volume increases and the concentrations decrease.

Example 5.16. A 50.00 mL sample of 0.100 M HCl is to be titrated with 0.100 M NH_3. Calculate the pH of the HCl solution:

   a) at the starting point
   b) after 15.00 mL of NH_3 is added

Chapter 5 Ionic Equilibria in Aqueous Solutions

    c) after 30.00 mL of NH_3 is added
    d) after 50.00 mL of NH_3 is added
    e) after 65.00 mL of NH_3 is added

a) At the starting point we have only a 0.100 M HCl solution.

$$[H^+] = 0.100 \text{ M}$$

$$pH = -\log(0.100) = 1.000$$

b) the balanced chemical equation is:

$$HCl + NH_3 \rightarrow NH_4Cl$$

The acid and base react in a 1:1 ratio. Let's calculate how many moles of each are present when 15.00 mL of the NH_3 has been added.

$$\frac{0.100 \text{ mol HCl}}{L} \times 0.05000 L$$

$$\frac{0.100 \text{ mol NH}_3}{L} \times 0.01500 \text{ L}$$

Obviously the NH_3 is the limiting reagent so only 1.50×10^{-3} mol NH_4Cl can be formed leaving 3.50×10^{-3} mol HCl. While NH_4Cl is an acidic salt, the stronger acid, HCl, will control the pH.

$$[H^+] = \frac{3.50 \times 10^{-3} \text{ mol}}{(0.05000+0.01500)L} = 0.0538 \text{ M}$$

$$pH = -\log[H^+] = 1.27$$

c) The same technique can be used as long as the HCl is the excess reagent, which is still the case when 30.00 mL of NH_3 has been added.

$$\frac{0.100 \text{ mol HCl}}{L} \times 0.05000 \text{ L} = 5.00 \times 10^{-3} \text{ mol HCl initially present}$$

$$\frac{0.100 \text{ mol NH}_3}{L} \times 0.03000 \text{ L} = 3.00 \times 10^{-3} \text{ mol NH}_3 \text{ initially present}$$

After reacting, there are 3.00×10^{-3} mol NH_4Cl formed and 2.00×10^{-3} mol HCl in

Ionic Equilibria in Aqueous Solutions Chapter 5

excess.

$$[H^+] = \frac{\text{mol HCl in excess}}{\text{total volume}}$$

$$[H^+] = \frac{2.00 \times 10^{-3} \text{ mol}}{(0.05000+0.03000)\text{L}} = 0.0250 \text{ M}$$

$$pH = -\log[H^+] = 1.60$$

d) When 50.00 mL of the NH_3 have been added, the reaction has reached the equivalence point.

$$\frac{0.100 \text{ mol HCl}}{\text{L}} \times 0.05000 \text{ L} = 5.00 \times 10^{-3} \text{ mol HCl initially present}$$

$$\frac{0.100 \text{ mol NH}_3}{\text{L}} \times 0.05000 \text{ L} = 5.00 \times 10^{-3} \text{ mol NH}_3 \text{ initially present}$$

After the acid and base react, there are 5.00×10^{-3} mol NH_4Cl formed and no excess reagent. Therefore the pH is controlled by the salt, which hydrolyzes water (the Cl^- will be a spectator ion as it is a very weak conjugate base).

$$NH_4^+ + H_2O \rightleftarrows NH_3 + H_3O^+$$

Due to the net production of H_3O^+ from this hydrolysis, the pH should be <7.

$$K_a \text{ for } NH_4^+ = \frac{K_w}{K_b \text{ for } NH_3} = \frac{1 \times 10^{-4}}{1.8 \times 10^{-5}}$$

$$K_a = 5.6 \times 10^{-10} = \frac{[H_3O^+][NH_3]}{[NH_4^+]}$$

$$\text{Let } x = [H_3O^+] = [NH_3]$$

$$[NH_4^+] = \frac{5.00 \times 10^{-3} \text{ mol}}{0.100 \text{ L}} - x$$

Chapter 5   Ionic Equilibria in Aqueous Solutions

$$5.6 \times 10^{-10} = \frac{x^2}{5.00 \times 10^{-2} - x}, \quad \text{assume } x \ll 5.00 \times 10^{-2}$$

$$5.6 \times 10^{-10} = \frac{x^2}{5.00 \times 10^{-2}}$$

$$5.29 \times 10^{-6} \text{ M} = x = [H_3O^+]$$

$$pH = -\log[H_3O^+] = 5.28$$

e) If we continue to add NH_3 past the equivalence point, we will have excess base and its salt - a buffer solution.

$$\frac{0.100 \text{ mol HCl}}{L} \times 0.05000 L = 5.00 \times 10^{-3} \text{ mol HCl initially present}$$

$$\frac{0.100 \text{ mol NH}_3}{L} \times 0.06500 \text{ L} = 6.50 \times 10^{-3} \text{ mol NH}_3 \text{ initially present}$$

After reaction there are 5.00×10^{-3} mol NH_4Cl formed and 1.50×10^{-3} mol NH_3 in excess. The pertinent equilibrium is:

$$NH_3 + H_2O \rightleftarrows NH_4^+ + OH^-$$

$$K_b = 1.8 \times 10^{-5} = \frac{[NH_4^+][OH^-]}{[NH_3]}$$

$$[NH_4^+] = \frac{5.00 \times 10^{-3} \text{ mol}}{(0.05000 + 0.06500)L} = 0.0435 \text{ M}$$

$$[NH_3] = \frac{1.50 \times 10^{-3} \text{ mol}}{(0.05000 + 0.06500)L} = 0.0130 \text{ M}$$

$$1.8 \times 10^{-5} = \frac{(0.0435)[OH^-]}{(0.0130)}$$

Ionic Equilibria in Aqueous Solutions   Chapter 5

$$5.38 \times 10^{-6} \text{ M} = [\text{OH}^-]$$

$$[\text{H}^+] = \frac{K_w}{[\text{OH}^-]} = \frac{1 \times 10^{-14}}{5.38 \times 10^{-6}} = 1.86 \times 10^{-9}$$

$$\text{pH} = -\log[\text{H}^+] = 8.73$$

A plot of pH *vs* volume of base added gives a curve called a titration curve from which the equivalence point can be easily detected. Usually an *indicator*, a substance that changes color near the endpoint pH, is added to the titrated solution. An indicator is itself a weak acid or base, but if added in only small amounts it will not affect the equilibrium of the acid and base being titrated.

Polyprotic acids are those acids with more than one ionizable H^+, such as H_2SO_4, H_2CO_3, or H_3PO_4. Each H^+ has its own equilibrium and K_a value. For instance for H_2SO_4,

$$\text{H}_2\text{SO}_4 + \text{H}_2\text{O} \rightleftarrows \text{HSO}_4^- + \text{H}_3\text{O}^+ \quad K_1 \gg 1$$

$$\text{HSO}_4^- + \text{H}_2\text{O} \rightleftarrows \text{SO}_4^{2-} + \text{H}_3\text{O}^+ \quad K_2 = 1.0 \times 10^{-2}$$

The first H^+ is completely transferred to H_2O since H_2SO_4 is a strong acid. However the second H^+ is transferred to a lesser extent, so HSO_4^- is a weak acid.

Example 5.17. What are the concentrations of H_2SO_4, HSO_4^-, H^+ and SO_4^{2-} in a bottle labelled 2.0 M H_2SO_4?

The bottle label is somewhat misleading since H_2SO_4 completely dissociates in water to H_3O^+ and HSO_4^-.

$$[\text{H}_2\text{SO}_4] = 0$$

$$[\text{HSO}_4^-] = 2.0 \text{ M}$$

The $[\text{H}_3\text{O}^+]$ also has a contribution from the second equilibrium.

	HSO_4^-	+ H_2O	$\rightleftarrows$	SO_4^{2-}	+ H_3O^+
start	2.0 M	-		0	2.0 M
change	-x	-		+x	+x
equilibrium	(2.0-x)	-		x	(2.0 + x)

-97-

Chapter 5   Ionic Equilibria in Aqueous Solutions

$$K_a = 1.0 \times 10^{-2} = \frac{[SO_4^{2-}][H_3O^+]}{[HSO_4^-]}$$

$$1.0 \times 10^{-2} = \frac{x(2.0+x)}{(2.0-x)}$$

if we assume $x \ll 2.0$, then $x = 1.0 \times 10^{-2}$.

$$x = [SO_4^{2-}] = 1.0 \times 10^{-2}$$

$$[H_3O^+] = 2.0 + x \simeq 2.0 \text{ M}$$

$$[HSO_4^-] = 2.0 - x \simeq 2.0 \text{ M}$$

$$pH = -\log[H_3O^+] = -0.30$$

Example 5.18. If a solution is saturated with H_2S (0.10 M) at 25° C, between what pH values can 0.10 M Zn^{2+} be separated from 0.10 M Cd^{2+}?

$$\begin{array}{llll} K_{sp} & ZnS & = & 2 \times 10^{-25} \quad \text{(assume only } \alpha\text{)} \\ K_{sp} & CdS & = & 8 \times 10^{-28} \\ K_1 & H_2S & = & 9.5 \times 10^{-8} \\ K_2 & H_2S & = & 1 \times 10^{-19} \end{array}$$

First let's calculate the critical $[S^{2-}]$ for each of the salts.

$$2 \times 10^{-25} = [Zn^{2+}][S^{2-}] = (0.10)[S^{2-}]$$

$$[S^{2-}] \geqslant 2 \times 10^{-24} \text{ M will cause ZnS precipitate}$$

$$8 \times 10^{-28} = [Cd^{2+}][S^{2-}] = (0.10)[S^{2-}]$$

Ionic Equilibria in Aqueous Solutions   Chapter 5

$$[S^{2-}] \geq 8 \times 10^{-27} \text{ M will cause CdS precipitate}$$

Using these two critical $[S^{2-}]$ values we can calculate the corresponding $[H^+]$ values in a saturated H_2S solution. The simplest way to do this is to combine K_1 and K_2 for H_2S.

H_2S	$\rightleftarrows$	H^+ + HS^-	K_1	=		9.5×10^{-8}
HS^-	$\rightleftarrows$	H^+ + S^{2-}	K_2	=		1×10^{-19}
H_2S	$\rightleftarrows$	$2H^+$ + S^{2-}	K	=	$K_1 \times K_2$ =	9.5×10^{-27}

$$9.5 \times 10^{-27} = \frac{[H^+]^2[S^{2-}]}{[[H_2S]}$$

For ZnS, the critical pH is:

$$9.5 \times 10^{-27} = \frac{[H^+]^2(2 \times 10^{-24})}{(0.10)}$$

$$2.18 \times 10^{-2} \text{ M} = [H^+]$$

pH = 1.66, which means a pH $\geq$ 1.66 will be needed to precipitate ZnS.

For CdS, the critical pH is:

$$9.5 \times 10^{-27} = \frac{[H^+]^2(8 \times 10^{-27})}{(0.10)}$$

$$0.345 \text{ M} = [H^+]$$

pH = -0.46, which means a pH ≥ -0.46 is needed to precipitate CdS.

If, as the question asks, we want to separate the Zn^{2+} from the Cd^{2+}, then the pH must be carefully controlled. A pH > -0.46 but < 1.66 will allow the CdS but not the ZnS to precipitate.

5.7. Complex Ion Equilibria

Many metal ions form complex ions in a stepwise fashion. These steps are usually combined into an overall *formation constant*. For instance, $Ni(CN)_4^{2-}$ forms in four steps in an aqueous solution.

Chapter 5
Ionic Equilibria in Aqueous Solutions

$$Ni^{2+} + CN^- \rightleftarrows Ni(CN)^+ \quad K_1$$

$$Ni(CN)^+ + CN^- \rightleftarrows Ni(CN)_2 \quad K_2$$

$$Ni(CN)_2 + CN^- \rightleftarrows Ni(CN)_3^- \quad K_3$$

$$Ni(CN)_3^- + CN^- \rightleftarrows Ni(CN)_4^{2-} \quad K_4$$

overall $\quad Ni^{2+} + 4CN^- \rightleftarrows Ni(CN)_4^{2-}$

$$K = K_1 \times K_2 \times K_3 \times K_4 = 10^{30} = \frac{[Ni(CN)_4^{2-}]}{[Ni^{2+}][CN^-]^4}$$

If necessary the concentrations of *all* the Ni^{2+}/CN^- complexes could be calculated, but usually we are interested in just the overall equilibrium.

Example 5.19. The complex ion $Ag(py)_2^+$, where py = pyridine, has a formation constant of 1×10^{10}. What concentration of pyridine will dissolve 0.15 mol AgCl in a liter of solution?

$$K_{sp} \; AgCl = 1.8 \times 10^{-10}$$

The overall balanced equation for the reaction described is:

$$AgCl_{(s)} + 2 \; py_{(aq)} \rightleftarrows Ag(py)_2^+{}_{(aq)} + Cl^-_{(aq)}$$

which is merely the sum of:

$$AgCl \rightleftarrows Ag^+ + Cl^- \qquad K_{sp} = 1.8 \times 10^{-10}$$

$$Ag^+ + 2 \; py \rightleftarrows Ag(py)_2^+ \qquad K = 1 \times 10^{10}$$

$$AgCl + 2py \rightleftarrows Ag(py)_2^+ + Cl^- \qquad K = 1.8$$

$$1.8 = \frac{[Ag(py)_2^+][Cl^-]}{[py]^2}$$

If $\dfrac{0.15 \; mol}{L}$ of AgCl dissolves:

$$[Cl^-] = [Ag(py)_2^+] = 0.15 \; M$$

This means that the initial [py] must be at least 0.30 M. In order to reach equilibrium with 0.15 M $Ag(py)_2^+$, there must be some pyridine remaining.

Ionic Equilibria in Aqueous Solutions   Chapter 5

$$1.8 = \frac{(0.15)(0.15)}{[py]^2}$$

[py] at equilibrium = 0.0125 M

So the solution must be 0.3125 M or greater in pyridine in order to dissolve 0.15 mol of AgCl.

Problems

1. The solubility of CaF_2 is 0.00021 mol/L. What is K_{sp} for CaF_2?

2. How many g/L of AgI will dissolve in pure water at 25° C?

3. What is the solubility of $BaCrO_4$ in a 1.15 M solution of Na_2CrO_4?

4. When 100 mL of a saturated MgF_2 solution is evaporated to dryness, 7.6×10^{-3} g of MgF_2 remain. What is K_{sp} for MgF_2?

5. When 60.00 mL of a 0.250 M BrO_3^- solution is added to 50.00 mL of a 0.100 M Ag^+ solution, $AgBrO_3$ precipitates. Assuming that volumes are additive, what is $[Ag^+]$ after equilibrium is reached? K_{sp} for $AgBrO_3 = 5.5 \times 10^{-5}$

6. A solution contains 2.0×10^{-4} M Ba^{2+} and 3.0×10^{-2} M Pb^{2+}. Sulfate ions (SO_4^{2-}) are added by dropwise addition of Na_2SO_4 solution. What salt, $BaSO_4$ or $PbSO_4$ will precipitate first?

7. What is the pH of a 0.20 M HNO_3 solution?

8. What is the pH of a saturated solution of $Mg(OH)_2$?

9. At body temperature, 37° C, $K_w = 2.1 \times 10^{-14}$. What is the pH of pure water at this temperature?

10. Rank the following solutions according to $[OH^-]$, from lowest to highest.

    a) pure water
    b) 0.1 M H_2SO_4
    c) 0.1 M HNO_3
    d) buffer at pH = 6
    e) buffer at pOH = 10

11. The following acids are listed from strongest to weakest:

    $$HClO_4 > HNO_3 > HNO_2 > HOCl$$

    List the conjugate bases from strongest to weakest.

Chapter 5 Ionic Equilibria in Aqueous Solutions

12. How many milliliters of 0.025 M H_3PO_4 would be required to completely neutralize 25 mL of 0.030 M $Ca(OH)_2$?

13. Rank the following solutions from lowest to highest pH.

    a) 1.0 M HCl
    b) 1.0 M HAc
    c) 1.0 M NaAc
    d) 1.0 M NaCl
    e) 1.0 M NH_4Cl

14. What is the pH of a solution made by dissolving 12.0 L of HCl gas, at STP, in enough water to make 1.0 liter of solution?

15. A 1.0 M solution of a weak acid, HA, has a pH = 4.0. What percent of HA is dissociated in solution? What is K_a for this acid?

16. Methylamine, CH_3NH_2, is a weak base with a $K_b = 4.3 \times 10^{-4}$. What is the pOH of a 0.250 M solution of CH_3NH_2?

17. Boric acid, H_3BO_3, is a weak acid commonly found in eye drops. It is monoprotic with $K_a = 5.5 \times 10^{-10}$. What is the pH of a 0.125 M solution of boric acid?

18. A 0.10 M solution of sodium formate, NaHCOO, has a pH of 8.34. What is the K_a of formic acid, HCOOH?

19. Calculate the pH of a 0.0010 M NH_4Cl solution.

20. How many grams of NH_4Cl are contained in 1.0 L of a 0.305 M NH_3 solution with a pH = 9.40?

21. What volume of 0.50 M NaAc must be added to 500 mL of a 0.01 M HAc solution to create at buffer of pH = 5.00?

22. What volume of 0.01 M NaOH solution will change the pH of the buffer in problem 21 by 1 pH unit?

23. What is the pH of a solution which is prepared by mixing 150 mL of 0.45 M NH_4Cl with 250 mL of 0.30 M NaOH?

24. Consider the titration of 40.00 mL of a 0.150 M HAc solution with 0.10 M NaOH. Calculate the pH for each of the following:

    a) the starting HAc solution
    b) after 25.00 mL of NaOH is added
    c) after 58.00 mL of NaOH is added
    d) at the equivalence point
    e) after 62.00 mL of NaOH is added

Ionic Equilibria in Aqueous Solutions   Chapter 5

25. A student takes an unknown quantity of an unknown weak monoprotic acid and dissolves it in an unknown amount of water. She then titrates the acid with a solution of NaOH of unknown concentration. After 15.00 mL of base is added the pH of the solution is 5.50 At the equivalence point she has added a total of 30.00 mL of base. What is the K_a of the acid?

26. What is the pH of a 0.01 M solution of sulfurous acid, H_2SO_3? What are $[HSO_3^-]$, $[SO_3^{2-}]$ and $[H_2SO_3]$?

27. The pH of a solution which contains H_2CO_3 is found to be 3.80. What was the initial $[H_2CO_3]$?

28. What is the pH at the *second* equivalence point for a 0.10 M $H_2C_2O_4$ solution titrated with 0.10 M KOH?

29. Phosphoric acid, H_3PO_4, and its sodium salts can be used to make many different buffers. Describe how to make a buffer at pH 2, 7.2, and 12 using only H_3PO_4, NaH_2PO_4, Na_2HPO_4 or Na_3PO_4 in water.

30. What is the solubility of $Cu(OH)_2$ in pure water? What is the solubility in a solution that is maintained at 2.0 M NH_3?

CHAPTER 6
VALENCE AND THE CHEMICAL BOND

This chapter will introduce bonding from the empirical point of view rather than from theory. This is exactly how our understanding of bonding evolved. Practical chemists devised models that adequately predicted bonds and the shapes of molecules without resorting to the complex mathematical derivations. We can learn much about bonding by using these empirical models. In later chapters we shall return to a discussion of bonding and find the theoretical basis of these models.

6.1. Radicals

In this course we shall use the term *radical* or *free radical* to indicate an atom or molecule with an odd number of electrons. Generally, any species with an odd number of electrons is unstable and reacts in order to pair all its electrons.

6.2. Valence

The word valence has a rather nebulous chemical meaning these days. Its simplest meaning is combining power. For example, carbon in CH_4 has a valence of 4 because it combines with 4 hydrogens. In H_2O, oxygen has a valence of 2 while in H_3O^+ it has a valence of 3.

More common is the term *valence electrons* which means those outermost electrons in an atom that are available for bonding. If atoms share their valence electrons, they form a *covalent bond*. If an atom transfers its valence electrons to another atom they form an *ionic bond*. We shall see in the next section how to determine the number of valence electrons that each atom has.

6.3. Lewis Electron-Dot Diagrams

G.N. Lewis' electron-dot diagrams were developed with little knowledge of the internal structure of atoms, but with an innate sense of their periodic behavior. Lewis' theory is often called octet theory, as it refers to the eight main groups of the periodic chart. Lewis' structures as described here will not include the transition elements in the middle of the periodic table. The bonding of these metals is best explained with another model.

The octet theory states that each atom (except for H, He, and a few others) will not be stable unless surrounded by eight outer electrons (four pairs of two). An atom could attain an octet by sharing valence electrons or transferring valence electrons. Usually when a metal combines with a nonmetal there is a transfer of electrons leading to an ionic bond. Each ion then has a completed octet or closed shell. The noble gases already have a complete octet and are generally unreactive. Other elements gain or lose valence electrons until they are *isoelectronic* with the noble gases. The number of valence electrons, and thus an indication of how many electrons must be gained or lost, is given by the group number. Take for instance, potassium chloride, KCl. K, a metal, is found in group I and has 1 valence electron. Cl, a nonmetal, is found in group VII and has 7 valence electrons. Both atoms become isoelectronic with Ar by forming an ionic bond. K will give up its valence electron to Cl. Each valence electron is symbolized by a dot around the element.

$$K\cdot \qquad \cdot \ddot{\underset{..}{Cl}}:$$

Valence and the Chemical Bond   Chapter 6

The compound KCl with its ionic bond is symbolized by the dot structure:

$$K^+, \left[:\!\ddot{\underset{..}{Cl}}\!:\right]^-$$

Notice that a complete octet is represented by four pairs of electrons in the valence shell of Cl^-. The K^+ also has a complete octet and is isoelectronic with Ar and Cl^-, but since this octet is in the *kernel* of electrons, not in the outer shell, we show no valence electrons around K^+.

When two nonmetals bond together there is a sharing of electrons in covalent bonds, each of which involves two electrons. Again, the number of valence electrons for each atom is given by the group number. The difference between eight and the group number indicates how many electrons must be shared by that atom. In the electron-dot structure a shared pair of electrons is placed between the two atoms bonded together. A line connecting the two atoms also indicates a shared pair of electrons. Let's take Cl_2 as an example. First we will draw the separate electron-dot structures of the atoms involved. Each Cl has 7 valence electrons (group VII) and so must share one electron with another atom.

$$:\!\ddot{\underset{..}{Cl}}\!\cdot \qquad \cdot\!\ddot{\underset{..}{Cl}}\!:$$

Each Cl contributes one electron to the shared pair, giving one covalent bond.

$$:\!\ddot{\underset{..}{Cl}}\!:\!\ddot{\underset{..}{Cl}}\!: \quad \text{or} \quad :\!\ddot{\underset{..}{Cl}}\!-\!\ddot{\underset{..}{Cl}}\!:$$

In this way each Cl can attain an octet. There is one shared pair of electrons and each Cl has three *lone pairs*, not used in bonding.

Frequently, more than one pair of electrons will be shared between two atoms resulting in a double or triple bond.

Example 6.1. Draw the Lewis electron-dot diagram for O_2.

Oxygen is in group VI and so each O will need to share two electrons with the other.

$$:\!\ddot{O}\!\cdot \qquad \cdot\!\ddot{O}\!:$$

The Lewis diagram for O_2 must have a double bond.

$$\ddot{O}\!:\!:\!\ddot{O} \quad \text{or} \quad \ddot{O}\!=\!\ddot{O}$$

Example 6.2. Draw the Lewis electron-dot diagram for CO.

$$\cdot\!\dot{C}\!\cdot \qquad \cdot\!\ddot{O}\!:$$
    4 valence electrons   6 valence electrons

Chapter 6 — Valence and the Chemical Bond

The carbon needs four more electrons while the oxygen needs only two more. How can these unequal demands be satisfied? There must be unequal contributions to the bonds.

$$:C::::O: \quad \text{or} \quad :C\equiv O:$$

The octet rule is satisfied for each atom, but notice that carbon only donated two electrons to the bonds while oxygen donated four. This leads to *formal charges*.

Let's use the above example to illustrate formal charges. Formal charges indicate nothing about actual electronic charge distribution but are only a formality for separating good and poor dot diagrams. To determine the formal charge on an atom in a covalent structure we assign half of all shared electrons to each atom, then find the difference between the assigned number of electrons and the valence electrons.

formal charge = # of valence electrons − # of electrons assigned from the diagram

---

Example 6.3. What are the formal charges on C and O in CO?

$$:C\equiv O:$$

Each atom is assigned 5 electrons in the structure - 1 lone pair plus 3 electrons for the shared triple bond.

O formal charge = 6 - 5 = 1
C formal charge = 4 - 5 = -1

The sum of all formal charges in CO must be zero since there is no net charge on the molecule.

We shall encounter polyatomic molecules or ions that have several possible Lewis electron-dot diagrams. The preferred diagram will have the lowest formal charges possible.

Constructing a Lewis electron-dot diagram for a polyatomic molecule or ion is not difficult providing your electron bookkeeping is good. If you are given a formula you must be able to decide which atoms are bonded together. In the case of H_2O we know that each H is bonded to the O, rather than together, but in a larger molecule like CH_3NO_2 it may not be clear initially how to attach the atoms. For this reason the first step in developing a dot diagram is to establish the *skeleton structure* arrangement of atoms. The formula can help with this. In CH_3NO_2 we have a CH_3 group attached to an NO_2 group. H atoms can only form one bond so they must be bonded to the C which must in turn be bonded to the N. The skeleton structure is below.

```
        H
    H   C   N   O
        H       O
```

Valence and the Chemical Bond   Chapter 6

Now count the total number of valence electrons in the molecule. This molecule has 24 valence electrons (1 per H, 6 per O, 4 per C, and 5 per N).

In the next step total the number of electrons needed to achieve an octet for each atom. (Remember H needs only two electrons to be isoelectronic with He.) For CH_3NO_2 this number is 38 electrons, 8 each for C, N, O and 2 each for H. The difference between the number of electrons needed for all atoms to have an isolated octet and the number of valence electrons equals the number of electrons that must be shared; that is, 38-24 = 14 shared electrons in CH_3NO_2. Since there are two electrons per bond, there must be seven covalent bonds in the molecule. If we first assign single bonds to the skeleton structure, we find only six are required. This means one of the bonds must be a double bond.

```
            H
            |
      H  -  C  -  N  -  O
            |     |
            H     O
```

Where does the double bond occur? An H atom can only have one bond. C already has an octet so the N must share two electron pairs with an O atom. Since both O atoms are the same, either one can be assigned the double bond.

```
            H
            |
      H  -  C  -  N  =  O
            |     |
            H     O
```

In the last step, we add the lone pairs necessary to complete all octets. In this molecule only the O atoms need lone pairs.

```
            H
            |
      H  -  C  -  N  =  Ö
            |     |
            H    :Ö:
```

Now double check that only 24 electrons were used in this diagram.

This same general technique can be used for most Lewis structures *if* they follow the octet rule. (Molecules with expanded octets cannot be handled this way.) Let's summarize the steps used for writing Lewis electron-dot diagrams following the octet rule:

1) Draw the skeleton structure from the formula.
2) Count the total number of valence electrons available, adjusting for any net positive or negative charge on the formula.
3) Count the total number of electrons needed if all atoms were to attain an isolated octet (8 for most atoms, 2 for H, 4 for Be and 6 for B).

Chapter 6 Valence and the Chemical Bond

4) Find the difference between the number of electrons. This is the number of electrons that must be shared. Dividing by 2 since there are two electrons per bond yields the number of bonds necessary in the molecule.
5) Assign all atoms single bonds in the skeleton structure. Any remaining bonds must be double or triple bonds. (These will usually involve N, O, C, S but never H).
6) Add lone pairs to complete any octets.
7) Double check octets to be sure formal charges are as low as possible.

Example 6.4. Draw the Lewis electron-dot formula for NCl_3.

```
Cl       N       Cl        # of valence electrons = 3(7) + 5 = 26
                           Total # of valence electrons needed = 4(8) = 32
         Cl                # of electrons to be shared = 32 - 26 = 6
                           # of bonds = 3
```

Obviously there must be only single bonds between each Cl and the N. Completing the octets gives the following structure.

$$:\!\ddot{\underset{..}{Cl}} - \underset{|}{\overset{..}{N}} - \ddot{\underset{..}{Cl}}\!:$$
$$:\!\ddot{\underset{..}{Cl}}\!:$$

Example 6.5. Draw the Lewis electron-dot formula for H_2CO_3.

The difficulty here is to get the skeleton structure correct. Logic dictates that if all the atoms follow the octet rule, H can have only 1 bond and C or O only 4 bonds, maximum. Therefore the H atoms must be terminal, and the C and O atoms attached to each other. As in this case, acidic H atoms are frequently bonded to an oxygen and rarely to a carbon.

```
        O   C   O   H
            O
            H
```

# of valence electrons = 3(6) + 4 + 2(1) = 24
total # of valence electrons needed = 4(8) + 2(2) = 36
# of electrons to be shared = 36 - 24 = 12
# of bonds = 6

Assigning all single bonds first, there is one bond remaining. This must be a double bond which exists between C and the lone O.

$$\ddot{\underset{..}{O}} = \underset{|}{C} - \ddot{\underset{..}{O}} - H$$
$$:\!\underset{..}{O}\!:$$
$$|$$
$$H$$

Valence and the Chemical Bond Chapter 6

For some molecules or ions there may be identical positions where double bonds could be placed in the diagram. This situation is called *resonance* (an unfortunate word). Let's use the carbonate ion, CO_3^{2-}, to illustrate resonance. The carbonate ion is simply H_2CO_3 (see example 6.5) after the loss of two protons.

$$\begin{bmatrix} O & C & O \\ & O & \end{bmatrix}^{2-}$$

\# of valence electrons = $3(6) + 4 + 2 = 24$
(Note that the two extra electrons that result in the -2 charge are included as available valence electrons)

$$\text{Total \# of valence electrons needed} = 4(8) = 32$$
$$\text{\# of electrons to be shared} = 32 - 24 = 8$$
$$\text{\# of bonds} = 4$$

Possible structures:

$$\begin{bmatrix} \ddot{\text{O}}-\text{C}-\ddot{\text{O}} \\ \| \\ :\text{O}: \end{bmatrix}^{2-} \text{ or } \begin{bmatrix} \ddot{\text{O}}=\text{C}-\ddot{\text{O}} \\ | \\ :\text{O}: \end{bmatrix}^{2-} \text{ or } \begin{bmatrix} \ddot{\text{O}}-\text{C}=\ddot{\text{O}} \\ | \\ :\text{O}: \end{bmatrix}^{2-}$$

All the oxygens are in fact equal. The three structures above are resonance structures. This doesn't mean that all three exist, rather that the true structure is an average of all three, and each oxygen has a $1\frac{1}{3}$ bond to the carbon.

Example 6.6. Draw the Lewis electron-dot diagram, with all resonance structures, for the oxalate ion, $C_2O_4^{2-}$.

Once again the most difficult part of this problem is deciding the skeleton structure. A good structure to try if you aren't sure is a symmetric one. For oxalate, it is the correct structure.

$$\begin{bmatrix} O & C & C & O \\ & O & O & \end{bmatrix}^{2-}$$

\# of valence electrons = $4(6) + 2(4) + 2 = 34$
total \# of valence electrons needed = $6(8) = 48$
\# of electrons shared = $48 - 34 = 14$
\# of bonds = 7

$$\begin{bmatrix} \ddot{\text{O}}=\text{C}-\text{C}=\ddot{\text{O}} \\ || \\ :\text{O}::\text{O}: \end{bmatrix}^{2-} \text{ or } \begin{bmatrix} \ddot{\text{O}}-\text{C}-\text{C}-\ddot{\text{O}} \\ \|\| \\ :\text{O}::\text{O}: \end{bmatrix}^{2-}$$

Chapter 6   Valence and the Chemical Bond

There are two double bonds which must be assigned one to each carbon (otherwise C would have 5 bonds). Because of the two-dimensional drawings, you may wonder why there aren't two more resonance structures, as below.

$$\left[\begin{array}{c}\ddot{\text{O}} = \text{C} - \text{C} - \ddot{\text{O}}\colon \\ | \quad\quad\; || \\ \colon\!\ddot{\text{O}}\colon \quad \colon\!\ddot{\text{O}}\colon \end{array}\right]^{2-} \text{or} \left[\begin{array}{c}\colon\!\ddot{\text{O}} - \text{C} - \text{C} = \ddot{\text{O}} \\ || \quad\quad\; | \\ \colon\!\ddot{\text{O}}\colon \quad \colon\!\ddot{\text{O}}\colon \end{array}\right]^{2-}$$

You will see from the later discussion of shapes and bond angles that the second set of structures is identical to the first set given.

We have already noted that some elements, like H, Be, and B, have incomplete octets (Li would be included in this group except that it usually forms ionic bonds). This may simply be due to the fact that these elements contain too few protons to attract and hold an octet of electrons.

Other elements with more than 14 protons (Si and beyond) frequently have expanded octets with 5 or 6 electron pairs in the valence shell. Usually it is obvious from the formula that an expanded octet is needed. For instance, BrF_5 or SF_6 must have more than four single bonds to the central atom. The steps to determine dot diagrams using the octet rule will not result in reasonable structures for these molecules.

Example 6.7. Draw the Lewis electron-dot diagram for PCl_5.

Phosphorous is the central atom surrounded by five chlorines. Begin by assigning only single bonds, then complete all octets around the terminal atoms, the chlorines.

$$\begin{array}{c} \colon\!\ddot{\text{Cl}}\colon \\ | \\ \colon\!\ddot{\text{Cl}} - \text{P} - \ddot{\text{Cl}}\colon \\ \diagup \quad \diagdown \\ \colon\!\ddot{\text{Cl}} \quad\quad \ddot{\text{Cl}}\colon \end{array}$$

Now count the valence electrons used thus far and compare with the number of valence electrons for phosphorous and five chlorines. If there are extra electrons yet to be assigned they belong in the valence shell of the central atom. For PCl_5 the diagram above is correct.

Example 6.8. Draw the Lewis electron-dot diagram for BrF_3.

In this instance there isn't any clue that Br may have an expanded octet but if you try and use the octet rule and the steps associated with it, you will immediately find a problem because only two bonds are called for. This is obviously wrong because a minimum of three bonds are required for the formula.

$$\begin{array}{c} \text{F} - \text{Br} - \text{F} \\ | \\ \text{F} \end{array}$$

Valence and the Chemical Bond — Chapter 6

The molecule has $(4 \times 7) = 28$ valence electrons. The following diagram shows the position of all the valence electrons:

$$:\ddot{\text{F}} - \ddot{\text{Br}} - \ddot{\text{F}}:$$
$$|$$
$$:\ddot{\text{F}}:$$

Bromine has two lone pairs, and an expanded octet surrounding its core.

By expanding the octet of a central atom of $Z > 14$, it is also possible to lower formal charges.

Example 6.9. Use an expanded octet for sulfur to find a reasonable alternate electron-dot diagram for H_2SO_4. Shown below is its octet form.

$$:\ddot{\text{O}}:$$
$$|$$
$$\text{H} - \ddot{\text{O}} - \text{S} - \ddot{\text{O}} - \text{H}$$
$$|$$
$$:\ddot{\text{O}}:$$

The formal charge on S is $+2$, while the O atoms with three lone pairs have a -1 formal charge. The remaining O and H atoms have no formal charge. We can remove all formal charges by expanding the octet of S.

$$:\text{O}:$$
$$\|$$
$$\text{H} - \ddot{\text{O}} - \text{S} - \ddot{\text{O}} - \text{H}$$
$$\|$$
$$:\text{O}:$$

6.4. Ionic and Polar Bonds and Dipole Moments

Between the two extremes of ionic bonding where electrons are completely transferred and pure covalent bonding where electrons are equally shared, lies a continuum of *polar covalent* bonding where the electrons between atoms are unequally shared. In an ionic bond there is complete charge separation but in a polar covalent bond there is only partial charge separation. This can be illustrated with examples of fluorine compounds. CsF is an ionic salt consisting of Cs^+ and F^- ions. F_2 contains a pure covalent bond of two electrons shared equally by the fluorine atoms. HF is a polar molecule because of the unequal electron sharing between H and F. The fluorine atom has a much stronger attraction for the shared pair of electrons than hydrogen. This difference in attraction is not large enough for hydrogen to totally relinquish the electron pair, so the fluorine has a partial negative charge while the hydrogen has a partial positive charge.

Chapter 6   Valence and the Chemical Bond

The measure of the overall polarity of a molecule is called *dipole moment,* usually measured in debyes (D). The dipole moment, u, for CsF is 7.9 D, for F_2 OD, and for HF 1.8 D. Some molecules have polar bonds yet have no net dipole moment. This would indicate a symmetry about the central atom such that the polarity vectors cancel. BF_3 is such an example. Certainly the B-F bonds are polar, yet there is no dipole moment for BF_3. This indicates that the F atoms must be symmetrically oriented about the B.

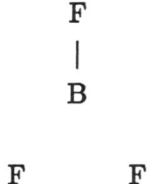

However, PF_3 does have a measurable dipole moment, indicating it has a different three-dimensional representation than BF_3. The electron-dot diagrams indicate why this is so. Boron only nesed six valence electrons while phosphorous requires eight (or more).

$$F - B - F \qquad F - \ddot{P} - F$$
$$\quad\; | \qquad\qquad\qquad | $$
$$\quad\; F \qquad\qquad\qquad F$$

PF_3 could not have the same shape as BF_3 and also have such a different u value. In fact PF_3 has a shape that shows how important the lone pair on a central atom can be. The P atom sits at the top of a pyramid whose base is comprised of F atoms.

$$\ddot{P}$$
$$F \diagup \; \diagdown F$$
$$F$$

6.5. Directed Valence and Molecular Geometry

Molecules are three-dimensional, and any two-dimensional drawing can only hint at the shape. Molecular models are the best way for a student to become familiar with the simple shapes, but even toothpicks and gumdrops or marshmallows--representing bonds and atoms--will provide good visualization. Remember that when chemists discuss the shape of a molecule they are referring to the position of the nuclei. Lone pair electrons around a central atom affect the geometry but aren't "seen" because they have virtually no mass. Thus the placement of the nuclei determine the shape. An empirical model for predicting molecular shape from a Lewis electron-dot diagram is explained next.

6.6. The Valence-Shell Electron-Pair Repulsion Model (VSEPR)

The VSEPR model, in its present form, was developed approximately 30 years ago by Gillespie and Nyholm. The fundamental concept behind the model is that electron pairs in the valence shell of a central atom will repel each other as much as possible in three-dimensions. Being able to construct the Lewis electron-dot diagram is the first step in using VSEPR. Then you must count the number of valence electron pairs around the central atom. This includes only lone pairs and single (or σ) bonds. Another way to count the valence pairs is to add the lone pairs and bonded atoms. In any case double (or π) bonds do not determine the basic molecular shape; they only modify it. The ideal geometries based on the number of valence electron pairs are found in Table 6.1.

Valence and the Chemical Bond   Chapter 6

Table 6.1. Ideal Geometry from Valence Shell Electron Pairs

# of Valence Electron Pairs	Ideal Geometry	Ideal Bond Angle	Example
2	linear	180°	BeH_2, CO_2
3	trigonal planar	120°	BF_3
4	tetrahedral	109.5°	CH_4
5	trigonal bipyramidal	90° & 120°	PF_5
6	octahedral	90°	SF_6

It is easy to make corrections for non-ideal geometry, the main kind being the lone pair effect.

Example 6.10. Predict the shape of a water molecule.

Lewis diagram   H−Ö−H

There are 4 valence pairs. The ideal geometry is tetrahedral with the 4 electron pairs 109.5° apart. Since we "see" only the atoms, not the electrons in the shape, H_2O appears as a bent molecule.

The bond angle between the hydrogen atoms is slightly less than 109.5° (104.5°), but predictably so. The lone pairs occupy a larger space than the bonded pairs which are directed between atoms, and repel each other and the bonded pairs more so than just the bonded pairs. The net effect is a smaller H−O−H bond angle.

Corrections for multiple bonds are also easily understood if you keep in mind that atoms with multiple bonds are much closer together than singly bonded atoms. The increased density of electrons in a double or triple bond repel all other bonded pairs more strongly than a single bond. This slightly affects the bond angles as can be seen in the next example.

Example 6.11. Predict the shape of a formaldehyde molecule, H_2CO.

Lewis diagram
$$\begin{array}{c} H \\ | \\ C=\ddot{\ddot{O}} \\ | \\ H \end{array}$$

Chapter 6 — Valence and the Chemical Bond

There are 3 valence electron pairs (remember to count only σ bonds and lone pairs around the carbon), so the ideal geometry is trigonal planar.

$$\begin{array}{c} H \\ \backslash \\ C = \ddot{\underset{..}{O}} \\ / \\ H \end{array}$$

The ideal bond angles would be 120°. Because of the double bond repulsion, we expect the H—C—H bond angle to be slightly less than 120°, the H—C—O bond angles will be slightly greater than 120°.

Example 6.12. Predict the shape of an SO_3^{2-} ion.

Lewis diagram
$$\left[\begin{array}{c} \ddot{\underset{..}{O}} - \ddot{S} - \ddot{\underset{..}{O}} \\ | \\ :\underset{..}{O}: \end{array} \right]^{2-}$$

There are 4 valence electron pairs but since one pair is a lone pair, the ion should have a trigonal pyramid shape, with O—S—O bond angles slightly less than 109.5°.

$$\left[\begin{array}{c} \ddot{S} \\ O \diagup \diagdown \\ O O \end{array} \right]^{2-}$$

Example 6.13. Predict the shape of an IF_5 molecule.

Lewis diagram
$$\begin{array}{c} F F \\ \backslash / \\ \ddot{I} \\ / | \backslash \\ F F F \end{array}$$

There are 6 valence electron pairs about the I. The ideal shape is an octahedron, where all positions are equivalent. The actual shape of the molecule should then be square pyramidal.

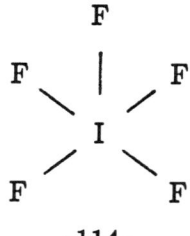

-114-

Valence and the Chemical Bond — Chapter 6

Example 6.14. Predict the shape of an NO_2 molecule.

Lewis diagram $:\ddot{O}-\ddot{N}=\ddot{O}$ or $\ddot{O}=\ddot{N}-\ddot{O}:$

This molecule is a radical because of the odd electron on the central atom. The ideal geometry would be that of 3 valence pairs, trigonal planar. The actual shape is bent.

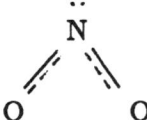

Both oxygen atoms have $1\frac{1}{2}$ bonds to nitrogen. The O—N—O bond angle is slightly greater than 120° because the lone electron does not repel to the same extent as a lone pair does.

6.7. Bond Energies and Distances

Bond energy is that energy required to break a bond or, conversely, the energy released when the bond forms. Bond energy has a strong dependence on the bond distance (or how far apart are the nuclei). The closer together the atoms are the stronger the bond and the greater the bond energy. Triple bonds are shorter and stronger than double bonds. In turn, double bonds are shorter and stronger than single bonds. Bond energies are generally listed for zero Kelvin.

Bond dissociation energies are measured at 298 K and vary slightly from bond energies.

Example 6.15. Using the bond dissociation energies provided in Table 6.7 of the main text, calculate $\Delta H°$ for the following reaction at 298 K:

$$H_2 + Br_2 \rightarrow 2HBr$$

We see that we must break the H—H and Br—Br bond and then form 2 H—Br bonds. Breaking bonds is *always* endothermic while making bonds is *always* exothermic. Therefore the correct sign must be given to the bond dissociation energies that are used.

	$\Delta H°$
$H_2 \rightarrow 2H$	436 kJ
$Br_2 \rightarrow 2Br$	193 kJ
$2H + 2Br \rightarrow 2HBr$	-2(366) kJ
$H_2 + Br_2 \rightarrow 2HBr$	-103 kJ

Chapter 6   Valence and the Chemical Bond

Problems

1. Draw the Lewis electron-dot diagrams for the following molecules.

    a) N_2

    b) I_2

    c) HCN

    d) NaCl

    e) BaO

    f) HBr

2. Draw the Lewis electron-dot diagrams for the following ions.

    a) H^+

    b) H^-

    c) OH^-

    d) O^{2-}

    e) Ca^{2+}

    f) Rb^+

3. For the following polyatomic molecules, choose an appropriate skeleton structure, then draw the Lewis structure. Be sure to include all resonance forms.

    a) SbF_3

    b) $HClO_2$

    c) HNO_2

    d) SO_2

    e) C_2H_2

    f) CH_2CCH_2

    g) HCF_3

    h) $SiCl_4$

    i) OCl_2

    j) N_2F_4

4. Draw the Lewis structure for the following polyatomic ions, including all resonance forms.

    a) NH_4^+

    b) NO^+

    c) H_3O^+

    d) CO_3^{2-}

    e) NO_2^+

    f) NO_2^-

    g) N_3^-

5. All the following molecules and ions violate the octet rule. Draw the appropriate Lewis structures.

    a) BCl_3

    b) BeH_2

    c) BO_2^-

    d) SF_4

    e) XeF_2

    f) XeO_4

    g) NO_2

    h) NO

    i) H_3BO_3

    j) BOH_6

Valence and the Chemical Bond Chapter 6

6. Mercury fulminate, $Hg(CNO)_2$, is commonly used to detonate TNT. Draw the Lewis structure of the fulminate ion, CNO^-. Include any resonance forms and indicate which form has the lowest formal charge.

7. The Lewis structure of the perchlorate ion, ClO_4^-, can be drawn according to the octet rule, but Cl also can have an expanded octet. Draw the electron-dot diagram of ClO_4^- in both instances and indicate all formal charges.

8. Decide whether each of the following molecules should have a dipole moment.

   a) HCF_3
   b) C_2F_4
   c) SCl_2
   d) AsH_3
   e) OBr_2
   f) $SiCl_4$
   g) BF_3
   h) $AsBr_5$

9. Predict the shape and the approximate bond angles in the following species using the VSEPR model.

   a) NH_4^+
   b) TeF_6
   c) SbF_5
   d) BCl_3
   e) SiH_4
   f) OF_2
   g) KrF_2
   h) IF_5
   i) Br_3^-
   j) PF_2^+
   k) ICl_4^-
   l) ClF_3

10. How do the shapes and H—N—H bond angles vary in the series NH_2^-, NH_3, NH_4^+?

11. How do the shapes and O—S—O bond angles vary in the series, SO_2^{2-}, SO_3^{2-}, SO_4^{2-}?

12. Calculate $\Delta H°$ at 298 K for the following reaction.

    $$2CO + O_2 \rightarrow 2CO_2$$

    Draw the Lewis structures of all products and reactants. Does the $\Delta H°$ value make sense, considering the general trend that triple bonds are stronger than double bonds.

13. Using bond dissociation energies, which of the following has the shortest bond length? Which has the longest bond length?

    $$N_2, O_2, F_2, Br_2, Cl_2$$

    Use Lewis structures to explain why.

CHAPTER 7
OXIDATION-REDUCTION REACTIONS

A substance is *oxidized* when it loses electrons, as with the sodium in the equation $Na \rightarrow Na^+ + 1e^-$. A substance is *reduced* when it gains electrons, as with the chlorine in the equation $Cl + 1e^- \rightarrow Cl^-$. Any complete, balanced reaction that includes an oxidation must necessarily include a reduction also because there is a *transfer of electrons*. This parallels the case of acid/base reactions where H^+ is transferred. If there is an acid there must also be a base involved.

Oxidation-reduction reactions are frequently called *redox* reactions. The substance that is oxidized is the *reducing agent*. The substance reduced is the *oxidizing agent*. The oxidizing and reducing agent can be identified by the changes that take place in the oxidation state.

7.1. Oxidation States

The *oxidation state* or *oxidation number* of an element is a formal assignment of charge to an atom within a molecule. Oxidation numbers are a bookkeeping technique which treats all molecules as though they were ionic in nature. Therefore we stress that oxidation numbers are a formality for covalent compounds.

The oxidation numbers of Na and Cl in NaCl are equal to the simple ionic charges. So +1 is the oxidation number of Na and -1 is the oxidation number of Cl. This is also expressed as Na^+ and Cl^-. These oxidation numbers indicate the electrical charge on the ions, but when a compound consists of covalently bonded atoms, like H_2O, the oxidation numbers do not give a true indication of charge distribution. In the case of covalent compounds, oxidation numbers only allow you to keep track of electron transfers, not to assign real charges. For H_2O, the oxygen is assigned -2 and the hydrogen is assigned +1 oxidation states. The total of all oxidation numbers must equal zero for a molecule or the net ionic charge for an ion.

Determining the oxidation states of elements within a molecule is rather simple if you know a few rules. You will find the periodic table to be a great help. All uncombined elements have an oxidation state equal to zero. This includes diatomic elements and all allotropic forms. Therefore Na, O_2, C (graphite), S_8, etc., all have an oxidation state of zero.

For simple monoatomic ions, the charge is the oxidation state. So Ca^{2+} has an oxidation state of +2 while S^{2-} has an oxidation state of -2. In a polyatomic ion all oxidation numbers must total the overall charge. This is straightforward in the case of HS^-, where S has a -2 and H a +1 oxidation state. With other more complex ions, you must know what oxidation numbers are feasible, and among those, which are most common. The alkali metals, group IA, have only one possible oxidation number (other than zero), which is +1. The alkaline earth metals, group IIA, are +2. The halogens, group VIIA are -1 except when combined with oxygen or another halogen. Hydrogen is +1 when directly bonded to a nonmetal and -1 when directly bonded to a metal. Oxygen is usually -2 except when bonded to fluorine or in a peroxide where it is O_2^{2-}. Almost all other oxidation states can be determined by difference. The transition metals and many nonmetals have several possible oxidation states.

Example 7.1. What are the oxidation states of the atoms in sodium oxalate, $Na_2C_2O_4$?

You should recognize this as the formula of a salt consisting of Na^+ ions and $C_2O_4^{2-}$ ions.

Oxidation-Reduction Reactions Chapter 7

If the O is assumed to be -2, then each C must be +3.

$$\underset{+1\ +3\ -2}{Na_2C_2O_4}$$

$$2(Na) + 2(C) + 4(O) = 0$$

$$2(+1) + 2(+3) + 4(-2) = 0$$

Note that while the oxidation states of Na and O are usually +1 and -2 respectively. C may have an oxidation state from +4 to -4, depending on the compound.

Example 7.2. Assign the oxidation numbers in the following compounds: Al_2S_3, SO_2, H_2SO_4.

Al is found in group IIIA and so usually is +3, therefore S is -2 in Al_2S_3.

$$\underset{+3\ -2}{Al_2S_3}$$

But in SO_2 and H_2SO_4 the S must have a positive oxidation state.

$$\underset{+4\ -2}{S\ O_2}$$

$$\underset{+1\ +6\ -2}{H_2\ S\ O_4}$$

These examples show the variety of oxidation states exhibited by S. There is a relationship between the number of valence electrons and the possible oxidation numbers. Sulfur has 6 valence electrons. In order to achieve a complete valence octet, S can gain two electrons (oxidation state -2) or lose up to six electrons (oxidation state +6). In fact these electrons are not gained or lost; they are shared. Nonetheless you can see that the number of valence electrons can be useful in predicting possible oxidation states.

Example 7.3. What are the upper and lower limits to the oxidation states of P? Give examples of compounds with these oxidation states.

Since P is in group VA and has 5 valence electrons, the limits to its possible oxidation states are -3 and +5. There are many examples of both these states.

P^{3-} in PH_3, Na_3P

P^{5+} in H_3PO_4, PF_5

Chapter 7   Oxidation-Reduction Reactions

A chemical reaction can be recognized as an oxidation-reduction if there are changes in oxidation states. For instance, $Na + \frac{1}{2}Cl_2 \rightarrow NaCl$ is easily seen to be a redox reaction because the Na is oxidized from 0 to +1 while the Cl is reduced from 0 to -1. Other reactions may be more complex, but it is still possible to tell if a reaction is redox in nature and, if so, what is being being oxidized and what is reduced.

Example 7.4. Decide whether the following balanced equation is an oxidation-reduction. What is the oxidizing agent?

$$14H^+ + Cr_2O_7^{2-} + 6Br^- \rightarrow 2Cr^{3+} + 3Br_2 + 7H_2O$$

Start by placing all oxidation numbers below each atom.

$$\underset{+1}{14H^+} + \underset{+6\ -2}{Cr_2O_7^{2-}} + \underset{-1}{6Br^-} \rightarrow \underset{+3}{2Cr^{3+}} + \underset{0}{3Br_2} + \underset{+1\ -2}{7H_2O}$$

The Cr is reduced from +6 to +3, while the Br is oxidized from -1 to 0. The oxidizing agent is the species containing the atom reduced.

$$\text{oxidizing agent} = Cr_2O_7^{2-}$$

$$\text{reducing agent} = Br^-$$

7.2. Half-Reactions Concept

Since any redox reaction contains both an oxidation and reduction, we frequently find it useful to separate the equation into its two component *half-reactions*. If a redox reaction is to be used to perform electrical work, like a battery, then the two half-reactions must be carried out in separate compartments that are electrically connected. Neither the oxidation half-reaction nor the reduction half-reaction can be carried out by itself. By making the distinction between the two half-reactions, however, we can better understand what happens and harness the available energy.

Example 7.5. Write the two half-reactions for the redox process

$$Na + \tfrac{1}{2}Cl_2 \rightarrow NaCl$$

$Na \rightarrow Na^+ + 1e^-$ oxidation

Oxidation-Reduction Reactions Chapter 7

$$\tfrac{1}{2}Cl_2 + 1e^- \to Cl^- \qquad \text{reduction}$$

Example 7.6. Write, in unbalanced form, the two half-reactions for the redox process

$$14H^+ + Cr_7O_2^{2-} + 6Br^- \to 2Cr^{3+} + 3Br_2 + 7H_2O$$

$$6Br^- \to 3Br_2 \qquad \text{oxidation}$$

$$Cr_2O_7^{2-} \to 2Cr^{3+} \qquad \text{reduction}$$

Note that the half-reactions are not balanced for mass or charge. Next we shall see how to balance half-reactions like this.

7.3. Balancing Oxidation-Reduction Reactions

We will use the popular half-reaction method of balancing redox reactions. The steps as presented in your text are:

1) Assign all oxidation numbers in the reaction and decide what is being oxidized and reduced.
2) Write separate half-reactions for the two processes.
3) Balance the mass first (use H^+ and H_2O in acid solution, and OH^- and H_2O in base solution) then balance charge by adding electrons where needed.
4) Adjust the number of electrons in the half-reactions to be equal, then combine the two half-reactions.
5) Double check the overall reaction for mass and charge.

Example 7.7. Balance the following redox reaction in acid solution.

$$H_2C_2O_4 + MnO_4^- \to Mn^{2+} + CO_2$$

We shall follow the steps in order:

1) $\underset{+1\ +3\ -2}{H_2C_2O_4} + \underset{+7\ \ -2}{MnO_4^-} \to \underset{+2}{Mn^{2+}} + \underset{+4\ -2}{CO_2}$

   the Mn is being reduced ($+7 \to +2$)
   the C is being oxidized ($+3 \to +4$)

2) $MnO_4^- \to Mn^{2+} \qquad \text{reduction}$

   $H_2C_2O_4 \to CO_2 \qquad \text{oxidation}$

Chapter 7 Oxidation-Reduction Reactions

3) To balance the reduction, we see the Mn is already mass-balanced, but to balance the O we must add water to the right side and then add H^+ to the left side.

$$8H^+ + MnO_4^- \to Mn^{2+} + 4H_2O$$

Now the mass is balanced but not the charge. The left side has a net +7 charge while the right side has a net +2 charge. The addition of 5 electrons to the left side will adjust this.

$$5e^- + 8H^+ + MnO_4^- \to Mn^{2+} + 4H_2O$$

The oxidation half-reaction can be balanced in a similar way. Note that the C must be balanced before the O and H.

$$H_2C_2O_4 \to 2CO_2$$

Then H_2O and H^+ and e^- can be added.

$$H_2C_2O_4 \to 2CO_2 + 2H^+ + 2e^-$$

In this case the O is balanced and only $2H^+$ and $2e^-$ are needed.

4) We can now add the two half-reactions. Since no electrons can appear in the net equation we must adjust the balanced half-reactions: the number of electrons generated in the oxidation must be exactly the same number of electrons required by the reduction. We generally use whole numbers and not fractions. Therefore we must find the lowest common multiple for the number of electrons in the two half-reactions. In this case it is 10.

$$2(5e^- + 8H^+ + MnO_4^- \to Mn^{2+} + 4H_2O)$$
$$5(H_2C_2O_4 \to 2CO_2 + 2H^+ + 2e^-)$$
$$\overline{10e^- + 16H^+ + 2MnO_4^- + 5H_2C_2O_4 \to 2Mn^{2+} + 8H_2O + 10CO_2 + 10H^+ + 10e^-}$$

Combining all like terms results in:

$$6H^+ + 2MnO_4^- + 5H_2C_2O_4 \to 2Mn^{2+} + 8H_2O + 10CO_2$$

5) Double checking, mass and charge seem OK.

Example 7.8. Balance the following redox reaction in base solution.

$$S_2O_3^{2-} + NO_3^- \to NO_2^- + SO_3^{2-}$$

1) $\underset{+2\ -2}{S_2O_3^{2-}} + \underset{+5\ -2}{NO_3^-} \to \underset{+3\ -2}{NO_2} + \underset{+4\ -2}{SO_3^{2-}}$

S is oxidized $(+2 \to +4)$
N is reduced $(+5 \to +3)$

Oxidation-Reduction Reactions

Chapter 7

2) $S_2O_3^{2-} \to SO_3^{2-}$ oxidation

    $NO_3^- \to NO_2^-$ reduction

3) Balancing the O and H using OH^- and H_2O can be frustrating, but trial and error is sometimes the best technique.

$$S_2O_3^{2-} \to 2SO_3^{2-}$$

We need 3 O atoms on the left, so we add twice as many OH^- ions as we need O atoms (this trick doesn't always work, but its a good place to start).

$$6OH^- + S_2O_3^{2-} \to 2SO_3^{2-}$$

Now we can add $3H_2O$ to the right to complete the mass balance.

$$6OH^- + S_2O_3^{2-} \to 2SO_3^{2-} + 3H_2O$$

With the electron balance we get a balanced oxidation.

$$6OH^- + S_2O_3^{2-} \to 2SO_3^{2-} + 3H_2O + 4e^-$$

The reduction half-reaction is balanced similarly.

$$NO_3^- \to NO_2^-$$

We need on O one the right, so we add $2OH^-$ to the right. That requires one H_2O on the left to balance the H. Two electrons balance the charge.

$$2e^- + H_2O + NO_3^- \to NO_2^- + 2OH^-$$

4)
$$\frac{\begin{array}{c}6OH^- + S_2O_3^{2-} \to 2SO_3^{2-} + 3H_2O + 4e^- \\ 2(2e^- + H_2O + NO_3^- \to NO_2^- + 2OH^-)\end{array}}{6OH^- + S_2O_3^{2-} + 4e^- + 2H_2O + 2NO_3^- \to 2SO_3^{2-} + 3H_2O + 4e^- + 2NO_2^- + 4OH^-}$$

Combining like terms:

$$2OH^- + S_2O_3^{2-} + 2NO_3^- \to 2SO_3^{2-} + H_2O + 2NO_2^-$$

5) Double check.

7.4. Galvanic Cells

As with all other chemical reactions, there is one direction in which the redox reaction proceeds spontaneously under standard conditions. The reverse reaction would then be non-

Chapter 7 — Oxidation-Reduction Reactions

spontaneous. A *galvanic cell* is based on a spontaneous redox reaction, while an *electrolytic cell* must be forced with electrical energy to proceed in a non-spontaneous direction. We shall first investigate galvanic cells. All cells capable of doing electrical work have the oxidation half-cell separated from the reduction half-cell by an external circuit that also includes two electrodes or terminals. The electrode where oxidation occurs is called the *anode*, and the electrode where reduction occurs is called the *cathode*. These electrodes may be active, which means they are part of the balanced reaction, or inert, which means they only serve as a conduit for electrons but are not changed by this transfer.

The standard state in electrochemistry is the same as we have previously encountered: 1 m for solutions (where we are actually using unit activities and $1\ m \simeq 1\ M$), 1 atm pressure for gases, and the most stable form of solids at 25°C.

The measurement of electrical force in a galvanic cell is in *volts*. We call this voltage the *standard cell potential* ($\Delta \epsilon$) if all substances involved are at standard state. The value of the cell potential depends on the two half-reactions involved as well as the concentrations and partial pressures of all products and reactants. Thus as a reaction proceeds, its cell potential changes. A spontaneous (left to right) reaction will always have a cell potential that is positive.

$\Delta \epsilon ° = +$ for spontaneous reactions, left to right

$\Delta \epsilon ° = -$ for spontaneous reactions in the reverse direction, right to left

$\Delta \epsilon °$ for a cell is simply the sum of the two half-cell voltages, but since we can't measure the potential of only one half-cell, we need to choose a standard half-cell and assign it some arbitrary value. Then all other half-cell potentials can be measured against this standard half-cell. The half-reaction chosen is that of hydrogen.

$$2H^+(1M) + 2e^- \rightleftarrows H_2\ (1\ atm) \qquad \epsilon° = 0\ \text{volts}$$

The voltage for this half-reaction, in either direction, is defined as zero volts. Now all other half-reactions can be measured against the hydrogen half-cell and each half-reaction can be assigned a voltage.

Example 7.9. The cell potential for a galvanic cell based on the following reaction was 0.236 volts under standard state conditions at 25°C. What is the half-cell potential assigned to $Ni^{2+} + 2e^- \rightarrow Ni$?

$$2H^+ + Ni \rightleftarrows H_2 + Ni^{2+}$$

Let us first separate this redox reaction into its two half-reactions.

	$\epsilon°$
$2H^+ + 2e^- \rightleftarrows H_2$	0 volts
$Ni \rightleftarrows Ni^{2+} + 2e^-$	?
$2H^+ + Ni \rightleftarrows H_2 + Ni^{2+}$	0.236 volts

Oxidation-Reduction Reactions   Chapter 7

Since the total potential of the cell is 0.236 volts and the H^+/H_2 cell has a potential of zero volts, the Ni/Ni^{2+} half-cell potential must be 0.236 volts.

$$Ni \rightleftarrows Ni^{2+} + 2e^- \qquad \epsilon° = +0.236 \text{ volts}$$

$$Ni^{2+} + 2e^- \rightleftarrows Ni \qquad \epsilon° = -0.236 \text{ volts}$$

Other half-cell potentials can be measured in the same way. This data is organized into tables of standard reduction potentials, where by convention all half-reactions are listed as reductions. Table 7.1 in the text contains about 60 of the most common half-reactions and their potentials. We can calculate the redox potential for any reaction by simply combining the appropriate half-reactions.

Example 7.10. Calculate the standard cell potential of the following reaction using Table 7.1.

$$Zn_{(s)} + Br_{2(l)} \rightleftarrows Zn^{2+}_{(aq)} + 2Br^-_{(aq)}$$

Again we should separate the reaction into its half-reactions.

$$Zn \rightleftarrows Zn^{2+} + 2e^-$$

$$2e^- + Br_2 \rightleftarrows 2Br^-$$

Now locate these half-reactions in Table 7.1 and tabulate the $\epsilon°$ values. Note that the Zn reaction is written as an oxidation above. All tabular values are reduction potentials. Therefore we must change the sign of the Zn potential for this problem. Since every redox reaction contains an oxidation, one half-cell potential must always be the reverse of that found in the reduction table and its potential must be of the opposite sign.

	$\epsilon°$
$Zn \rightleftarrows Zn^{2+} + 2e^-$	+0.762
$Br_2 + 2e^- \rightleftarrows 2Br^-$	+1.078
$Zn + Br_2 \rightleftarrows Zn^{2+} + 2Br^-$	+1.830 volts

So the reaction is spontaneous.

The potential of a cell does not depend on the coefficients of the balanced equation. The reaction $H^+ + Cl^- \rightleftarrows \frac{1}{2}H_2 + \frac{1}{2}Cl_2$ has the same cell potential as the reaction $2H^+ + 2Cl^- \rightleftarrows H_2 + Cl_2$. At first this may not make sense, but remember that cell potential is

Chapter 7 Oxidation-Reduction Reactions

not the same as work or heat. Think about the following case. A standard 9 volt battery can be made larger. The voltage will still be 9 volts but the battery will simply run longer. The total available work is greater from the larger battery but the voltage is the same because the redox reaction is identical in both batteries.

Example 7.11. Calculate the standard cell potential for the spontaneous reaction below, which is unbalanced.

$$Ni^{2+} + Al \rightleftarrows Ni + Al^{3+}$$

	$\epsilon°$
$3(Ni^{2+} + 2e^- \rightleftarrows Ni)$	-0.236 volts
$2(Al \rightleftarrows Al^{3+} + 3e^-)$	1.68 volts
$3Ni^{2+} + 2Al \rightleftarrows 3Ni + 2Al^{3+}$	1.44 volts

Although it is necessary for 3 moles of Ni^{2+} to be reduced for every 2 moles of Al that are oxidized, the potential is not adjusted by these factors. This means that it is unnecessary to balance an equation in order to calculate its potential.

Spend a few minutes perusing Table 7.1 in the text. Where are the oxidizing agents? Where are the reducing agents? Where are the strongest or weakest oxidizing agents? With a little practice you can become adept at predicting either the spontaneous direction of a reaction or whether one substance can oxidize another. Below is a skeleton reproduction of the standard reduction potential table with some trends denoted.

	Half-Reaction (In Acid)	$\epsilon°$ (v)
↑ increasing strength of oxidizing agents	$F_2 + 2e^- \rightleftarrows 2F^-$	+2.890
		increasing strength of reducing agents
	$Li^+ + e^- \rightleftarrows Li$ ↓	-3.040

The strongest oxidizing agent listed in Table 7.1 is F_2. The strongest reducing agent is Li. Conversely the weakest oxidizing agent is Li^+ and the weakest reducing agent is F^-.

Example 7.12. Refer to Table 7.1 to answer the following. Which is the better oxidizing agent, H^+ or Zn^{2+}? Because H^+ is above Zn^{2+}, and has a higher reduction potential it is a better oxidizing agent than Zn^{2+} or any other oxidizing agent below it on the table, such as Pb^{2+}, Ni^{2+}, etc.

Example 7.13. Again referring to Table 7.1, decide whether manganese metal will dissolve in a 1 M H^+ solution.

In order to answer this question you must write the pertinent chemical equation and calculate its cell potential. If $\Delta\epsilon°$ is positive the manganese dissolves.

	$\epsilon°$
$Mn \rightleftarrows Mn^{2+} + 2e^-$	1.182
$2H^+ + 2e^- \rightleftarrows H_2$	0
$Mn + 2H^+ \rightleftarrows Mn^{2+} + H_2$	1.182 volts

Yes, 1 M H^+ will dissolve Mn as well as any other metal with a reduction potential less than zero, that is below H^+ on the table.

7.5. Nernst Equation

Normally the conditions with which chemists work are not standard state. In these instances we would like to know $\Delta\epsilon$ as well as $\Delta\epsilon°$. Qualitatively we can use Le Chatelier's principle to predict how $\Delta\epsilon$ compares with $\Delta\epsilon°$. Let's illustrate this with the reaction below.

$$Zn + 2H^+ \rightleftarrows Zn^{2+} + H_2$$

When $[Zn^{2+}] = [H^+] = 1$ M and $P_{H_2} = 1$ atm, $\Delta\epsilon° = +0.76$ volts. But if $[Zn^{2+}]$ falls below 1 M or P_{H_2} falls below 1 atm then Le Chatelier's principle predicts the system will be even farther away from equilibrium than at standard state. This means the reaction driving force will be even greater and $\Delta\epsilon > \Delta\epsilon°$. Conversely if the $[H^+]$ falls below 1 M (i.e., pH > 0) then the reaction driving force will be less than at standard state and $\Delta\epsilon < \Delta\epsilon°$. Changing the amount of $Zn_{(s)}$ has no effect on the cell potential.

The mathematical relationship between $\Delta\epsilon$ and $\Delta\epsilon°$ is given by the Nernst equation.

$$\Delta\epsilon = \Delta\epsilon° - \frac{0.059}{n} \log Q \qquad \text{common form at 25°C}$$

$$\Delta\epsilon = \Delta\epsilon° - \frac{RT}{nF} \ln Q \qquad \text{general form}$$

Chapter 7 　　　　　　　　　　　　　　　　　　　　　　Oxidation-Reduction Reactions

where R = ideal gas constant

T = temperature (K)

F = Faraday's constant

In both equations N= number of electrons transferred in the balanced redox equation. Q is the reaction quotient in the same form we previously used in Chapter 4. Therefore voltage depends on the concentrations of all products and reactions.

Example 7.14. What is the cell potential of the reaction given below when $[Zn^{2+}] = 0.50$ M and $P_{H_2} = 0.15$ atm?

$$Zn_{(s)} + 2H^+_{(aq)} \rightarrow H_{2(g)} + Zn^{2+}_{(aq)}$$

$$Q = \frac{[Zn^{2+}]P_{H_2}}{[H^+]^2} = \frac{(0.50)(0.15)}{(1)^2} = 0.075$$

Note that concentrations and pressures are mixed in Q (and in K also). This is because the standard phase for Zn^{2+} is aqueous but for H_2 it is gaseous. The values used are activities and have no units. It is therefore legitimate and necessary to mix phases in the reaction quotient.

$$\Delta\varepsilon = \Delta\varepsilon° - \frac{0.059}{n} \log Q$$

$$\Delta\varepsilon = 0.76 - \frac{0.059}{2} \log (0.075)$$

$$\Delta\varepsilon = 0.79 \text{ volts}$$

The value of n is 2 because the Zn transfers two electrons to H^+.

Since cell potential varies with the relative concentrations of the oxidized and reduced species we ought to be able to construct a cell based on the same half-reaction, but with differing concentrations. This is called a *concentration cell*.

Example 7.15. What is $\Delta\varepsilon$ for a cell consisting of two Zn electrodes connected to a voltmeter, one electrode in 0.07 M Zn^{2+} solution, the other in 1.0 M Zn^{2+} solution? If both Zn^{2+} solutions were the same concentration there would be no driving force and no reaction. So $\Delta\varepsilon° = 0$. But since the concentrations are different, there will be a reaction until the two

Oxidation-Reduction Reactions — Chapter 7

concentrations are equal. In order for the concentrations to become equal, the half-cells must be:

$Zn \rightleftarrows Zn^{2+} + 2e^-$ oxidation on the dilute side

$Zn^{2+} + 2e^- \rightleftarrows Zn$ reduction on the concentrated side

The overall reaction is+ Zn^{2+} (1 M) $\rightleftarrows Zn^{2+}$ (0.07 M)

$$Q = \frac{[Zn^{2+}\text{dilute}]}{[Zn^{2+}\text{conc.}]} = \frac{0.07}{1} = 0.07$$

$$\Delta\epsilon = \Delta\epsilon^\circ - \frac{0.059}{n} \log Q$$

$$\Delta\epsilon = 0 - \frac{0.059}{2} \log (0.07) = 0.034 \text{ volts}$$

The greater the difference in the two concentrations, the larger the voltage will be.

Certainly $\Delta\epsilon^\circ$ and ΔG° for a reaction must be related since both indicate whether or not a reaction is spontaneous. ΔG° values, however, depend on stoichiometry and moles of substance while $\Delta\epsilon^\circ$ values are independent of stoichiometry. Therefore, besides converting volts to joules, the relationship between ΔG° and $\Delta\epsilon^\circ$ must contain a mole dependence

$$\Delta G^\circ = -nF\epsilon^\circ$$

where   n = moles of electrons transferred in the redox reaction

F = Faraday's constant

Faraday's constant gives the quantity of charge per mole of electrons.

$$F = \frac{96,5000 \text{ coulombs}}{\text{mol e}^-}$$

$$\text{Since 1 coulomb} = \frac{1 \text{ Joule}}{\text{volt}}$$

$$F = \frac{96,500 \text{ J}}{\text{mol e}^- \cdot \text{volt}}$$

Chapter 7 Oxidation-Reduction Reactions

Under non-standard state conditions, ΔG and $\Delta \epsilon$ have a similar relationship:

$$\Delta G = -nF\Delta\epsilon$$

Substitution for $\Delta\epsilon$ and $\Delta\epsilon°$ in the Nernst equation gives:

$$\Delta G = \Delta G° + RT \ln Q$$

When a system is at equilibrium it can do no more electrical work nor release more free energy, and $Q = K$. This gives us a mathematical relationship between K and $\Delta\epsilon°$.

$$\Delta\epsilon° = \frac{RT}{nF} \ln K$$

or

$$K = e^{nF\Delta\epsilon°/RT}$$

Example 7.16. Calculate $\Delta\epsilon°$, $\Delta G°$ and K for the following unbalanced redox reaction.

$$Cr + Ni^{2+} \rightleftarrows Cr^{3+} + Ni$$

The $\epsilon°$ value can easily be found from Table 7.1.

	$\epsilon°$
$2(Cr \rightleftarrows Cr^{3+} + 3e^-)$	0.74 volts
$3(Ni^{2+} + 2e^- \rightleftarrows Ni)$	-0.236 volts
$2Cr + 3Ni^{2+} \rightleftarrows 2Cr^{3+} + 3Ni$	0.50 volts

$$\Delta G° = -nF\Delta\epsilon° = -(6 \text{ mol } e^-)\left(96{,}500 \frac{J}{\text{mole}^- \cdot v}\right)(0.50 \text{ v})$$

$$\Delta G° = -2.90 \times 10^5 \text{ J} = -290 \text{ kJ}$$

$$K = e^{-\Delta G°/RT} \quad \text{or} \quad e^{nF\Delta\epsilon°/RT}$$

$$K = e^{290{,}000 J / (8.314 J/ \text{mol·K})(298 K)} = 6.8 \times 10^{50}$$

Oxidation-Reduction Reactions   Chapter 7

Although a cell potential of 0.50 volts doesn't seem to be very large, the corresponding equilibrium constant value indicates that very little reactants remain after equilibrium is reached.

We can even calculate or measure a cell potential for a reaction that doesn't appear to be an oxidation-reduction.

Example 7.17. Given that the K_{sp} of CuI is 5.1×10^{-12} at 25°C, calculate $\Delta \epsilon °$ for the following half-reaction.

$$CuI + e^- \rightleftarrows Cu + I^-$$

The voltage for this half-reaction can be found in many reduction potential tables. Since it isn't listed in our Table 7.1 we shall have to calculate the potential. We know K_{sp} of CuI from which we can calculate an $\Delta \epsilon °$.

$$CuI \rightleftarrows Cu^+ + I^- \qquad K_{sp} = 5.1 \times 10^{-12}$$

$$\Delta \epsilon ° = \frac{RT}{nF} \ln K_{sp} = \frac{(8.314 \frac{J}{mol \cdot K})(298K)}{1 \text{ mol } e^- (96,500 \frac{J}{mol \ e^- \cdot v})} \ln(5 \times 10^{-12})$$

$$\Delta \epsilon ° = -0.67 \text{ volts}$$

Although the dissolving of solid CuI doesn't appear to be a redox process, by allowing one of the half-cells to consist of the unknown reaction $CuI + e^- \rightleftarrows Cu + I^-$ and the other half-cell to be $Cu \rightleftarrows Cu^+ + e^-$, we can get electrical work from the system.

	$\epsilon °$
$CuI + e^- \rightleftarrows Cu + I-$	?
$Cu \rightleftarrows Cu^+ + I-$	-0.518 volts
$CuI \rightleftarrows Cu^+ + I^-$	-0.67 volts

The half-cell potential for the reduction half-reaction is directly found from the difference.

$$-0.67 = -0.518 + ?$$

$$-0.15 \text{ volts} = ? = \epsilon ° \text{ for CuI reduction}$$

-131-

Chapter 7 — Oxidation-Reduction Reactions

Example 7.18. Using only data from Table 7.1, calculate the cell reduction potential for $Fe^{3+} + 3e^- \to Fe$.

This half-reaction is not listed in the table but is the sum of two other half-reactions.

		ϵ°
(1)	$Fe^{2+} + 2e^- \rightleftarrows Fe$	-0.44 volts
(2)	$Fe^{3+} + e^- \rightleftarrows Fe^{2+}$	-0.77 volts
(3)	$Fe^{3+} + 3e^- \to Fe$	0.33 volts

Although the reaction is simply the sum of two others, the ϵ° value is not. This problem is different from others we've solved because the electrons do not drop out of the equations when added. We do not have a redox process here, only two reductions adding up to a third reduction. Whenever electrons remain in the equation, cell potentials cannot be added. Rather, ΔG° values must be added.

$$\Delta G_3^0 = \Delta G_2^0 + \Delta G_1^0$$

$$-n_3 F \epsilon_3^0 = -n_2 F \epsilon_2^0 - n_1 F \epsilon_1^0$$

which rearranges to:

$$\epsilon_3^0 = \frac{n_1 \epsilon_1^0 + n_2 \epsilon_2^0}{n_3}$$

or in a more generalized form:

$$\epsilon_T^0 = \frac{n_1 \epsilon_1^0 + n_2 \epsilon_2^0 + \cdots}{n_T}$$

For this Fe example:

$$\epsilon_3^0 = \frac{2(-0.44) + 1(0.77)}{3} = -0.037 \text{ volts}$$

Example 7.19. Does Fe^{2+} spontaneously disproportionate under standard state conditions?

Disproportionation means the same substance, in this case Fe^{2+}, is both the oxidizing agent and the reducing agent.

Oxidation-Reduction Reactions Chapter 7

$$\begin{array}{ll} & \epsilon^\circ \\ Fe^{2+} + 2e^- \to Fe & -0.44 \text{ volts} \\ 3Fe^{2+} \to Fe + 2Fe^{3+} & -0.77 \text{ volts} \\ \hline 3Fe^{2+} \to Fe + 2Fe^{3+} & -1.21 \text{ volts} \end{array}$$

Now we are adding an oxidation and a reduction so the ϵ° values are additive. Since a negative ϵ° is the result, Fe^{2+} does not disproportionate.

7.6. Oxidation-Reduction Titrations

Example 7.20. The iron in a 1.000 g rock sample is dissolved and converted to all Fe^{2+} which is titrated with $C_2O_7^{2-}$. If the titration requires 45.30 ml of 5.00×10^{-3} M $Cr_2O_7^{2-}$ to be complete, what was the original percentage of iron in the rock sample?

The stoichiometry portion of such a problem should be familiar to you. But what is the chemical reaction that takes place? Dichromate, $Cr_2O_7^{2-}$, cannot be oxidized, rather it is a strong oxidizing agent. Therefore the Fe^{2+} must be oxidized to Fe^{3+}.

$$\begin{array}{ll} & \epsilon^\circ \\ Cr_2O_7^{2-} + 14H^+ + 6e^- \rightleftarrows 2Cr^{3+} + 7H_2O & 1.36 \text{ volts} \\ 6(Fe^{2+} \rightleftarrows Fe^{3+} + e^-) & -0.77 \text{ volts} \\ \hline Cr_2O_7^{2-} + 14H^+ + 6Fe^{2+} \rightleftarrows 2Cr^{3+} + 6Fe^{3+} + 7H_2O & -0.59 \text{ volts} \end{array}$$

$$0.04530 \text{ L} \times \frac{5.00 \times 10^{-3} \text{ mol } Cr_2O_7^{2-}}{L} = 2.27 \times 10^{-4} \text{ mol } Cr_2O_7^{2-}$$

$$\text{mol } Fe^{2+} = 6 \times \text{mol } Cr_2O_7^{2-}$$

$$\text{mol } Fe^{2+} = 6(2.27 \times 10^{-4}) = 1.36 \times 10^{-3} \text{ mol } Fe^{2+}$$

$$\text{mol Fe in rock} = \text{mol } Fe^{2+} \text{ titrated} = 1.36 \times 10^{-3}$$

$$1.36 \times 10^{-3} \text{ mol Fe} \times \frac{55.8 \text{ g Fe}}{\text{mol}} = 0.760 \text{ g Fe}$$

$$\% \text{ Fe in rock} = \frac{0.760}{1.000} \times 100 = 76.0\%$$

Chapter 7 Oxidation-Reduction Reactions

7.7. Electrolysis

In the early 19th century the British scientist Michael Faraday discovered certain quantitative relationships while studying electrolytic cells. *Electrolysis* is the use of an external source of electricity to carry out a chemical reaction (the reverse of a galvanic cell). Some definitions are in order at this point.

> current - electrical charge flowing past a certain point per unit time. The SI unit of current is the ampere (A).

The SI unit of charge is the coulomb (c).

$$1A = \frac{1c}{s}$$

The charge on 1 mole of electrons is Faraday's constant

$$F = \frac{96,485 \text{ coulombs}}{\text{mol } e^-} \simeq \frac{96,500 c}{\text{mol } e^-}$$

With these few relationships many calculations can be made. Though both an oxidation and a reduction must be occurring, normally only one of the half-reactions is pertinent to the calculation.

Example 7.21. Calculate how many coulombs of charge are necessary to produce 10.5 L of H_2 (g) at STP from the electrolysis of H_2O.

A balanced half-reaction is necessary to start.

$$2e^- + 2H_2O \rightleftarrows H_2 + 2OH^-$$

Then we can convert 10.5 L of H_2 to moles

$$\frac{10.5 \text{ L } H_2}{22.4 \text{ L/ mol@STP}} = 0.469 \text{ mol } H_2$$

$$\text{mol } e^- = 2 \times \text{mol } H_2 = 2 \times 0.469$$

$$\text{mol } e^- = 0.938$$

$$0.938 \text{ mol } e^- \times \frac{96,500 \text{ c}}{\text{mol } e^-} = 90,500 \text{ c}$$

Example 7.22. A current of 12.3 A is used for the electrolysis of molten NaCl. How long will it take for 1.0 kg of sodium metal to be deposited?

Again, a balanced half-reaction is required.

$$Na^+ + e^- \rightarrow Na$$

$$\frac{1,000 \text{ g}}{23.0 \text{ mol}} = 43.4 \text{ mol Na}$$

$$\text{mol } e^- = \text{mol Na} = 43.4$$

$$43.4 \text{ mol } e^- \times \frac{96,500 \text{ c}}{\text{mol } e^-} = 4.19 \times 10^6 \text{ c}$$

$$\frac{4.19 \times 10^6 \text{ c}}{12.3 \text{ c/s}} = 3.4 \times 10^5 \text{ s} \quad \text{or} \quad 94.6 \text{ hrs}$$

7.8. Electrochemical Applications

Anyone who has experienced biting a piece of aluminum foil so that it comes in contact with a metal dental filling, has experienced an electrochemical cell. Saliva makes an excellent salt bridge between the Al and the filling, which is usually a Ag/Hg or Sn/Hg mixture. The tarnish that forms on silver is Ag_2S produced in the redox reaction between Ag and SO_2 in the air.

Many everyday redox reactions are not as simple as those we have studied thus far. Frequently an assortment of products result, as in the case of rust forming. Fe is first oxidized to Fe^{2+} while O_2 is reduced to OH^-. Iron (II) oxide or iron (II) hydroxide may form depending on the pH. The Fe^{2+} also can be further oxidized to Fe^{3+} which in turn can form iron (III) oxide or iron (III) hydroxide. It is impossible to write a single net redox reaction showing all these possible products, whose ratios vary with the pH. Paint or absence of moisture may retard the rusting process. Another common method of protecting iron and steel is to coat it with a thin layer of Zn. This process is called *galvanizing*. The Zn corrodes but the $Zn(OH)_2$ or $ZnCO_3$ formed adheres tightly to the metal surface blocking out the oxygen and water.

A *battery* is nothing more than a galvanic cell in a neat little package. Your text describes alkaline dry cells, lead storage batteries and the new NiCad batteries. Advances in battery technology are made constantly, showing up in longer lasting, more reliable batteries for consumer use. Fuel cells, another type of battery, while not currently practical for mass production, may soon find their way into everyday applications.

Chapter 7　　　　　　　　　　　　　　　　　　　　　　　　Oxidation-Reduction Reactions

Problems

1. Assign oxidation numbers for all elements in the following molecules.

    a) N_2H_4

    b) $KBrO_4$

    c) LiH

    d) Ga_2S_3

    e) XeO_2

    f) Fe_3O_4

    g) $C_{12}H_{22}O_{11}$

2. Assign oxidation numbers for all elements in the following ions.

    a) FeO_4^{2-}

    b) VO^{2+}

    c) S_2O_3

    d) AlH_4^-

    e) $P_2O_7^{4-}$

3. Complete and balance the following redox equations in acid solution. Identify the oxidizing agent and the reducing agent.

    a) $H_2O_2 + Cu \rightleftarrows Cu^{2+} + H_2O$

    b) $Cr^{3+} + HNO_2 \rightleftarrows Cr + NO_3^-$

    c) $Cr_2O_7^2 + HClO_2 \rightleftarrows Cr^{3+} + ClO_3^-$

    d) $HNO_3 + As_4 \rightleftarrows H_3AsO_4 + NO$

    e) $Mn^{2+} + NaBiO_3 \rightleftarrows MnO_4^- + Na^+ + Bi^{3+}$

    f) $NCS^- + O_2 \rightleftarrows NO_2 + CO_2 + SO_2$

4. Complete and balance the following redox equations in basic solution. Identify the oxidizing agent and the reducing agent.

    a) $Be_2O_3^{2-} + ClO_2^- \rightleftarrows Be + ClO_4^-$

    b) $P_4 + IO_3^- \rightleftarrows HPO_4^{2-} + I^-$

    c) $NH_3 + HO_2^- \rightleftarrows N_2H_4 + OH^-$

    d) $C_8H_{18} + O_2 \rightleftarrows CO_3^{2-} + H_2O$

5. List the following species in order of increasing strength as oxidizing agents in 1 M acid solution.

    $Na^+, Cl_2, Zn^{2+}, H^+, MnO_4^-$

Oxidation-Reduction Reactions   Chapter 7

6. List the following species in order of increasing strength as reducing agents in 1 M base solution.

$$Mn^{2+}, Cl^-, H_2, Na, F^-, I^-$$

7. Which of the following metals should dissolve to an appreciable extent in a solution of pH = 0?

$$Zn, Pt, Mn, Fe$$

8. Refer to Table 7.1 and list at least 3 substances that will reduce Ag^+ but not Mn^{2+}.

9. Copper will dissolve in nitric acid but not in hydrochloric acid. Explain why.

10. Silver tarnish, Ag_2S, can be electrochemically removed from silverware by wrapping the object loosely in aluminum foil and placing it in warm salt water. Explain why the tarnish is removed.

11. Calculate the standard cell potential at 25 °C for the following redox reactions.

   a) $2Hg + 2Ag^+ \rightleftarrows Hg_2^{2+} + 2Ag$

   b) $Zn + Cl_2 \rightleftarrows Zn^{2+} + 2Cl^-$

   c) $Al + 3Fe^{3+} \rightleftarrows Al^{3+} + 3Fe^{2+}$

   d) $Ca + 2H_2O + OH^- \rightleftarrows H_2 + Ca(OH)_2$

12. A lead storage battery is based on the following reaction:

$$Pb + PbO_2 + 4H^+ + 2SO_4^{2-} \rightleftarrows 2PbSO_4 + 2H_2O$$

$$\Delta \epsilon° = 1.81 \text{ volts}$$

   How does the value of $\Delta \epsilon$ compare to $\Delta \epsilon°$ in each of the following circumstances:

   a) more Pb and PbO_2 are added
   b) the pH is increased
   c) water is added
   d) when the system reaches equilibrium, what is $\Delta \epsilon°$?

13. The following half-reaction has a standard reduction potential of 0.94 volts. What is ϵ at pH = 7?

$$NO_3^- + 3H^+ + 2e^- \rightleftarrows HNO_2 + H_2O$$

Chapter 7	Oxidation-Reduction Reactions

14. The following reaction can take place wherever air and water are in contact.

$$2N_2 + 5O_2 + 2H_2O \rightleftarrows 4H^+ + 4NO_3^-$$

$$\Delta\epsilon° = -0.02 \text{ volts}$$

What will the voltage be under normal ocean conditions, $P_{N_2} = 0.80$ atm, $P_{O_2} = 0.20$ atm, pH = 8.0, $[NO_3^-] = 3\times10^{-5}$ M?

15. Show that the half-reaction $O_2 + 4H^+ + 4e^- \rightleftarrows 2H_2O$ with $\epsilon° = 1.229$ volts in standard acid has $\epsilon° = 0.414$ volts in standard base.

16. What is the voltage expected from a concentration cell with 0.007 M Fe^{2+} in one cell, and 1 M Fe^{2+} in the other?

17. At 298K the fuel cell below releases free energy.

$$2H_2 + O_2 \rightleftarrows 2H_2O$$

$$\Delta G° = -237 \text{ kJ/mol}$$

What is $\Delta\epsilon°$? What is K?

18. What is the standard free energy change at 25°C for:

$$2H^+ + H_2O_2 + Cu \rightleftarrows Cu^{2+} + 2H_2O?$$

19. The K_{sp} for AgI is 8.3×10^{-17}. Using data from Table 7.1, what is $\epsilon°$ for AgI + $e^- \rightleftarrows Ag + I^-$?

20. Use the following information to find $\epsilon°$ for $Sn^{4+} + 4e^- \rightleftarrows Sn$.

$Sn^{2+} + 2e^- \rightleftarrows Sn$ $\epsilon° = -0.14$ volts

$Sn^{4+} + 2e^- \rightleftarrows Sn^{2+}$ $\epsilon° = 0.15$ volts

Will Sn^{2+} disproportionate under standard state conditions?

21. a) What is $\Delta\epsilon°$ for the disproportionation of Cr^{3+} to $Cr_2O_7^{2-}$ and Cr?

b) What is $\Delta\epsilon°$ for the disproportionation of Cr^{2+} to Cr^{3+} and Cr?

c) What is the potential for $Cr_2O_7^{2-} \rightleftarrows Cr$ in 1 M H^+?

22. In the reaction between $FeSO_4$ and MnO_2, Fe^{3+} and Mn^{2+} are formed in 1M acid solution. What weight of MnO_2 is required to react with 0.850 g of $FeSO_4$?

-138-

Oxidation-Reduction Reactions	Chapter 7

23. A solution of $KMnO_4$ has an unknown concentration but it can be standardized against a piece of pure iron. The iron is dissolved in acid and reduced to Fe^{2+} then titrated with $KMnO_4$ according to the following equation. If a 0.3265 g piece of pure iron requires 53.32 mL of the $KMnO_4$ solution to reach the equivalence point, what is the molarity of the $KMnO_4$ solution?

$$5Fe^{2+} + MnO_4^- + 8H^+ \rightleftarrows 5Fe^{3+} + Mn^{2+} + 4H_2O$$

24. Electrolysis is one method of determining atomic mass. If a current of 0.600 A deposits 2.38 g of a certain metal in 59 minutes, what are the possible atomic masses of this metal? What metal and what oxidation state are most likely?

25. How long will it take for a current of 3.0 A to decompose one mole of water electrolytically?

26. An alkaline dry cell containing 0.85 g of Zn is used to power a hearing aid that draws 2.5 milliamps of current. How long will the battery last?

CHAPTER 8
CHEMICAL THERMODYNAMICS

Thermodynamics is the study of the relationship between heat and other forms of energy. Using thermodynamics we shall learn how to calculate equilibrium constants from energy values of the reacting molecules. So thermodynamics will tell us about equilibrium; which side of a reaction is energetically favored at a given temperature, but *not* about how long the process takes.

8.1. Systems, States, and State Functions

A *system* is whatever part of the universe we are considering. Everything else in the universe constitutes the *surroundings*. A chemical system may frequently be a reaction vessel and its contents, but a system can be defined in many ways, so you must be alert as to what is considered part of the system and what is considered surroundings. Plenty of examples will be forthcoming.

A *state function* is a property that depends only on the state of the system and not on its history. State functions are usually denoted by capital letters. Some state functions that we have encountered are pressure (P), volume (V), temperature (T), enthalpy (H), free energy (G) and entropy (S).

Any property that *does* depend on the history of the system is called a *path function*. Path functions are usually denoted by lower-case letters, like heat (q) and work (w), which are discussed in the next section.

An analogy can be developed that may make state functions and path functions clearer. Consider travelling from the Pacific Ocean at Los Angeles to the Atlantic Ocean at New York City. Many different paths are possible and many different modes of transportation may be used. The change in altitude from L.A. to N.Y., however, is the same regardless of whether you travel by plane at 30,000 ft. or walk the entire distance going up and down the mountains. The change in altitude depends only on where you start and where you finish and *not* on any point in between.

$$\Delta \text{altitude} = (\text{final altitude} - \text{initial altitude})$$

So the change in altitude is a state function. On the other hand, the work and time involved in travelling by air versus walking are obviously path dependent functions.

8.2. Work and Heat

There are many different kinds of *work* (electrical, magnetic, etc.) but for the discussion at hand we shall use mechanical or displacement work.

$$\text{work} = \text{force} \times \text{distance}$$

For gases, displacement work occurs due to a change in volume.

$$\text{work} = -P_{\text{external}} \Delta V$$

Chemical Thermodynamics — Chapter 8

In the more general form:

$$\text{work} = -\int_{V_1}^{V_2} P_{ext}\, dV$$

Since w = work done on the system by the surroundings, the sign conventions are that w = + when the surroundings do work on the system as in the mechanical compression of a gas in the downstroke of a piston, and w = − when the system does work on the surroundings as when an expanding gas forces the piston in an upstroke.

Work is not the only way for a system and its surroundings to exchange energy. *Heat* describes the flow of energy due to a change in temperature. Heat and work are both interconvertible. We can generate heat by doing work (a person sweats during exercise) and we can do work by heating (boiling water). Both heat and work are path functions due to energy transfer between system and surroundings. They are related by the first law of thermodynamics. The sign conventions for heat (q) are that q = + when the system absorbs heat from the surroundings, and q = − when the system gives off heat to the surroundings.

8.3. First Law of Thermodynamics

Simply stated, the first law of thermodynamics is that energy can be neither created nor destroyed, but may change form. Mathematically stated, the internal energy (E) of a system can change only due to heat or work being exchanged with the surroundings.

$$\Delta E = q + w$$

where ΔE = change in internal energy of the system

q = heat absorbed by the system

w = work done to the system

Internal energy, E, is a state function that depends on two path functions: heat and work. The units for all three functions must be the same, usually joules or kilojoules. Since w = −PΔV, this involves a unit conversion.

$$w = -P\Delta V = \text{atm} - L$$

The relationship between atm-L and J can be easily found using the ideal gas constant, R, expressed in two different units.

$$R = 0.08206 \frac{\text{atm}-L}{\text{mol}-K} = 8.314 \frac{J}{\text{mol}-K}$$

So 1 atm-L = 101.3 J.

The most straightforward way to measure ΔE is to carry out the reaction in question in a sealed container so that volume cannot change $\Delta V = 0$. Any heat absorbed or released, as measured by a temperature change, gives ΔE.

Chapter 8   Chemical Thermodynamics

$$\Delta E = q + w + q - P\Delta V$$

$$\Delta E = q_v \text{ at constant volume since } \Delta V = 0$$

The heat exchanged at constant volume, q_v, is equal to the change in internal energy.

If, however, the reaction is carried out so that no heat can be exchanged between the system and the surroundings, as in a thermos bottle then $q = 0$, and $\Delta E = w$. This is called an *adiabatic* process.

By far, the most common method is to study a reaction open to the atmosphere. The heat measured at constant pressure, q_p, is equal to the change in enthalpy, ΔH.

$$\Delta H = q_P$$

This equation is easily derived from the relationship between ΔH and ΔE.

Enthalpy and internal energy are closely related.

$$\Delta H = \Delta E + \Delta(PV)$$

Under constant pressure conditions:

$$\Delta H = q + w + P\Delta V$$

and since $w = -P\Delta V$, $\Delta H = q_P$.

A common mistake that students make is assuming heat and enthalpy are the same. Obviously that is only true under constant pressure conditions.

When a process involves only solids and liquids, the relationship between ΔH and ΔE can be simplified. The $\Delta(PV)$ term approaches zero because there are little changes in the pressure or volume of liquids and solids. Thus:

$$\Delta H \simeq \Delta E \quad \text{for liquid and solid reactions}$$

When gases are involved, the $\Delta(PV)$ term can be replaced with $\Delta(nRT)$ from the ideal gas law.

$$\Delta H = \Delta E + \Delta(PV) = \Delta E + \Delta(nRT)$$

If the temperature is constant then RT will not change and the equation becomes:

$$\Delta H = \Delta E + \Delta n(RT)$$

where Δn = change in *gas* moles.

Chemical Thermodynamics Chapter 8

Example 8.1. An ideal gas expands against a constant external pressure of 1.05 atm, from 2.3 liters to 6.5 liters. In the process, 820 J of heat are evolved. Calculate q, w, ΔE and ΔH for the system.

The value for q is directly given in the problem. Since it is heat *evolved* by the system the sign is negative.

$$q = -820 \text{ J}$$

Since the external pressure remains constant, $\Delta H = q_P = -820$ J.

In order to find ΔE, we must first calculate the work involved.

$$w = -P\Delta V = -(1.05 \text{ atm})(6.5\text{L} - 2.3\text{L}) = -4.41 \text{ atm} - \text{L}$$

$$w = -(4.41 \text{ atm} - \text{L})\left(101.3 \frac{\text{J}}{\text{atm} - \text{L}}\right) = -447 \text{ J}$$

$$\Delta E = q + w = -820 \text{ J} - 447 \text{ J} = -1267 \text{ J}$$

Example 8.2. One mole of propane, C_3H_8, was combusted under constant volume conditions at 300 K, and 62.8 kJ of heat were released. Calculate ΔE and ΔH for the combustion.

Since this is a *chemical* process we are investigating, the balanced chemical equation may be needed.

$$C_3H_{8(g)} + 5O_{2(g)} \rightleftarrows 3CO_{2(g)} + 4H_2O_{(l)}$$

Since it is a constant volume process, $\Delta E = q_v = -6.28$ kJ. Then ΔH can be found.

$$\Delta H = \Delta E + \Delta(PV) = \Delta E + \Delta n(RT)$$

Remember that Δn refers only to gas mole changes.

$$\Delta n = (3-6) = -3 \text{ moles}$$

$$\Delta H = -62.8 \text{ kJ} + (-3 \text{ mol})\left(8.314 \times 10^{-3} \frac{\text{kJ}}{\text{mol-k}}\right)(300 \text{ K})$$

Chapter 8 Chemical Thermodynamics

$$\Delta H = -62.8 \text{ kJ} - 7.5 \text{ kJ} = -70.3 \text{ kJ}$$

8.4. Thermochemistry

Thermochemistry is the bookkeeping, or stoichiometry, of thermodynamics. Many aspects will already be familiar to you. In Chapter 3.2 we introduced $\Delta H°$, the standard state enthalpy. $\Delta H_f°$ at 298 K was defined as zero for all elements in their most stable form. Tabulated values of $\Delta H°$ for elements and molecules can be found in the *Handbook of Chemistry and Physics* or in Chapter 3 of the text. The $\Delta H°$ value for a balanced reaction could be easily calculated because enthalpy can be treated stoichiometrically.

$$\Delta H_{rxn}^0 = \sum \Delta H_f°(\text{products}) - \sum \Delta H_f°(\text{reactants})$$

This treatment can be extended to include addition of equations or other manipulations.

Example 8.3. Pure iron can be formed by passing CO over mixed iron (II) and iron (III) oxides according to the equation below.

$$Fe_3O_4 + 4CO \rightleftarrows 3Fe + 4CO_2 \quad \Delta H° = -10.9 \text{ kJ}$$

If we know the enthalpy of CO oxidation:

$$CO + \tfrac{1}{2}O_2 \rightleftarrows CO_2 \quad \Delta H° = -283.0 \text{ kJ}$$

then what is $\Delta H°$ for the following reaction?

$$Fe_3O_4 \rightleftarrows 3Fe + 2O_2$$

Although it may not be immediately apparent to you, the top two can be rearranged and added to produce the equation in question. The corresponding $\Delta H°$ values are treated accordingly.

$$\begin{array}{ll}
Fe_3O_4 + 4CO \rightleftarrows 3Fe + 4CO_2 & \Delta H° = -10.9 \text{ kJ} \\
4(CO_2 \rightleftarrows CO + \tfrac{1}{2}O_2) & \Delta H° = 4(+283.0 \text{ kJ}) \\
\hline
Fe_3O_4 \rightleftarrows 3Fe + 2O_2 & \Delta H° = +1,121.1 \text{ kJ}
\end{array}$$

Example 8.4. Use Table 8.1 in the text to calculate $\Delta H°$ for the following reaction.

$$3C_2H_{2(g)} \rightleftarrows C_6H_{6(l)}$$

$$\Delta H° = \sum \Delta H_f°(C_6H_6) - \sum 3\Delta H_f°(C_2H_2)$$

Chemical Thermodynamics   Chapter 8

$$\Delta H° = 49.08 \text{ kJ} - 3(227.48 \text{ kJ}) = -6.33.36 \text{ kJ}$$

Remember that the coefficients of all products and reactants must be included in such a calculation.

Hess's Law states that the enthalpy of a reaction is the same regardless of how the reaction is carried out. Therefore a process may be completed in one step or in 20 steps, but as long as the starting materials and the products are the same the enthalpy change will be the same. Hess' findings for enthalpies now apply to all state functions. Compare Hess's Law with our earlier definition of a state function - they are the same!

Each substance absorbs heat differently, depending on its phase and the number and types of bonds. The amount of heat needed to raise the temperature of one mole of a substance by 1°C or 1 K is called the *molar heat capacity*. Heat capacity measured at constant pressure is C_p while heat capacity measured at constant volume is C_v. The units of molar heat capacity are J/mol-K. The general relationships between heat and heat capacities are shown below.

$$q_v = n \int C_v dT \qquad q_p = n \int C_p dT$$

Over normal temperature ranges, C_v and C_p are usually constant. Then the equations can be simplified.

$$q_v = nC_v \Delta T = \Delta E \qquad q_p = nC_p \Delta T = \Delta H$$

For a given solid or liquid substance, the C_p value is approximately the same as the C_v value since temperature effects on the pressure and volume of a solid or liquid are quite small. For an ideal gas, $C_p = C_v + R$ where R = 8.314 J/mol-K.

Example 8.5. How much heat is needed to raise the temperature of 1000 g of H_2O from 298 K to 308 K at constant pressure? (C_p $H_2O_{(l)}$ = 75.3 J/mol-K.) Under the same conditions, how much heat would be required for 1000 g of mercury? ($C_p Hg_{(l)}$ = 27.9 J/mol-k.)

$$q_p = nC_p\Delta T = \left(\frac{1000 \text{ g}}{18 \text{ g/mol}}\right)(75.3 \text{ J/mol-K})(308-298 \text{ K})$$

$$q_p = 41,800 \text{ J for } H_2O$$

$$q_p = nC_p\Delta T = \left(\frac{1000 \text{ g}}{200.6 \text{ g/mol}}\right)(27.9 \text{ J/mol-K})(308-298 \text{ K})$$

$$q_p = 1,390 \text{ J for Hg}$$

We can use heat capacities to calculate how ΔH changes with temperature.

$$\Delta H^0_{T_2} = \Delta H^0_{T_1} + \int_{T_1}^{T_2} \Delta C_p^0 \, dT$$

where $\Delta C_p^0 = \sum C_p(\text{products}) - \sum C_p(\text{reactants})$. Usually ΔC_p^0 changes little with temperature and can be brought outside the integral leading to the simpler equation below.

$$\Delta H^0_{T_2} = \Delta H^0_{T_1} + \Delta C_p^0 (T_2 - T_1)$$

Example 8.6. Use the following balanced reaction and the tabulated data to calculate ΔH° at 298 K and at 400 K assuming DC_p^0 to be a constant over that temperature range.

	$N_{2(g)}$ +	$3H_{2(g)}$ ⇌	$2NH_{3(g)}$	
ΔH_f^0 (kJ/mol)	0	0	-46.11	@298K
C_p^0 (J/mol-K)	29.125	28.824	35.564	@298K

$$\Delta H^0_{298} = [2(-46.11)] - [0 + 3(0)] = -92.22 \text{ kJ}$$

$$\Delta C_p^0 = [2(35.564)] - [29.125 + 3(28.824)] = -44.469 \frac{\text{J}}{\text{K}}$$

$$\Delta H^0_{400} = \Delta H^0_{298} + \Delta C_p (400 - 298 \text{ K})$$

$$\Delta H^0_{400} = -92{,}220 \text{ J} + (-44.469 \text{ J/K})(400 - 298 \text{ K})$$

$$\Delta H^0_{400} = -96{,}760 \text{ J} = -96.76 \text{ kJ}$$

8.5. Criteria for Spontaneous Change

A *reversible process* is one carried so that the system and surroundings are seemingly at equilibrium. Any small change in conditions will shift the process, thus it is reversible. The state functions of the system and the surroundings never differ by more than an infinitesimal amount. But many of these very small changes can add up to a large change. A process carried out reversibly always requires the least amount of work done on the system or, said in another way, the system can do a maximum amount of work on the surroundings. Thus reversible processes are the most efficient.

Chemical Thermodynamics Chapter 8

An *irreversible process* is not the most efficient as the changes that occur are larger and cannot be reversed by a simple change of direction. Since $w_{rev} < w_{irrev}$, the first law of thermodynamics predicts that $q_{rev} > q_{irrev}$. This inequality will become very important to the second law of thermodynamics.

8.6. Entropy and the Second Law of Thermodynamics

Entropy was defined in Chapter 3 as a measure of randomness or disorder. Now we shall look at how the entropy change of a system, ΔS, relates to other thermodynamic functions. By definition,

$$\Delta S = \int \frac{dq_{rev}}{T}$$

We must always use the *reversible* heat to calculate ΔS, regardless of whether or not the process is reversible or irreversible.

Entropy is central to the second law of thermodynamics which states that the total entropy of a system and its surroundings (ie. the universe) must increase in a spontaneous process. Stated in another way, the second law modifies the first law; all the heat exchanged between a system and its surroundings cannot be converted to useful work. There is always some heat lost to entropy in a spontaneous process.

Under isothermal conditions:

$$\Delta S = \frac{q_{rev}}{T}$$

If the isothermal process is a physical change, like the expansion or contraction of a gas (as opposed to a chemical change) then $\Delta E = 0$ because the internal energy of a gas can't change unless the temperature does.

$$\Delta E = 0 = q + w$$

$$q_{rev} = -w_{rev} = \int P dV = \int \frac{nRT}{V} dV = nRT \ln \frac{V_2}{V_1}$$

$$\text{So} \quad \Delta S = \frac{q_{rev}}{T} = nR \ln \frac{V_2}{V_1}.$$

This result makes qualitative sense since $\Delta S = +$ for an increase in disorder which accompanies an increase in volume, and $\Delta S = -$ for a decrease in disorder which accompanies a contraction in volume.

Example 8.7. Predict the sign of ΔS for the following processes.

a) melting a solid
b) expanding a gas at constant temperature
c) dissolving a solid in a liquid
d) dissolving a gas in a liquid
e) $H_{2(g)} + \frac{1}{2}O_{2(g)} \rightarrow H_2O_{(l)}$

Answers:

a) a liquid has more disorder or freedom than a solid, so ΔS = +
b) greater volume means greater disorder so ΔS = +
c) two pure substances are mixed which increases disorder, ΔS = +
d) this is tricky: gases have much more entropy than liquids and solids, so although a mixture is made here, the loss of a gas means ΔS = -
e) using similar reasoning, $1\frac{1}{2}$ moles of gas are converted to 1 mole of liquid, the loss of gas moles means ΔS = -

---

The ability to predict the sign of ΔS for a system (or any other function) is useful and usually not too difficult. The second law of thermodynamics can be used to predict $DS_{universe}$.

$$\Delta S_{universe} = + \text{ for spontaneous processes}$$

When a process is reversible, $\Delta S_{system} = -\Delta S_{surroundings}$

$$\Delta S_{universe} = \Delta S_{system} + \Delta S_{surroundings}$$

$$\Delta S_{universe} = 0 \text{ for reversible processes}$$

Of course $\Delta S_{universe}$ is never negative or else the second law of thermodynamics would not be valid. For calculation purposes we shall usually confine ourselves to ΔS_{system}.

Entropy, like enthalpy or internal energy, varies with temperature and depends on heat capacity.

$$\Delta S = \int \frac{dq_{rev}}{T} = \int \frac{nC_p dT}{T} \quad \text{at constant pressure}$$

$$\Delta S = \int \frac{nC_v dT}{T} \quad \text{at constant volume}$$

If C_v and C_p are constant with temperature:

Chemical Thermodynamics   Chapter 8

$$\Delta S = nC_p \ln\frac{T_2}{T_1} \quad , \quad \Delta S = n\, C_v \ln\frac{T_2}{T_1}$$

8.7. The Molecular Interpretation of Entropy

The bulk property of disorder also can be understood on the molecular scale. Statistically, entropy is a measure of how many different ways something can be organized internally to yield the same external condition. Let's use five coins to illustrate how different internal arrangements (microstates) can lead to the same external arrangement (macrostate). The coins are identical. Let's label them 1 through 5 in order to keep our microstates straight. The macrostate of all coins with heads showing has only one way to form - one microstate.

```
Coin:  1  2  3  4  5
       H  H  H  H  H
```

This is a very ordered, low entropy condition, but its probability is also low. Now let's look at the macrostate of 4 heads and 1 tail. This can be arrived at by 5 different microstates shown below:

```
Coin:  1  2  3  4  5
       H  H  H  H  T
       H  H  H  T  H
       H  H  T  H  H
       H  T  H  H  H
       T  H  H  H  H
```

This macrostate is more random and thus has higher entropy than the macrostate with all heads. There is a higher probability of forming this macrostate because there are more possible microstates.

Although molecules aren't like coins, every mole of a substance has 10^{23} atoms that populate closely spaced energy levels in a statistical manner. There are more microstates than you could possibly count. Molar entropy is a measure of these.

Example 8.8. For each of the following pairs, choose the condition of greater entropy:

  a) two poker hands, one with two pairs and one with a full house
  b) five coins showing 3 heads and 2 tails and five coins showing 1 head and 4 tails
  c) a bookshelf with one book on each of six shelves and a bookshelf with all six books on the same shelf

Answer: The first choice in all three instances is more probable, has more microstates and has greater entropy.

The entropy of a phase change, done at constant pressure, is both simple to calculate and easy to understand qualitatively. Melting a solid involves breaking down the crystal structure. A liquid has less rigid molecular interactions and so more molecular freedom and $DS_{fusion} = +$. Boiling a liquid should have a large increase in entropy since all molecular interactions must be

Chapter 8 — Chemical Thermodynamics

broken. So $\Delta S_{vap} = +$.

$$\text{Since} \quad \Delta S° = \frac{q_{req}}{T}, \quad \text{and} \quad q_p = \Delta H°,$$

$$\Delta S°_{\text{phase change}} = \frac{\Delta H°_{\text{phase change}}}{T}$$

Example 8.9. The enthalpy of vaporization of ethanol is 42.3 kJ/mol and the entropy of vaporization is 106 J/mol-K. What is the normal boiling point of ethanol?

$$\Delta S°_{vap} = \frac{\Delta H°_{vap}}{T}$$

$$106 \text{ J/mol-K} = \frac{42{,}300 \text{ J/mol}}{T}$$

$$T = 300 \text{ K}$$

8.8. Absolute Entropies and the Third Law of Thermodynamics

As long as a molecule or atom is moving it should have some disorder. If all movement could be stopped, a molecule would have no entropy. Theoretically, at absolute zero (zero K), a perfect crystal (one with no defects) would have no entropy. This is the third law of thermodynamics and it gives us a non-arbitrary starting point for entropy values.

$$S_0 = 0 \quad \text{for all substances}$$

$$\text{So} \quad S_T^0 = \int_0^T \frac{C_p}{T} \, dT$$

Entropies for substances at 298 K are found by measuring C_p from 0 to 298 K and then plotting C_p/T vs T. S_T^0 is the area under such a curve. S_{298}^0 values are tabulated for easy use (see Table 8.3 in text).

Because entropy is a state function $\Delta S°$ for a reaction can be found if all the S^0 values are known for the products and reactants. The method is the same one used for $\Delta H°$ of a reaction.

$$\Delta S° = \sum S^0(\text{products}) - \sum S^0(\text{reactants})$$

Chemical Thermodynamics   Chapter 8

Example 8.10. First predict the sign of ΔS^0_{298}, then calculate the value for the combustion of propane.

$$C_3H_{8(g)} + 5O_{2(g)} \rightleftarrows 3CO_{2(g)} + 4H_2O_{(l)}$$

We predict $\Delta S^0_{298} = -$ because 6 gas moles of reactants are converted to only 3 gas moles of products.

$$\Delta S° = \sum S^0(\text{products}) - \sum S^0(\text{reactants})$$

$$\Delta S° = [3(213.63)+4(69.91)] - [(198.41 + 5(205.029)]$$

$$\Delta S° = -303.03 \text{ J/mol-K}$$

8.9. Free Energy

Thus far in thermodynamics our only criterion for spontaneity has been $\Delta S_{\text{universe}}$, which is positive for spontaneous processes. We would rather have a system property that predicts spontaneity. This state function is Gibbs free energy, G. At constant pressure and temperature,

$$\Delta G_{\text{system}} = \Delta H_{\text{system}} - T\Delta S_{\text{system}}$$

$\Delta G = 0$ for reversible processes or a system at equilibrium

$\Delta G = -$ for a spontaneous, irreversible process

$\Delta G = +$ for a non-spontaneous process, the reverse reaction is spontaneous

The equation for ΔG shows how enthalpy and entropy both contribute to the overall determination of spontaneity. Table 8.1 below shows all permutations. A negative ΔH value is favorable and so is a positive ΔS value. But neither alone is sufficient for predicting spontaneity.

Chapter 8   Chemical Thermodynamics

Table 8.1 Contribution of ΔH and ΔS Values to Spontaneity

ΔH	ΔS	ΔG	Comments
−	+	−	The process is spontaneous at all temperatures.
+	−	+	The process is non-spontaneous at all temperatures.
+	+	?	The process is spontaneous only at high temperatures.
−	−	?	The process is spontaneous only at low temperatures.

ΔG^0_{298} values for a reaction can be easily calculated by using ΔH_f° values and S^0_{298} values.

Example 8.11. Limestone, $CaCO_3$, can form quicklime, CaO, by evolving CO_2. Is this process spontaneous at 298 K? Assuming ΔS° and ΔH° do not change with temperature (not always a good assumption) at what temperature does the process become spontaneous?

$$CaCO_{3(s)\text{calcite}} \rightleftarrows CaO_{(s)} + CO_{2(g)}$$

$$\Delta H^\circ = [(-635.09) + (-393.509)] - [-1206.09] = 177.91 \text{ kJ/mol}$$

$$\Delta S^\circ = [39.75 + 213.63] - [92.9] = 160.48 \text{ J/mol-K}$$

$$\Delta G^0_{298} = \Delta H^\circ - T\Delta S^\circ = 177{,}910 \text{ kJ/mol} - 298 \text{ K}(160.48 \text{ J/mol-K})$$

$$\Delta G^0_{298} = 130{,}000 \text{ J/mol} = 130 \text{ kJ/mol}$$

So the reaction is not spontaneous at 298 K, but since ΔH = + and ΔS = +, it should become spontaneous if the temperature is raised high enough. To find out what temperature this is we can solve for T when $\Delta G^\circ = 0$. At this point ΔG° is no longer positive, so any higher T will give a negative value.

$$\Delta G^\circ = 0 = \Delta H^\circ - T\Delta S^\circ$$

$$0 = 177{,}910 \text{ kJ/mol} - T(160.48 \text{ J/mol-K})$$

$$T = 1108 \text{ K}$$

Chemical Thermodynamics　　　　　　　　　　　　　　　　　　　　　　　　　　　　Chapter 8

This means that when T > 1108 K the reaction will be spontaneous, providing ΔS° and ΔH° are the same at 298 K as at 1108 K.

8.10. Free Energy and Equilibrium Constants

In addition to tabulated ΔH_f° and $S°$ values, it is frequently convenient to use standard molar free energy values (ΔG_f°). The arbitrary zero point for free energies of formation is the same as used for enthalpies.

$$\Delta G_f^\circ \equiv 0 \text{ for all elements in their standard phase at 298 K}$$

Thus we have a second method at 298 K to find ΔG° for a reaction because, as with other state functions, it can be found stoichiometrically.

$$\Delta G_{rxn}^\circ = \sum \Delta G_f^\circ(\text{products}) - \sum \Delta G_f^\circ(\text{reactants})$$

Example 8.12. Calculate ΔG° for the following reaction at 298 K using DG_f° values in Table 8.4 of the text. Compare with the answer to example 8.11.

$$CaCO_{3(s)\text{calcite}} \rightleftarrows CaO_{(s)} + CO_{2(g)}$$

$$\Delta G° = [(-604.05) + (-394.359)] - [1128.84]$$

$$\Delta G° = 130.43 \text{ kJ/ mol}$$

We must remember that ΔG° (ΔH° and ΔS° too) indicate which direction a reaction will proceed spontaneously under *standard state conditions* only. But rarely do we start with 1.0 atm of each gas and 1.0 M of each solution. So while standard state conditions, are convenient for calculations they are not normally encountered in everyday life.

Another misconception that may have developed during our treatment of free energy and spontaneity also needs to be dispelled. *Just because a reaction is spontaneous does not mean that all the reactants are consumed and only products remain.* Not all reactions go to virtual completion. Any spontaneous reaction stops when equilibrium is attained. So ΔG° indicates how much free energy is released or absorbed, starting with standard state conditions and ending with equilibrium conditions. Therefore we need to know the relationship between ΔG° and K, and between ΔG° and ΔG (the non-standard state free energy).

$$\Delta G = \Delta G° + RT \ln Q$$

where Q = reaction quotient

Chapter 8   Chemical Thermodynamics

From this general relationship, several other special equations can be derived. Under standard state conditions, all concentrations are 1 M and all partial pressures are 1 atm. That means Q = 1.

$$\text{At standard state } Q = 1$$

$$\Delta G = \Delta G°$$

When equilibrium conditions prevail, there is no more free energy available from a process and $\Delta G = 0$.

$$\text{at equilibrium} \quad \Delta G = 0$$

$$\text{and} \quad Q = K$$

$$\Delta G = 0 = \Delta G° + RT \ln K$$

$$\Delta G° = -RT \ln K$$
$$\text{or}$$
$$K = e^{-\Delta G°/RT}$$

Recall that the form of Q and K use *activities* and so are dimensionless.

Example 8.13. Find the free energy absorbed or released by the following reaction when 1.0 atm N_2, 3.0 atm H_2, and 1.0 atm NH_3 are mixed at 25 °C. Calculate K.

$$N_{2(g)} + 3H_{2(g)} \rightleftarrows 2NH_{3(g)} \quad \Delta G°_{298} = -32.96 \text{ kJ}$$

$$\Delta G = \Delta G° + RT \ln \frac{(P_{NH_3})^2}{P_{N_2}(P_{H_2})^3}$$

$$\Delta G = -32.96 \text{ kJ} + (8.314 \times 10^{-3} \frac{\text{kJ}}{\text{mol-K}})(298 \text{ K}) \ln \frac{1.0^2}{1.0(3.0)^3}$$

Chemical Thermodynamics   Chapter 8

$$\Delta G = -41.13 \text{ kJ}$$

The equilibrium constant is calculated from $\Delta G°$ at the appropriate temperature.

$$K = e^{-\Delta G°/RT}$$

$$K = e^{32.96/(8.314\times 10^{-3})298}$$

$$K = 5.99 \times 10^5$$

Example 8.14. Assuming that $\Delta H°$ and $\Delta S°$ are invariant with temperature, calculate the equilibrium constant for the following reaction at 298 K and at 600 K. Use Tables 8.1 and 8.2 in the text.

$$Ag_{(s)} + \tfrac{1}{2}Cl_{2(g)} \rightleftarrows AgCl_{(s)}$$

$$\Delta H° = [-127.068] - [0 + \tfrac{1}{2}(0)] = -127.068 \text{ kJ}$$

$$\Delta S° = [96.2] - [42.55 + \tfrac{1}{2}(222.957)] = -57.83 \text{ J/K}$$

$$\Delta G° = \Delta H° - T\Delta S°$$

$$\Delta G°_{298} = -127.068 \text{ kJ} - (298 \text{ K})(-0.05783 \text{ J/K})$$

$$\Delta G°_{298} = -109.8 \text{ kJ}$$

$$K_{298} = e^{-\Delta G°/RT} = e^{109.8/(8.314\times 10^{-3})298} = 1.79 \times 10^{19}$$

$$\Delta G°_{600} = -127.068 \text{ kJ} - (600 \text{ K})(-0.05783 \text{ J/K})$$

$$\Delta G°_{600} = -92.37 \text{ kJ}$$

Chapter 8 — Chemical Thermodynamics

$$K_{600} = e^{92.37/(8.314\times 10^{-3})600} = 1.10 \times 10^8$$

8.11. Electrochemical Cells

We've seen that free energy is related to the equilibrium constant. We also looked at the relationship between electrical potential and the equilibrium constant. Now we can interrelate K, $\epsilon°$, $\Delta H°$, and $\Delta S°$. The Nernst equation can be derived very easily.

$$\Delta G = \Delta G° + RT \ln Q$$

$$\Delta G = -nF\epsilon \;,\; \Delta G° = -nF\epsilon°$$

$$-nF\epsilon = -nF\epsilon° + RT \ln Q$$

which rearranges to give the Nernst equation

$$\epsilon = \epsilon° - \frac{RT}{nF} \ln Q$$

Example 8.15. Refer to example 8.14. Calculate $\epsilon°$ for $Ag + \tfrac{1}{2}Cl_2 \rightleftarrows AgCl$ at 298 K.

$$\Delta G°_{298} = -109.8 \text{ kJ} \;,\; K = 1.79 \times 10^{19}$$

$$\Delta G° = -nF\epsilon°$$

$$-109.8 \text{ kJ} = -1 \text{ mol } e^- \left(\frac{96.5 \text{ kJ}}{\text{mol } e^-\cdot v}\right)\epsilon°$$

$$\epsilon° = 1.14 \text{ v}$$

8.12. Temperature Dependence of Equilibria

You may think that chemists delight in deriving as many mathematical formulas as possible just so students have to memorize them. Rather, we advocate using as few basic relationships as possible, and deriving the particular equation that is needed. An exception to this is the equation to calculate how K changes with temperature.

Chemical Thermodynamics   Chapter 8

Qualitatively, Le Chatelier's principle predicts that if a reaction is endothermic, K will increase with temperature, and if a reaction is exothermic, K will decrease with temperature.

$$\text{heat} + A \rightleftarrows B \quad \text{as T increases, K increases}$$

$$C \rightleftarrows D + \text{heat} \quad \text{as T increases, K decreases}$$

The exact dependence is easily derived (see text) and leads to:

$$\ln\left(\frac{K_2}{K_1}\right) = -\frac{\Delta H°}{R}\left(\frac{1}{T_2} - \frac{1}{T_1}\right)$$

assuming that $\Delta H°$ doesn't change with temperature.

Example 8.16. In example 8.14 we calculated the equilibrium constant at 298 K and 600 K for $Ag + \frac{1}{2}Cl_2 \rightleftarrows AgCl$. Use only $\Delta H° = -127.068$ kJ and $K_{298} = 1.79 \times 10^{19}$ to show that $K_{600} = 1.10 \times 10^8$.

$$\ln\left(\frac{K_2}{K_1}\right) = -\frac{\Delta H°}{R}\left(\frac{1}{T_2} - \frac{1}{T_1}\right)$$

$$\ln\left(\frac{K_2}{1.79 \times 10^{19}}\right) = \frac{127.068 \text{ kJ}}{8.314 \times 10^{-3} \text{ kJ/mol-K}}\left(\frac{1}{600 \text{ K}} - \frac{1}{298 \text{ K}}\right)$$

$$K_2 = 1.10 \times 10^8$$

Example 8.17. What is $\Delta H°$ for a reaction whose equilibrium constant triples in value when the temperature is raised from 27°C to 110°C?

$$\ln\left(\frac{K_2}{K_1}\right) = -\frac{\Delta H°}{R}\left(\frac{1}{T_2} - \frac{1}{T_1}\right)$$

$$\ln\left(\frac{3}{1}\right) = \frac{-\Delta H°}{8.314 \times 10^{-3} \text{ kJ/mol-K}}\left(\frac{1}{383 \text{ K}} - \frac{1}{300 \text{ K}}\right)$$

$$\Delta H° = 12.6 \text{ kJ}$$

Chapter 8 — Chemical Thermodynamics

8.13. Colligative Properties

The equation showing how K varies with T can be used to derive the expressions for boiling point elevation and freezing point depression. The results are:

$$\Delta T_b = K_b m \quad \text{where} \quad K_b = \frac{RT_b^2 MW_1}{1000 \Delta H_{vap}^o}$$

$$\Delta T_F = K_f m \quad \text{where} \quad K_f = \frac{RT_f^2 MW_1}{1000 \Delta H_{fus}^0}$$

Example 8.18. Estimate the boiling point constant for water. Why is this number different than the actual K_b for water (0.51 K/molal)?

$$H_2O_{(l)} \rightleftarrows H_2O_{(g)}$$

$$\Delta H_{vap}^0 = [-241.818] - [-285.830] = 44.01 \text{ kJ/mol}$$

$$K_b = \frac{RT_b^2 MW}{1000 \Delta H_{vap}^o}$$

$$K_b = \frac{(8.314 \times 10^{-3} \text{kJ/mol-K})(373 \text{ K})^2 (18.0 \text{ g/mol})}{(1000 \text{g/kg})(44.01 \text{ kJ/mol})}$$

$$K_b = 0.473 \text{ K-kg/mol} = 0.473 \text{ K/molal}$$

Although this value is close to the actual number, the difference is more than negligible. One of our basic assumptions is at fault. ΔH_{vap}^o at 298 K is not identical to ΔH_{vap}^o at 373 K. There is a change in ΔH° with temperature due to DC_p°. At its normal boiling point, H_2O has a vaporization enthalpy of 40.7 kJ/mol. Substituting this value in the above calculation results in $K_b = 0.512$ K/molal, the correct number.

8.14. Heat Energies

Scientists, conservationists, and the general public continue to dream of energy processes that are 100% efficient. After all, a coal fired generating plant operating at only 40% efficiency

Chemical Thermodynamics   Chapter 8

would seem to be wasting 60% of its energy. By following the reasoning behind the Carnot cycle outlined in the text you can see that efficiency (η) can only approach 100% when the heat sink temperature (T_c) approaches zero K or when the heat reservoir temperature (T_h) approaches infinity. Now 40% efficiency doesn't sound so bad.

$$1 - \frac{T_c}{T_h} = \eta$$

Problems

1. A gas expands against a constant external pressure of 2.00 atm, increasing in volume to 3.40 L. Simultaneously the system absorbs 400 J of heat from the surroundings. What is ΔE for the gas in joules? in calories?

2. The temperature of 2.0 moles of helium drops, at a constant pressure of 1.0 atm, from 298 K to 250 K. What are ΔE, q, and w of the gas?

3. An ideal gas is isothermally and freely expanded (free expansion means into a vacuum) from 2 L to 10 L. What are ΔE, q, and w of the gas?

4. An ideal gas absorbs 1100 J of heat without undergoing any change of temperature. How much work is done on the gas?

5. A small swimming pool containing 175 M^3 of water is heated from 15.5 °C to 23.0 °C. Use the heat capacity of water (C_p) to calculate how much heat is required for the pool, assuming no heat loss to the environment.

6. How much heat is required to raise the temperature of 0.500 mol of argon gas from 20 °C to 85 °C? (C_p of Ar = 20.786 J/mol-K).

7. If exactly 4.00 g of aluminum is burned in excess O_2 to form Al_2O_3 at a constant pressure and at 298 K, 123.7 kJ of heat is evolved. What is ΔH_f° for Al_2O_3?

8. When 5000 J of heat were added to 100.0 g of CH_3OH there was a temperature increase of 19.6 °C. What is the molar heat capacity of CH_3OH under this constant pressure condition?

9. The reaction below was carried out under constant volume conditions at 298 K, where 627.0 kJ of heat were evolved. What are ΔE and ΔH?

$$Ba_{(s)} + O_{2(g)} \rightleftarrows BaO_{2(s)}$$

10. Calculate $\Delta H°$ for the reaction below at 298 K. Assuming DC_p does not change, calculate $\Delta H°$ at 500 K. (ΔC_p = 1.02 J/ mol−K)

$$4NH_{3(g)} + 7O_{2(g)} \rightleftarrows 4NO_{2(g)} + 6H_2O_{(g)}$$

11. Given that $\Delta H°$ for the combustion of one mole of glucose is -1259.8 kJ, what is ΔH_f° for glucose?

Chapter 8   Chemical Thermodynamics

$$C_6H_{12}O_{6(s)} + 6O_{2(g)} \rightleftarrows 6CO_{2(g)} + 6H_2O_{(l)}$$

12. Find ΔH_f° for $CuCl_2$ using the following equations.

$$2Cu_{(s)} + Cl_{2(g)} \rightleftarrows 2CuCl_{(s)} \quad \Delta H^\circ = -274 \text{ kJ}$$

$$2CuCl_{2(s)} \rightarrow 2CuCl_{(s)} + Cl_{2(g)} \quad \Delta H^\circ = 165 \text{ kJ}$$

13. For each reaction below, decide whether the entropy will increase, decrease, or stay the same.

   a) $2CO_{(g)} + O_{2(g)} \rightarrow 2CO_{2(g)}$

   b) $Na_2O_{(s)} + 2H_{2(g)} \rightarrow 2NaH_{(s)} + H_2O_{(g)}$

   c) $CH_{4(g)} + Cl_{2(g)} \rightarrow CH_3Cl_{(g)} + HCl_{(g)}$

   d) $PbCl_{2(s)} + H_2O_{(l)} \rightarrow Pb^{2+}_{(aq)} + 2Cl^{-}_{(aq)}$

14. What are ΔS°_{298} and ΔS°_{400} for the following equation, assuming ΔC_p to be constant? ($\Delta C_p = 31.3$ J/mol–K)

$$4NH_{3(g)} + 5O_{2(g)} \rightleftarrows 4NO_{(g)} + 6H_2O_{(g)}$$

15. The heat of vaporization of water at a constant pressure of 1.0 atm is 40.7 kJ/mol at 100°C. What is ΔS° for this process?

16. Carbon tetrachloride, CCl_4, has $\Delta H^\circ_{vap} = 32.5$ kJ/mol and $\Delta S^\circ_{vap} = 93$ J/mol–K at 25°C. Assuming ΔH° and ΔS° don't change with temperature, estimate the boiling point of CCl_4.

17. Trouton's rule states that $\Delta S^\circ_{vap} \simeq 88$ J/mol–K for many substances. If ΔH°_{vap} for benzene is 33.85 kJ/mol and we assume benzene behaves according to Trouton's rule, what is the estimated normal boiling point of benzene?

18. What is the standard free energy change at 298 K and at 350 K for the following reaction.

$$4NH_{3(g)} + 5O_{2(g)} \rightleftarrows 4NO_{(g)} + 6H_2O_{(g)}$$

19. Calculate K for the reaction above at 25°C.

20. The K_a for acetic acid is 1.8×10^{-5}. What is the standard free energy of dissociation of CH_3COOH in water at 25°C?

21. The K_a for chloroacetic acid, $ClCH_2COOH$, is 1.53×10^{-3} at 0°C at and 1.23×10^{-3} at 40°C. What is ΔH° of dissociation for the acid in water?

Chemical Thermodynamics   Chapter 8

22. The standard free energy change of the reaction below at 500 K is 35.4 kJ. What is the partial pressure of N_2O_4 at equilibrium with 2.0 atm of NO_2 at 500 K?

$$2NO_{2(g)} \rightleftarrows N_2O_{4(g)}$$

23. If the standard free change of the following reaction is 30.4 kJ at 298 K, what is the equilibrium partial pressure of CO_2?

$$CaCO_{3(s)} \rightleftarrows CaO_{(s)} + CO_{2(g)}$$

24. Consider the dissolving of $BaSO_4$ in water.

$$BaSO_{4(s)} \rightleftarrows Ba^{2+}_{(aq)} + SO^{2-}_{4(aq)}$$

$$K_{sp} @ 18°C = 8.7 \times 10^{-11}$$

$$K_{sp} @ 50°C = 2.0 \times 10^{-10}$$

   a) Is $BaSO_4$ more or less soluble at 50°C than at 18°C?
   b) What are the values of $\Delta G°$ at both temperatures?
   c) Is this reaction endothermic or exothermic?

25. Suppose a chemist wished to synthesize N_2O using the following reaction. Estimate the temperature needed to obtain an equilibrium mixture with all partial pressures equal to 1.0 atm.

$$2N_{2(g)} + O_{2(g)} \rightleftarrows 2N_2O_{(g)}$$

$$\Delta H° = 164 \text{ kJ} \qquad \Delta S° = -149 \text{ J/K}$$

26. Consider the following redox reaction

$$Cr_2O_7^{2-} + 14H^+ + 6Br^- \rightarrow 2Cr^{3+} + 7H_2O + 3Br_2$$

$$\epsilon° = 0.28 \text{ v}$$

   a) What is $\Delta G°$ of this cell at 298 K?
   b) What is the equilibrium constant at 298 K?
   c) What are ϵ and ΔG when $[Cr_2O_7^{2-}] = 0.100$ M, pH = 4.00 and $[Cr^{3+}] = 3.0 \times 10^{-2}$ M?

Chapter 8 — Chemical Thermodynamics

27. Use Table 8.1 in the text to calculate $\Delta H°$ for boiling of benzene, C_6H_6. Use this $\Delta H°$ value to calculate K_b for benzene. Compare this to the actual value, $K_b = 2.53°C/m$.

CHAPTER 9
CHEMICAL KINETICS

9.1. Concentration Effect

Stoichiometry and thermochemistry allow us to calculate the direction in which a reaction proceeds and how much of each product and reactant there will be at equilibrium. Neither tells us how long it takes to reach equilibrium. *Kinetics* is the study of rates of reaction. Not only can we learn about reaction speed from kinetics but frequently we can gain an understanding of the steps involved in a reaction.

In the simple reaction below, let's consider what happens if we start only with A and B.

$$A + 2B \rightarrow C$$

As time passes, the concentrations of A and B decrease and the concentration of C increases until equilibrium is reached. The reaction rate can be described in terms of the disappearance of A or B, or the appearance of C. In differential terms:

$$\frac{d[C]}{dt} = \text{the rate of appearance of C with time}$$

$$\frac{d[A]}{dt} = \text{rate of disappearance of A with time}$$

$$\frac{d[B]}{dt} = \text{rate of disappearance of B with time}$$

Reaction rates are always positive values so a disappearance will be preceded by a negative sign. This means that if we want to equate the rate of appearance of A we must include a sign change.

$$\frac{d[C]}{dt} = -\frac{d[A]}{dt} = \text{rate of reaction}$$

Because A and C have the same stoichiometric coefficient in the balanced equation, the equation above is logical. But what about B? Reactant B is used up at twice the rate as A, or A is used up half as fast as B.

$$\text{rate} = -\frac{d[A]}{dt} = \frac{d[C]}{dt} = -\frac{1}{2}\frac{d[B]}{dt}$$

The units of rate are variable. Commonly we will use molarity/second for solutions and atmospheres/second for gas phase reactions. Many reactions occur more slowly than can be measured in seconds; so M/minute or even M/year may be appropriate.

Chapter 9 — Chemical Kinetics

How the rate of a reaction depends on the concentrations (or pressures) of all products and reactants is given by the *differential rate law*. For our example reaction, $A + 2B \rightarrow C$, the general form of the differential rate law is:

$$\text{rate} = -\frac{d[A]}{dt} = -\frac{1}{2}\frac{d[B]}{dt} = \frac{d[C]}{dt} = k[A]^x[B]^y[C]^z$$

The numbers x, y, and z are called *orders* and are usually whole numbers or zero, showing the exact nature of dependence. If an order is positive, increasing the concentration of that substance leads to an increase in rate. If an order is negative, increasing the concentration of that substance slows down the reaction. Frequently an order will be zero, indicating there is no effect on rate when that substance is increased or decreased in concentration. The sum $x + y + z$ = the overall order of the reaction. *The values of x, y and z can be found only by experimentation and do not necessarily correspond to the coefficients in the balanced equation.* The number represented by k in the rate law is the *rate constant*. Its value is specific to the reaction being considered and at the temperature being considered.

Example 9.1. A reaction is studied in which H_2 gas is consumed. If, at a constant temperature, the concentration of H_2 changes from 0.0280 M to 0.0236 M in 15.0 minutes, what is the rate of reaction based on H_2?

$$-\frac{d[H_2]}{dt} = -\frac{(0.0236-0.0280)M}{15.0 \text{ min}} = 2.93 \times 10^{-4} \text{ M/min or } 1.76 \times 10^{-2} \text{ M/hr}$$

A useful procedure for experimentally determining a rate law is the *initial rate method*. By starting a reaction with only pure reactants and no products, a chemist can monitor the progress of the reaction by measuring the formation of a product. There will not be sufficient quantities of the product to expect a dependence on its concentration. This method is illustrated in the next example.

Example 9.2. The oxidation of manganate ion by periodate ion in basic solution has the following stoichiometry:

$$2MnO_4^{2-} + H_3IO_6^{2-} \rightarrow 2MnO_4^- + IO_3^- + 3OH^-$$

The product is permanganate ion and has an easily observable color. Four separate experiments are carried out varying the starting quantities of the two reactants. Determine the experimental rate law.

Chemical Kinetics Chapter 9

Experiment	Initial [MnO_4^{2-}]	Initial [$H_3IO_6^{2-}$]	Initial Rate $d[MnO_4^-]/dt$
1	1.6×10^{-4} M	3.1×10^{-4} M	2.6×10^{-6} Ms^{-1}
2	1.6×10^{-4}	6.2×10^{-4}	2.6×10^{-6}
3	3.2×10^{-4}	3.1×10^{-4}	1.0×10^{-5}
4	6.4×10^{-4}	3.1×10^{-4}	4.1×10^{-5}

In order to determine the order with respect to each of the reactants, we must see how the rate changes when the concentration of each reactant changes. Consider experiments 1 and 2: [MnO_4^{2-}] remains constant but [$H_3IO_6^{2-}$] doubles with no corresponding rate change. This indicates that the rate does not depend on [$H_3IO_6^{2-}$] and that the order is zero.

$$\text{rate } \alpha \ [H_3IO_6^{2-}]^0$$

Now consider experiments 1 and 3: the [$H_3IO_6^{2-}$] is held constant and the [MnO_4^{2-}] is doubled with a corresponding four-fold increase in the rate. The same observation is made when comparing experiments 3 and 4. Doubling [MnO_4^{2-}] quadruples the rate. Since the dependence must be to a power we need only decide what power relates 2 and 4. Obviously it is a square dependence.

$$\text{rate } \alpha \ [MnO_4^{2-}]^2$$

The rate law therefore is:

$$\text{rate} = \frac{d[MnO_4^-]}{dt} = k \ [MnO_4^{2-}]^2 [H_3IO_6^{2-}]^0$$

or

$$\frac{d[MnO_4^-]}{dt} = k \ [MnO_4^{2-}]^2$$

Once the rate law is established it is a simple matter to calculate k. Use the rate law and any set of data to find k. An average of the four possible values is best.

From experiment 1:

$$\text{rate} = k\,[MnO_4^{2-}]^2$$

$$2.6 \times 10^{-6} \text{ M s}^{-1} = k(1.6 \times 10^{-4} \text{ M})^2$$

$$1.02 \times 10^2 \text{ M}^{-1} \text{ s}^{-1} = k$$

From experiments 2, 3 and 4: $k = 1.02 \times 10^2$, 9.77×10^1, and 1.00×10^2 M^{-1} s^{-1}, respectively. The *average* value is:

$$k = 1.00 \times 10^2 \text{ M}^{-1} \text{ s}^{-1}$$

The differential rate law is used to show how rate varies with concentration of reactants or products. Chemists also are interested in knowing how concentration varies with time. For instance, if two reactants are mixed together at a certain temperature we would like to know how to calculate the quantity of reactants that remains after 15 minutes. The differential rate law is not much help for this calculation. But the *integrated rate law* directly relates time and concentration. The calculus is basic and the general forms of the common rate laws are derived in the text. Below is a list of the differential and integral rate law forms for zero, first and second orders.

Experimentally, directly measuring concentration vs. time allows k to be calculated *if* the order of the reaction is known.

Order	Rate Law Differential Form	Rate Law Integrated Form
0	rate = $\dfrac{dc}{dt} = kC^0$	$C_0 - C = kt$
1	rate = $\dfrac{dc}{dt} = kC$	$\ln\dfrac{C}{C_0} = -kt$
2	rate = $\dfrac{dc}{dt} = kC^2$	$\dfrac{1}{C} - \dfrac{1}{C_0} = kt$

A first order reaction requires that lnC vs t be plotted so that the slope yields -k, but for a second order reaction 1/C vs t would be plotted and the slope would be k. Therefore it is important both to discover the order of the reaction (differential rate law) and to find k (via integrated rate law).

Chemical Kinetics   Chapter 9

A *half-life* is the time it takes for half of a reactant to be consumed; that is, the time required for C to equal $\frac{1}{2}$ C.

$$t_{\frac{1}{2}} = \text{half-life}$$

zero order $\quad t_{\frac{1}{2}} = \dfrac{C_0}{2k}$

first order $\quad t_{\frac{1}{2}} = \dfrac{0.693}{k}$

second order $\quad t_{\frac{1}{2}} = \dfrac{1}{kC_0}$

Note that the half-life of a first-order reaction depends only on k and not on C. This means that a first order reaction has a consistent half-life, while a zero-order or second-order reaction has a changing half-life that depends on the remaining concentration of reactant. This can be useful for determining the reaction order.

Example 9.3. Determine the reaction order with respect to reactant A in the following experiment.

$$A \rightarrow B$$

time (hr)	[A]
0.0	1.00 M
2.0	0.50 M
4.0	0.33 M
6.0	0.25 M

Your should observe that the half-life is changing. The first $t_{\frac{1}{2}}$ is 2 hours while the second $t_{\frac{1}{2}}$ is 4 hours. This doubling of successive half-lives indicates a second-order reaction. This inverse relationship between half-life and concentration is present only in second order, or higher, reactions.

$$t_{\frac{1}{2}} = \dfrac{1}{kA_0}$$

$$\text{rate} = k[A]^2$$

Chapter 9 — Chemical Kinetics

Example 9.4. The first-order disproportionation of hydrogen peroxide, at some temperature, has a rate constant of 0.0410 min^{-1}. If the [H$_2$O$_2$] is originally 0.650 M, how much peroxide remains after 12.0 minutes? What is the half-life of H$_2$O$_2$ at this temperature? The balanced equation, while not needed to solve the problem, is given below.

$$2H_2O_{2(aq)} \rightarrow 2H_2O_{(l)} + O_{2(g)}$$

Since the reaction is first-order (even if it was stated above, the units of k, t^{-1}, are for a first-order reaction) we can use the integrated rate law to solve the problem.

$$\ln \frac{[H_2O_2]}{[H_2O_2]} = -kt$$

$$\ln \frac{[H_2O_2]}{0.650} = -(0.0410 \text{ min}^{-1})(12.0 \text{ min})$$

$$[H_2O_2] = 0.397 \text{ M after 12.0 minutes}$$

The half-life is found from k.

$$t_{\frac{1}{2}} = \frac{0.693}{k} = \frac{0.693}{0.0410 \text{ min}^{-1}} = 16.9 \text{ min}$$

What if a reaction is more complex than those described thus far? Although the initial rate method can be used, it is not practical for reactions that are very fast or very slow. A popular kinetic technique for studying complex reactions involves *flooding* the system with one or more reactants so that the concentrations don't change during the course of the reaction. Other reactants can be monitored with time. This method simplifies the data interpretation and is best shown with an example.

Example 9.5. A hypothetical reaction involving A and B as reactants was found to be first-order in A when B was in 100-fold excess ([B]$_0$ = 1.5 M), and had a rate constant of 8.3 × 10^{-3} sec^{-1}. When A was in excess, the reaction was found to be second order in B. What is the rate law? What is the true value of k?

Since the reaction is first-order in A and second-order in B, the rate law is:

$$\text{rate} = k[A][B]^2$$

But the observed rate constant in the first experiment is not the true rate constant because [B] was so much greater than [A] and thus didn't change. Therefore [B] was a constant in the first

Chemical Kinetics　　Chapter 9

experiment and was included in the observed rate constant, k_{obs}.

$$k_{obs} = k_{real}[B]_0^2$$

$$8.3 \times 10^{-3} \text{ sec}^{-1} = k_{real}(1.5 \text{ M})^2$$

$$k_{real} = 3.69 \times 10^{-3} \text{ M}^{-2} \text{ sec}^{-1}$$

Sometimes these flooding experiments are called pseudo-rate experiments.

9.2. Reaction Mechanisms

Chemically similar reactions do not necessarily have similar rate laws. For example the gas-phase reactions below have very different rate laws.

$$H_2 + I_2 \rightleftarrows 2HI \qquad H_2 + Br_2 \rightleftarrows 2HBr$$

$$\text{rate} = k[H_2][I_2] \qquad \text{rate} = k[H_2][Br_2]^{\frac{1}{2}}$$

In this example the rate laws are an indication that, although the balanced equations appear similar, the path of the reaction, called the *mechanism*, must be different in each case. A mechanism consists of a series of steps or elementary processes. Each elementary process involves the collision of the reactant molecules. A bimolecular elementary process involves a two-molecule collision, termolecular involves a three-molecule collision and a unimolecular process has only one reactant molecule. A collision of more than three molecules at the same time is improbable enough that it needn't be considered. *The order and the molecularity of an elementary process are the same.* This means that a rate law for an elementary process can be written from its balanced equation.

An *overall* balanced equation is frequently the sum of two or more elementary processes. The slow step in this mechanism is always the rate-determining step since the overall process can occur no faster than its slowest step. An experienced kineticist can usually suggest at least one mechanism, and usually several, that coincide with the observed rate law for a reaction. A new student of kinetics probably will have difficulty working from experimental data to theoretical mechanism. The logic involved may become clearer if you can work from the opposite direction; from possible mechanism to consistent rate law. This is accomplished in many cases by following two steps. If a plausible mechanism is presented: 1) the rate of the overall reaction can be equated with the rate of the slow step, which can be written from the stoichiometry, then 2) this rate law may only contain a dependence on product or reactant concentrations and not on any intermediate or activated-complex species.

Example 9.6. The reaction $I^-_{(aq)} + ClO^-_{(aq)} \xrightarrow{OH^-} IO^-_{(aq)} + Cl^-_{(aq)}$ is thought to occur via the following three-step mechanism. If the mechanism is plausible, it must be consistent with the observed rate law below. Show this to be true.

$$\frac{d[Cl^-]}{dt} = \frac{k[I^-][ClO^-]}{[OH^-]}$$

Mechanism: $ClO^- + H_2O \underset{k_{-1}}{\overset{k_1}{\rightleftarrows}} HClO + OH^-$ (fast equilibrium)

$$I^- + HClO \overset{k_2}{\rightarrow} HIO + Cl^- \quad \text{(slow)}$$

$$OH^- + HIO \overset{k_3}{\rightarrow} H2O + IO^- \quad \text{(fast)}$$

Since the second step is the slow one, we can write directly from the stoichiometry

overall rate = rate of step 2

$$\frac{d[Cl^-]}{dt} = k_2[I^-][HClO]$$

This is not an acceptable rate law form however because HClO is neither a product nor a reactant. It is an intermediate formed in the fast equilibrium step 1. Since this equilibrium controls the [HClO], we can write the equilibrium constant for step 1 and solve for [HClO].

$$K_1 = \frac{[HClO][OH^-]}{[ClO^-]}$$

$$[HClO] = \frac{K_1[ClO^-]}{[OH^-]}$$

Now this can be substituted for [HClO] in the rate law.

$$\frac{d[Cl^-]}{dt} = k_2[I^-][HClO]$$

$$\frac{d[Cl^-]}{dt} = k_2 K_1 \frac{[I^-][ClO^-]}{[OH^-]}$$

This is the observed rate law, where

$$k_2 K_1 = k_{observed}$$

Chemical Kinetics Chapter 9

and OH^- is an inhibitor because it slows the reactions.

Sometimes all steps in a mechanism take place at comparable speeds, with no single step being significantly slower than any other. In such a case we use the *steady-state approximation*, where an intermediate species is formed at approximately the same rate as it is consumed. Derivation of the relationships as well as an excellent example can be found in chapter 9 of the text.

Another class of reactions is called *chain reactions,* where an intermediate, usually a radical, is used up and regenerated quickly. A chain reaction is different from the steady-state condition discussed above. Once the *initiation reaction* produces the radical, the subsequent *propagation* steps can repeat many times before termination or re-initiation occurs.

9.3. Reaction Rates and Equilibria

When equilibrium is reached, the rate of the the forward reaction is equal to the rate of the reverse reaction. Thus equilibrium is dynamic. In the text the relationship $K_1 = k_1/k_{-1}$ is derived for an *elementary* process but can be shown to be true for any overall process.

At equilibrium:

$$\text{rate forward} = \text{rate back}$$

$$K = \frac{\text{all forward rate constants, multiplied}}{\text{all reverse rate constants, multiplied}}$$

Example 9.7. For the reaction $NO_{2(g)} + F_2 \rightleftarrows 2NO_2F_{(g)}$, the rate law for forward reaction near equilibrium is:

$$\text{rate} = -\frac{d[F_2]}{dt} = k[NO_2][F_2]$$

Determine a consistent rate law for the reverse reaction near equilibrium.

$$K = \frac{[NO_2F]^2}{[NO_2]^2[F_2]} = \frac{k_{forward}}{k_{back}}$$

At, or near, equilibrium the rate forward = rate back. We can rearrange the equilibrium expression so that the measured rate forward is on one side of the equation.

$$\frac{k_f}{k_b} = \frac{[NO_2F]^2}{[NO_2]^2[F_2]}$$

Chapter 9   Chemical Kinetics

$$k_f[NO_2][F_2] = k_b \frac{[NO_2F]^2}{[NO_2]}$$

rate forward = rate back

$$\text{rate back} = k_b \frac{[NO_2F]^2}{[NO_2]}$$

9.4. Collision Theory of Gaseous Reactions

Collision theory presents a molecular interpretation of the rate constant, k. A reaction can be represented as a series of elementary processes. The reactants of an elementary process must collide in order to react, according to this theory. At room temperature and 1 atmosphere pressure there would be a collision rate of $\sim 10^{29}$ molecules/cm^3–s or $\sim 10^8$ M/s. Obviously most reactions don't take place as quickly as the molecules collide. This means that many collisions do not lead to products. Bimolecular values of k as large as 10^8 M/s are rare. (Generally the value of k is indicative of reaction speed. The larger the value of k, the faster the reaction.) According to collision theory there are two other factors that contribute to k besides the number of collisions.

Not only must reacting molecules collide, they must have the proper orientation during collision and have sufficient energy to react. A crude but effective analogy can be drawn from pocket billiards. In order to get a ball in the pocket there must first be a collision with the cue ball, at the proper angle and with enough momentum. All three criteria must be met before the desired result is attained. Similarly, the reaction rate, and thus k of, a bimolecular process depends on these criteria. In mathematical form, as presented in your text:

$$k_{\text{bimolecular}} = \left[\left(\frac{8\pi k_B T}{\mu}\right)^{\frac{1}{2}} \rho^2 \right] \times p \times e^{-E_a/RT}$$

$\left(\dfrac{8\pi kT}{\mu}\right)^{\frac{1}{2}} \rho^2$ = # of bimolecular collisions where k_B is the Boltzmann constant, μ is reduced mass and ρ is internuclear distance of the colliding species

p = steric factor or orientation factor

$e^{-E_a/RT}$ = fraction of molecules with sufficient energy to react--this critical energy is called the activation energy, E_a

While the collision number depends on temperature, the mass of colliding species and the internuclear distance during collision, the value of the collision number does not vary enough to account for the wide variety of rate constants observed by chemists. Most reactions speed up when the temperature is raised but this cannot be totally accounted for by an increase in the

Chemical Kinetics — Chapter 9

number of collisions. The dominant factor is the term which includes the activation energy. While E_a does not change with temperature, the fraction of molecules that contain that critical energy rises (see the Maxwell-Boltzmann distribution curve in figure 9.4 of the text). This temperature sensitivity explains why rates and k increase with temperature. The activation energy is always ≥ 0 since it is essentially an energy barrier that reactants must surmount.

Both forward and reverse reactions have E_a values and are related to the overall difference in energy between the products and reactants.

$$\Delta E = E_a \text{ forward} - E_a \text{ back}$$

Remember that $\Delta E \simeq \Delta H$ for reactions where $\Delta n_{gas} = 0$.

9.5. Temperature Effects

We already have discussed the qualitative effect of raising the temperature. Due to more molecules having the necessary energy to react, reactions speed up and k values increase as the temperature is raised.

If the collision number and the steric factor are combined into one term that is essentially independent of temperature (not strictly true) we have the common Arrhenius form of k.

$$k = Ae^{-E_a/RT}$$

where A = Arrhenius pre-exponential factor

Since A remains constant, or nearly constant, with temperature changes, the relationship between k and T becomes:

$$\ln \frac{k_2}{k_1} = -\frac{E_a}{R}\left(\frac{1}{T_2} - \frac{1}{T_1}\right)$$

This equation form should look familiar to you because it is similar to the relationship between K and T and $\Delta H°$.

Example 9.8. The reaction $2N_2O_5 \rightarrow 4NO_2 + O_2$ has a rate constant at 300 K equal to $2.77 \times 10^{-5}\ s^{-1}$ and $E_a = 103.3$ kJ/mol. What is the rate constant at 425 K?

$$\ln \frac{k_2}{k_1} = -\frac{E_a}{R}\left(\frac{1}{T_2} - \frac{1}{T_1}\right)$$

$$\ln \frac{k_2}{2.77 \times 10^{-5}} = \frac{-103.3\ \text{kJ/mol}}{8.314 \times 10^{-3}\ \text{kJ/mol-K}}\left(\frac{1}{425\ \text{K}} - \frac{1}{300\ \text{K}}\right)$$

Chapter 9　　　　　　　　　　　　　　　　　　　　　　　　　　　　　　　　　Chemical Kinetics

$$k_2 = 5.40 \text{ s}^{-1}$$

In other words the reaction is 20,000 times faster at 425 than at 300 K.

Example 9.9. What is the activation energy of a reaction that proceeds 4.5 times faster at 35° C than at 20° C?

$$\ln \frac{4.5}{1} = -\frac{E_a}{8.314 \times 10^{-3} \text{kJ/mol-K}} \left(\frac{1}{333 \text{ K}} - \frac{1}{293 \text{ K}} \right)$$

$$E_a = 7.41 \times 10^{-2} \text{ kJ/mol} = 74.1 \text{ J/mol}$$

9.6. Rates of Reactions in Solution

While the experimental determination of the rate laws of gas and liquid phase reactions are similar, understanding the rates and rate constants on a molecular level is much more complicated for solution reactions than for gas phase reactions.

Several additional factors must be considered in solutions; the close range intermolecular forces between solvent and solute, rate of diffusion in the solvent, and any charges or polarity of species. Thus, we can consider *qualitatively* how a reaction rate in solution may depend on solvent viscosity, ionic strength, pH, etc. Quantitative treatment is beyond the scope of an introductory text.

9.7. Activated Complex Theory

An *activated complex* or *transition state* is a high-energy, loose but definite, association of reactant species that exists fleetingly. An intermediate differs from an activated complex because it has some stability and is lower in energy than an activated complex. An activated complex is symbolized by a superscript ‡.

What we can learn by investigating activated complex theory is that the rate constant, k, depends on the free energy of activation, $\Delta G°^{\ddagger}$.

$$k = \frac{k_B T}{h} e^{-\Delta G°^{\ddagger}/RT}$$

Since $\Delta G°^{\ddagger} = \Delta H°^{\ddagger} - T \Delta S°^{\ddagger}$,

$$k = \frac{k_B T}{h} e^{\Delta S°^{\ddagger}/R} e^{-\Delta H°^{\ddagger}/RT}$$

where k_B = Boltzmann's constant, h = Planck's constant, $\Delta S°^{\ddagger}$ = entropy of activation and $\Delta H°^{\ddagger}$ = enthalpy of activation.

Chemical Kinetics   Chapter 9

In many cases this correlates nicely with collision theory. The Arrhenius A factor corresponds to the $\frac{k_B T}{h} e^{\Delta S^{\circ \ddagger}/R}$ term and $e^{-E_a/RT}$ corresponds to $e^{-\Delta H^{\circ \ddagger}/RT}$ term. Although E_a and $\Delta H^{\circ \ddagger}$ are not identical, it does make sense that the two terms are similar.

The entropy of activation is a useful though complex term that can give an indication of reaction type. A negative $\Delta S^{\ddagger}$ value indicates that the activated complex has less freedom than the reactants, possibly due to the formation of new bonds. A positive $\Delta S^{\circ \ddagger}$ value indicates that the activated complex has more freedom than the reactants possibly due to bonds being broken, as in the case of one molecule dissociating into two parts.

This theory makes use of experimental rate constants to infer the structural properties of an activated complex, and provides an alternate viewpoint of reaction paths. Collision theory and activated complex theory each provide a unique perspective.

9.8. Catalysis

A *catalyst* is a substance that speeds up a reaction but is not required by the overall stoichiometry of the reaction. An *inhibitor* slows down a reaction.

It is important to remember that a catalyst operates by providing a pathway with a lower activation energy for *both* the forward and the reverse reaction. In other words, a catalyst helps the reaction get to equilibrium faster but a catalyst does not change the position of equilibrium.

A catalyst might be recognized in a reaction in several ways. Frequently a catalyst will be written over the arrows of a reaction to show that it is not part of the stoichiometry. In the equation below

$$C_2H_4 + H_2 \xrightarrow{Ni} C_2H_6$$

nickel metal is the catalyst. Since H_2 and C_2H_4 are gases and Ni is a solid, this is a case of *heterogeneous* catalysis. In aqueous systems OH^- and H^+ are common catalysts but since they exist in the same phase as reactions they speed up, they are *homogeneous catalysts*.

A catalyst can also be recognized from a mechanism. A catalyst is used up in one or more steps but regenerated in subsequent steps. For example, in the general mechanism of enzyme catalysis, E is the enzyme, S is the reactant substrate, P is the product, and ES are intermediate.

$$E + S \underset{k_{-1}}{\overset{k_1}{\rightleftarrows}} ES$$

$$ES \xrightarrow{k_2} E+P$$

The enzyme is regenerated and ready to catalyze another substrate molecule.

Most enzyme reactions follow that simple mechanism given above. Biochemists commonly refer to certain mathematically rearranged equations that deal with this mechanism. Two of these, the Michaelis-Menten treatment and the Lineweaver-Burke treatment, are just different ways of expressing the same kinetic equations.

$$\text{rate} = \frac{d[P]}{dt} = V$$

Michaelis-Menten equation $\quad V = \dfrac{k_2[E_0][S]}{[K_m+[S]]}$

where $\quad K_m = \dfrac{k_{-1}+k_2}{k_1}$

Lineweaver–Burke equation $\quad \dfrac{1}{v} = \dfrac{1}{V_{max}} + \dfrac{K_m}{V_{max}} \cdot \dfrac{1}{[S]}$

where $\quad \nu = $ initial rate

$V_{max} = $ maximum rate

An important result of investigating the limits of these equations is that increasing the [S] only increases the rate of reaction up to a certain point, beyond which the enzyme is being completely used at all times and the substrate molecules are competing for available enzyme molecules.

Problems

1. Consider the following reaction and rate law:

$$2NO_{(g)} + H_{2(g)} \rightleftarrows N_2O_{(g)} + H_2O_{(g)}$$

$$\text{rate} = \frac{d[H_2O]}{dt} = k[NO]^2[H_2]$$

a) If the initial rate of formation of H_2O is 6.0×10^{-3} M s^{-1}, what is the initial rate of disappearance of NO?

b) If concentration is expressed in molarity, and time in seconds, what are the units of the rate constant?

c) What is the order of the reaction with respect to NO? With respect to H_2?

d) If the initial [NO] is decreased to one-half and the initial [H_2] is increased by a factor of three, how does the initial rate change?

2. The concentration of reactant A was followed over several hours of reaction. Some data points are presented below.

$$A \rightarrow B$$

[A]	Time	[A]	Time
0.150	0	0.114 M	20 min
0.148 M	60 s	0.077 M	1.00 hr
0.141 M	4.0 min	0.052 M	2.00 hr
		0.039 M	3.00 hr

a) What is the initial rate over the first 4.0 minutes?
b) What is the rate over the first 3.0 hours?
c) Plot the data a [A] vs t, ln[A] vs t, and 1/[A] vs t to determine the reaction order. Which plot gives a straight line?
d) What is $t_{\frac{1}{2}}$?
e) What is k (expressed in seconds, minutes, and hours)?

3. The following are experimental data for the reaction below, at some constant temperature.

$$H_3AsO_4 + 3I^- + 2H^+ \rightarrow H_3AsO_3 + I_3^- + H_2O$$

$[H_3AsO_4]_0$	$[I^-]_0$	$[H^+]_0$	Initial Rate $d[I_3^-]/dt$
0.010 M	0.20 M	0.10 M	9.33×10^{-8} M min^{-1}
0.010 M	0.40 M	0.10 M	1.87×10^{-7} M min^{-1}
0.020 M	0.20 M	0.10 M	1.87×10^{-7} M min^{-1}
0.020 M	0.20 M	0.05 M	4.67×10^{-8} M min^{-1}

a) What is the rate law?
b) What is the value of k?

4. The rate constant of a pure substance that reacts in a one-step unimolecular process is 4.6×10^{-3} s^{-1}. What is the reaction rate when the reactant concentration is 5.0×10^{-2} M?

Chapter 9 — Chemical Kinetics

5. The gas phase reaction $2N_2O_5 \rightleftarrows 4NO_2 + O_2$ has the rate law: $d[O_2]/dt = k[N_2O_5]$. If the volume of the reaction vessel is suddenly reduced to one-half its original volume, how does the new rate compare with the original rate?

6. A compound decomposes by a first order reaction and the concentration falls from 0.0963 M to 0.0796 M in 5.1 minutes. What is the concentration of the compound after 9.0 minutes?

7. The decomposition of 0.50 M SO_2Cl_2 has a rate constant of 2.2×10^{-5} s^{-1} at some temperature. What is the half-life of SO_2Cl_2 at this temperature?

8. The decomposition of N_2O_5 is a first order process with a $t_{\frac{1}{2}} = 5.71$ hr at 25° C. How long will it take for 35% of the N_2O_5 to decompose?

9. The decomposition of HI to H_2 and I_2 is second order at 500° C and an initial [HI] = 0.10 M, $k = 4.74 \times 10^{-2}$ M^{-1}min^{-1}. What is the half-life of this HI? When the [HI] falls to 0.05 M, what is k and what is the half-life?

10. A reaction found to have a rate law $-dA/dt = k[A][B]$ was studied under conditions where $[A]_0 = 1.00 \times 10^{-3}$ M and $[B]_0 = 2.35$ M. The half-life of A under these conditions was 75 s. What are $k_{observed}$ and k_{real}?

11. The following mechanism is a major pathway in the hydrolysis of organic esters.

$$CH_3CH_2COOCH_3 + H^+ \underset{k_{-1}}{\overset{k_{-1}}{\rightleftarrows}} CH_3CH_2COHOCH_3^+ \quad \text{(fast)}$$

$$CH_3CH_2COHOCH_3^+ + H_2O \overset{k_2}{\rightarrow} CH_3CH_2C(OH)_2^+ + CH_3OH \quad \text{(slow)}$$

$$CH_3CH_2C(OH)_2^+ \overset{k_3}{\rightarrow} CH_3CH_2COOH + H^+ \quad \text{(fast)}$$

    a) What is the overall reaction?
    b) What substance is a catalyst?
    c) What substance is an intermediate?
    d) What is the rate law?

12. The following is a proposed mechanism for the reaction
$5Br^- + BrO_3^- + 6H^+ \rightarrow 3Br_2 + 3H_2O$.

$$BrO_3^- + 2H^+ \underset{k_{-1}}{\overset{k_1}{\rightleftarrows}} H_2BrO_3^+$$

$$Br^- + H_2BrO_3^+ \overset{k_2}{\rightarrow} Br_2O_2 + H_2O$$

Chemical Kinetics Chapter 9

$$Br_2O_2 + 4H^+ + 4Br^- \rightarrow 3Br_2 + 2H_2O \quad \text{(series of very fast reactions)}$$

a) If step 2 is the slowest, what should be the rate law?
b) If step 1 is the slowest what should be the rate law?

13. The reaction $H_2 + 2ICl \rightarrow 2HCl + I_2$ has an experimental rate law: $d[I_2]/dt = k[H_2][ICl]$. Which of the following mechanisms is the most reasonable?

mechanism 1	$2ICl + H_2 \rightarrow 2HCl + I_2$	(slow)
mechanism 2	$ICl \rightarrow I + Cl$	(slow)
	$I + ICl \rightarrow I_2 + Cl$	(fast)
	$2Cl + H_2 \rightarrow 2HCl$	(fast)
mechanism 3	$H_2 \rightarrow 2H$	(slow)
	$H + ICl \rightarrow HCl + I$	(fast)
	$I + ICl \rightarrow I_2 + Cl$	(fast)
	$H + Cl \rightarrow HCl$	(fast)
mechanism 4	$H_2 + ICl \rightarrow HI + HCl$	(slow)
	$HI + ICl \rightarrow HCl + I_2$	(fast)

14. Consider the following redox reaction and its potential.

$$V(H_2O)_6^{2+} + Ru(NH_3)_6^{3+} \underset{k_{-1}}{\overset{k_1}{\rightleftarrows}} V(H_2O)_6^{3+} + Ru(NH_3)_6^{2+}$$

$$\epsilon° = 0.35 \text{ volts}$$

If $k_1 = 1.5 \times 10^3 \text{ M}^{-1}\text{s}^{-1}$, what is k_{-1} at 298 K?

15. The reaction below has two parallel pathways.

$$H_2O_2 + 3I^- + 2H^+ \rightleftarrows 2H_2O + I_3^-$$

pathway 1, rate $= k_1[H_2O_2][I^-]$

pathway 2, rate $= k_2[H^+][H_2O_2][I^-]$

What are the rate laws for the reverse pathways, -1 and -2?

Chapter 9 — Chemical Kinetics

16. The termination step of a chain reaction mechanism may involve two radicals colliding. For example

$$H\cdot + H\cdot \rightarrow H_2$$

Radical recombinations occur at nearly the same rate as the radicals collide. What can you infer from this about p and E_a?

17. Assuming $\Delta E° \simeq \Delta H°$ for the following reaction, what is the calculated ΔH_f^0 for SiH_2? (ΔH_f^0 SiH_4 = 34.3 kJ/mol)

$$SiH_4 \rightleftarrows SiH_2 + H_2 \quad E_f = 249.4 \text{ kJ}$$

$$E_b = 8.4 \text{ kJ}$$

18. What is E_a for a reaction that proceeds 50 times faster at 350 K than at 275 K?

19. A reaction has k = 6.2 × 10^{-4} M^{-1}s^{-1} at 310 K and an activation energy of 324 kJ. At what temperature is k = 0.17 M^{-1}s^{-1}?

20. Sometimes the glass surface of a reaction vessel can act as a heterogeneous catalyst for a gas phase reaction. Can you devise a simple set of experiments to test for this?

CHAPTER 10
THE ELECTRONIC STRUCTURE OF ATOMS

10.1. Electrical Nature of Matter

Faraday's observations combined with Thomson's and Millikan's experiments, using equipment now considered obsolete, led to the discovery and characterization of the electron. The properties of the electron are:

electron charge = e = 1.60×10^{-19} C

electron mass = m_e = 9.1×10^{-31} kg

Example 10.1. Using conversion factors given in the text (Chapter 10 and appendices) calculate the mass of an electron in g and amu. Calculate the mass of a mole of electrons in kg, g, and amu.

$$m_e = 9.1 \times 10^{-31} \text{kg} = 9.1 \times 10^{-28} \text{g}$$

since 1 amu = 1.66×10^{-24} g, $m_e = 5.48 \times 10^{-4}$ amu

For one mole of electrons, simple multiply each mass by Avogadro's number.

$$m_e \times N_A = 5.48 \times 10^{-7} \text{kg} = 5.48 \times 10^{-4} \text{g} = 3.30 \times 10^{20} \text{amu}$$

Example 10.2. Convert the charge of an electron from coulombs per electron to electrostatic units per electron and to c/mol and esu/mol.

e = 1.60×10^{-19} C per electron = 96,500 C/mol (if you use 1.6022×10^{-19} C and $N_A = 6.022 \times 10^{23}$ you will get the more precise number, 96,484 C/mol--Faraday's Constant).

Since 1 esu = 3.34×10^{10} C

e = 4.8×10^{-10} esu per electron = 2.9×10^{14} esu/mol

10.2. The Structure of the Atom

Rutherford's experiment in which α-particles were scattered by gold foil shows the importance of maintaining an open mind in science. Rutherford would hardly have suspected the far-reaching effects his experiments would have. The positive charges in an atom were compressed into a very small space and the light-weight electrons occupied most of the atomic volume. This unexpected development won the Nobel Prize in Chemistry in 1908.

Example 10.3. The volume of a silver atom can be calculated from its solid density, d = 10.5 g/cm^3. Find this volume and calculate what percentage of the atomic volume is occupied by the silver nucleus which has a radius of 6.33×10^{13} cm.

Chapter 10 — The Electronic Structure of Atoms

Using the molecular weight of Ag and Avogadro's number, the volume per atom can be found.

$$\frac{10.5 \text{g/cm}^3}{108 \text{g/mol}} = 0.0972 \text{ mol/cm}^3$$

0.0972 mol/cm^3 x 6.02 x 10^{23} atoms/mol = 5.85 x 10^{22} atoms/cm^3 or 1.71 x 10^{-23} cm^3/atom

The volume occupied by the nucleus of a Ag atom is:

$$V = \frac{4}{3}\pi r^3 = \frac{4}{3}\pi (6.33 \times 10^{-13} \text{cm})^3$$

$$V = 1.06 \times^{-36} \text{ cm}^3/\text{nucleus}$$

This leads to the percentage calculation.

$$\frac{1.06 \times 10^{-36} \text{ cm}^3}{1.71 \times 10^{-23} \text{ cm}^3} \times 100 = 0.00000000000620\% \text{ of the atomic volume is occupied by the nucleus.}$$

10.3. The Origins of the Quantum Theory

Rutherford's model of an atom generated more questions than it answered. Difficulties arose because experimental evidence could not be reconciled with classical mechanics. Modifications of modifications became necessary until the current atomic theory was developed. We shall see how classical views of waves and particles eventually were brought in line with the experimental evidence and with each other.

We know that light is a type of electromagnetic radiation that travels in waves and as a wavelength (λ) and a frequency (ν). All electromagnetic radiation, in a vacuum, travels at the same speed (c). The relationship between these quantities is

$$\nu\lambda = c = 3 \times 10^8 \text{ ms}^{-1}$$

The SI unit of wavelength is meters, and of frequency is seconds^{-1} or hertz. But the common units of wavelength used by chemists and physicists are nanometers (nm), angstroms (Å) and picometers (pm).

The energy associated with electromagnetic radiation is given by the equation:

$$E = h\nu = \frac{hc}{\lambda}$$

where h = Planck's Constant

The Electronic Structure of Atoms   Chapter 10

$$= 6.626 \times 10^{-34} \text{ J·S}$$

But in order to explain the non-continuous absorption or emission of light by metal surfaces, the wave model of light had to be modified. Planck said energy occurred in discrete packets or *quanta*. Einstein called these quanta of light *photons*. So light was not only described as a wave, but on a more fundamental level was composed of particles, photons.

Example 10.4. A sodium vapor lamp of the type used in most street lights emits light with a wavelength of 589 nm. What is the frequency and energy of this light?

$$\nu = \frac{c}{\lambda} = \frac{2.998 \times 10^8 \text{ms}^{-1}}{589 \times 10^{-9} \text{ m}} = 5.09 \times 10^{14} \text{s}^{-1} \text{ or Hertz}$$

$$E = h\nu = (6.626 \times 10^{-34} \text{ J·s})(5.09 \times 10^{14} \text{ s}^{-1})$$

or

$$E = 203 \text{ kJ/ mol of photons}$$

Niels Bohr applied quantum theory to the emission spectra of atoms and succeeded in deriving the relationship below to account for the quantized light emission spectra of hydrogen, called the Balmer Series.

$$\nu = \left(\frac{1}{n_2^2} - \frac{1}{n_1^2}\right) \frac{e^2}{8h\pi\epsilon_o a_o}$$

where

e = charge of the H nucleus

ϵ_o = constant = $8.854 \times 10^{-12} c^2 J^{-1} M^{-1}$

a_o = Bohr radius of the electron in hydrogen = 0.529 Å

Qualitatively, Bohr was saying that the electron in an H atom was "orbiting" the nucleus at a set distance, a_o. This distance could be changed by the absorption of energy which would allow the electron to move from its home orbit (n = 1) to a new, more distant orbit (n = 2, 3, 4, etc.). When this energetically excited electron falls back down to its home orbit (the one of lowest energy) it emits the excess energy as light. Because only certain orbits are permitted, only certain wavelengths of light can be absorbed or emitted. The collection of constants, $e^2/8h\pi\epsilon_o a_o$, divided by c, have come to be known as the Rydberg Constant, R_H.

$$R_H = \frac{e^2}{8h\pi\epsilon_o a_o c} = 1.097 \times 10^7 M^{-1}$$

Bohr's introduction of quantized orbits for electrons had no theoretical basis at the time and unfortunately the equation did not work for any atom except those with only one electron.

Chapter 10　　　　　　　　　　　　　　　　　　　　　　　The Electronic Structure of Atoms

The modification of Bohr's model was the next step.

Example 10.5. For an H atom calculate the frequency, wavelength and energy for the emission line n = 3 to n = 1.

$$\frac{1}{\lambda} = \frac{\nu}{c} = \left(\frac{1}{n_2^2} - \frac{1}{n_1^2}\right) R_H$$

$$\frac{1}{\lambda} = \left(\frac{1}{1^2} - \frac{1}{3^2}\right) 1.097 \times 10^7 M^{-1} = 9.75 \times 10^6 M^{-1}$$

$$\lambda = 1.03 \times 10^{-7} M = 103 nm$$

$$\Delta E = \frac{hc}{\lambda} = \frac{(6.626 \times 10^{-34} J \cdot s)(2.998 \times 10^8 ms^{-1})}{1.03 \times 10^{-7} m} = 1.93 \times 10^{-20} J \text{ per photon}$$

$$\Delta E = 11.6 \text{ kJ/ mol}$$

It quickly became obvious that for multielectron atoms, the nuclear charge, Z, was different for each electron. This is logical because the net attraction that an electron experiences should depend on its distance from the nucleus as well as how many inner electrons have orbits in this intervening space. Thus an outer electron is "screened" from the nucleus by inner electrons. The *effective nuclear charge*, Z_{eff}, must always be less than or equal to the real nuclear charge, Z.

10.4. Quantum Mechanics

By the 1920's, the wave and particle nature of light was well demonstrated. If light could behave as both waves and particles, why couldn't matter exhibit some wave characteristics? Louis de Broglie in his two page doctoral thesis of 1924 said that matter could have a wavelength given by the equation below:

$$\lambda = \frac{h}{mv}$$

Where mv = mass x velocity = momentum for heavy objects like a baseball or a car, the wavelength is very short and so the wave motion is indiscernible. But small particles like electrons should have important wave properties that must be considered.

Example 10.6. An electron in the n = 1 Bohr orbit has a velocity of 2.2 x $10^6 ms^{-1}$. What

The Electronic Structure of Atoms — Chapter 10

is its de Broglie wavelength?

$$\lambda = \frac{h}{mv} = \frac{6.626 \times 10^{-34} \text{J·s}}{(9.1 \times 10^{31}\text{kg})(2.2 \times 10^{6}\text{ms}^{-1})}$$

$$\lambda = 3.3 \times 10^{-10}\text{m} = 0.33\text{nm} = 3.3\text{Å}$$

A direct consequence of this wave particle duality is that very small particles will be fuzzy. Imagine viewing a object under a microscope that is illuminated with light of $\lambda = 500$ nm. You won't be able to distinguish details smaller than 500 nm. Therefore particle details at a distance $\leqslant 500$ nm will be uncertain. If $\Delta\chi$ is this uncertainty position:

$$\Delta\chi = \lambda$$

and since

$$\Delta\chi \Delta(mv) = h$$

In more polished form this is *Heisenberg's Uncertainty Principle:*

$$\Delta\chi \, \Delta(mv) \geqslant \frac{h}{4\pi}$$

The more accurately the position of a particle is known, the more uncertain is the momentum and vice versa.

Physicist Erwin Scrödinger developed wave equations which took this wave-particle duality into account. Very simply stated, he used de Broglie's wave expression, $\lambda = h/mv$ to replace λ in the wave equations. The mathematics involved are rigorous. Of primary importance to chemists are the constraints and physical descriptions of electron behavior that result from the Schrödinger equation.

10.5. The Hydrogen Atom

The Schrödinger equation correctly predicts that electrons are found in a three-dimensional space surrounding a nucleus, not in a planetary Bohr orbit. Bohr's quantum number, n, does have meaning however. It is one of four quantum numbers that describe the space occupied by an electron (hereafter called an orbital).

The four quantum numbers and to what they refer, physically, are given below.

The principal quantum number, n, defines the energy level, or shell, where the electron can be found; n can have whole numbers values starting with 1.

The angular momentum quantum number, l, defines the sub-shell that an electron may occupy. The angular momentum defines the shape of this subshell. l can have values of positive whole numbers ranging from zero up to n-1. Thus shell number 2 (n = 2) can have two subshells because l = 0 and l = 1. The subshells have more common chemical references. When l = 0 we are referring to an s subshell, l = 1 is a p subshell, l = 2 is a d subshell and l = 3 is an f subshell.

The *magnetic quantum number*, m_l, depends on the angular momentum quantum number, and can have integer values between $-l$ and $+l$ including zero. The number of m_l values indicates how many energetically identical orbitals are within each subshell. For example, in a p subshell, $l = 1$ so $m_l = -1, 0, +1$ indicating there are three orbitals within a p subshell. An s subshell, $l = 0$, has only one orbital, $m_l = 0$. A d subshell, $l = 2$, has five orbitals, $m_l = -2, -1, 0, +1, +2$.

The last quantum number is m_s, the *spin quantum number*. Each electron is a charged particle spinning about its own axis, which generates a magnetic field perpendicular to its spin direction. The spin quantum number can have only two possible values, $m_s = +1/2$ or $-1/2$ depending on the direction of electron spin.

When the single electron of an H atom is in its most stable configuration or *ground state*, it must be as close to the nucleus as possible. The nearest orbital to the nucleus is found in the $n = 1$ shell. The electron will have the following quantum numbers describing its position: $n = 1$, $l = 0$, $m_l = 0$, possible values of m_s, each orbital can hold two electrons each with a different spin quantum number.

We will use an additional shorthand notation when referring to the orbitals occupied by electrons. In the case of a ground state H atom, the electron is said to be in the 1s orbital (from $n = 1$ and $l = 0$, the s orbital). Since there is only one electron in the orbital, the configuration is written $1s^1$. The H electron can be promoted or excited to a higher energy state such as $2s^1$ or $2p^1$ or $3s^1$, depending on the energy absorbed by the electron. Figures 10.13, 10.14 and 10.16 show the shapes of the s, p, and d orbitals respectively. Although the 1s, 2s, 3s, 4s, etc. orbitals all have a similar shape, their radial probability density functions are different. For instance, an electron in a 1s orbital is more likely to be found closer to the nucleus than a 2s electron.

Keep in mind that the orbital diagrams are a three-dimensional picture of the probability, $|\psi|^2$, of finding an electron. Electrons do not orbit the nucleus as Bohr thought, but rather exist in a cloud of probability whose shape defines where the electron may be found 99% of the time.

10.6. Multielectron Atoms

We would like to be able to construct a "picture" or electron configuration for an atom with many protons and electrons. The techniques we shall use to arrive at a ground state configuration are based on several rules, some empirical and some theoretical. The *Pauli exclusion principle* states that no two electrons in the same atom can have the same four quantum numbers. When ranking the shells and subshells from low to high energy it is generally true that $n = 1 < n = 2 < n = 3$, ... etc. and that $s < p < d < f$ for any given n value. There is some overlap in these generalities that occurs because of two interacting factors. Usually the further away the orbital is from the nucleus, the higher its energy, but an s orbital penetrates closer to the nucleus than a p or d orbital. As a result, the 4s orbital is slightly lower in energy than the 3d orbitals. The overall order of filling for orbitals can be found in figure 10.18 of the text.

The electron configuration of H, Li, and Na are shown below.

H $1s^1$

Li $1s^2 2s^1$

Na $1s^2 2s^2 2p^6 3s^1$

The Electronic Structure of Atoms   Chapter 10

Notice that these three elements from group IA all have an ns^1 outer shell configuration. This similarity in electron configurations, and thus chemical behavior, is at the heart of the structure of the periodic table.

Example 10.8. Write the ground state electron configuration of P. P has 15 electrons: $1s^22s^22p^63s^23p^3$

Example 10.9. Write the ground state electron configuration of Ni.

Ni has 28 electrons: $1s^22s^22p^63s^23p^64s^23d^8$

The form of the modern periodic table was developed from similarities in chemical behavior. We now know that the underlying reason for these similarities is that the electron configuration of the outer electrons is the same. Elements in the first group (vertical column) of the periodic table all have an ns^1 configuration of their valence electrons. The group IIA elements have an ns^2 configuration, where n = principal quantum number and the period number (horizontal row) of the chart. Group IIIA elements have an ns^2np^1 configuration. Groups IVA through the noble gases (group VIIIA) have configurations from ns^2np^2 to ns^2np^6. The transition metals in the midsection of the chart all have partially filled d orbitals. The d orbitals are available starting with the 3rd quantum shell, yet the first transition metals are in period 4 because the 4s orbital is lower in energy than 3d. The lanthanide and actinide series elements have partially filled f orbitals. With some practice you should be able to determine the configuration of elements with only the periodic table.

Example 10.10. Write the electron configuration of Ar.

Argon is in group VIII, period 3 so the valence electrons are $3s^23p^6$ and all inner orbitals are filled.

Ar has 18 electrons: $1s^22ssp22p^63s^23p^6$

Example 10.11. Using argon as a base, write the electron configuration of Sc.

Scandium has 21 electrons, 3 more than the filled subshells of argon.

  Sc $[Ar]4s^23d^1$

         or

       $1s^22s^22p^63s^23p^64s^23d^1$

In order to write the electron configuration of an ion, begin with the ground state configuration of the neutral atom. For a positive ion, remove electrons beginning with the outermost quantum shell. For a negative ion, add electrons to the lowest energy orbitals available.

Example 10.12. Write the electron configuration of Se^{2-}.

Se has 34 electrons: $1s^2 2s^2 2p^6 3s^2 3p^6 4s^2 3d^{10} 4p^4$

Se^{2-} has 36 electrons: $1s^2 2s^2 2p^6 3s^2 3p^6 4s^2 3d^{10} 4p^6$

or [Kr]

Example 10.13. Write the electron configuration of Fe^{3+}.

Fe has 26 electrons: $1s^2 2s^2 2p^6 3s^2 3p^6 4s^2 3d^6$

Fe^{3+} has 23 electrons: $1s^2 2s^2 2p^6 3s^2 3p^6 3d^5$. Notice that when three electrons are removed from Fe, the first to leave are the $4s^2$ electrons because they reside in the outermost quantum shell. The 3d electrons then can be removed.

In a shell that is not completely filled, how are the electrons arranged? If we consider a nitrogen atom we see there are several possibilities, but only one corresponds to the ground state. Nitrogen has an unfilled p subshell ($1s^2 2s^2 2p^3$). There are three energetically equivalent p orbitals for the three electrons. Do the electrons pair up or spread out? *Hund's rule* states that electrons, whenever possible, will occupy as many orbitals as possible and have parallel spins. A more descriptive depiction is shown below where an arrow represents an electron. Two electrons in the same orbit must have opposing spins.

N ground state $\underline{\uparrow\downarrow}\;\underline{\uparrow\downarrow}\;\underline{\uparrow}\;\underline{\uparrow}\;\underline{\uparrow}$
$\phantom{N\ \text{ground state}\quad}$1s$$2s$$2p

N an excited state $\underline{\uparrow\downarrow}\;\underline{\uparrow\downarrow}\;\underline{\uparrow\downarrow}\;\underline{\uparrow}\;\underline{}$
$\phantom{N\ \text{an excited state}\quad}$1s$$2s$$2p

A *paramagnetic* substance has one or more unpaired electrons. A *diamagnetic* substance has all paired electrons.

Ionization energy is the amount of energy needed to remove the highest energy electron from an atom in the gas phase, I_1.

$$M_{(g)} \rightarrow M^+_{(g)} + e^-_{(g)}$$

The equation above describes the first ionization energy. The following equation would be that describing the second ionization energy.

$$M^+_{(g)} \rightarrow M^{2+}_{(g)} + e^-_{(g)}$$

I_1, I_2, I_3, etc. values are positive values because energy must be absorbed in order to remove the electron. There are periodic trends observable that correlate well with electron configurations.

The Electronic Structure of Atoms — Chapter 10

Comparing the first ionization energies of the period 3 atoms, Na through Ar, we see there is a general increase in I_1, which means that the outer electrons are held more tightly from left to right in the period. This is explained by screening. A Na atom has its outer electron in a 3s orbital while a Cl atom has its outer electron in a 3P orbital. The Cl nucleus with 17 protons is not screened as well as the sodium nucleus with 11 protons. Electrons in the same shell are less effective at screening each other.

Although there is a *general* increase in I_1 across a period, there are some explainable interruptions. In period 3, Al has a smaller I_1 than Mg, and S has a smaller I_1 than P. A look at the configurations shows why.

$$Mg(3s^2) \rightarrow Mg^+(3s^1) + e^-$$

$$Al(3s^2 3p^1) \rightarrow Al^+(3s^2) + e^-$$

$$P(3s^2 3p^3) \rightarrow P^+(3s^2 3p^2) + e^-$$

$$S(3s^2 3p^4) \rightarrow S^+(3s^2 3p^3) + e^-$$

It would seem that a filled subshell or a half-filled subshell is more stable than other configurations. Conversely, if a filled or half-filled shell results from a one electron loss it will be easier than expected. Thus P is less likely to give up an electron and S is more likely to give one up due to the half-filled configuration. Similar predictable trends are found for I_2 and I_3 values.

Ionization energies also vary for the elements in a group. For example, group IIA, Be to Ra, shows a decrease in I values from top to bottom even though all the elements have the same outer shell configuration, ns^2. This means that each quantum shell is more effective at screening than the previous one. It is easier to ionize Ra or Ba than Be or Mg.

We expect, and find, that I_2 values are considerably higher than I_1 values, and that I_3 values are higher still. It is more difficult to remove an electron from a positive species than from a neutral species.

Example 10.15. Rank the following atoms according to their expected I_1 values. Use only the periodic table.

$$O, P, Ge$$

The correct order is Ge < P < O. This can be deduced from placement on the periodic table.

Example 10.16. Which of the following species should have the smallest I value?

$$Cl^+, Cl, Cl^-$$

Since all the above species have the same nucleus (17 protons), the one with the most electrons should have the lowest I value.

$$Cl^- < Cl < Cl^+$$

Electron affinity, A, is the amount of energy required, or given off, when an electron is removed from a negative ion to form a neutral atom.

$$M^-_{(g)} \rightarrow M_{(g)} + e^-_{(g)}$$

Substances such as Mg or the noble gases have a filled shell or subshell and have low electron affinities (A < 0). Other species such as the halogens have a strong attraction for an additional electron and have a high electron affinity. The periodic trends are that A values decrease for group members from top to bottom (the same trend as for I values, for the same reasons). A values *generally* increase from left to right across a period. There are predictable interruptions as with I values; Mg has a lower A than Na and P has a lower A than Si. This is easily explained by their electron configurations.

$$Na^-(3s^2) \rightarrow Na(3s^1) + e^- \qquad Mg^-(3s^2 3p^1) \rightarrow Mg(3s^2) + e^-$$

$$Si^-(3s^2 3p^3) \rightarrow Si(3s^2 3p^2) + e^- \qquad P^-(3s^2 3p^4) \rightarrow P(3s^2 3p^3) + e^-$$

By losing an electron, Mg^- and P^- can gain a filled or half-filled shell, while Na^- and Si^- already have such a situation.

10.7. Quantum Mechanical Calculations of Atomic Properties

Calculations of electron energies in multielectron atoms must necessarily be more complex than for the simple H atom. Now, the nucleus-electron attractions must be modified by the electron-electron repulsions in the orbitals. *Hartree-Fock* calculations, which include the repulsions, accurately predict the size of atomic orbitals for multielectron atoms and give a reasonable estimate of their energy levels.

Problems

1. In Millikan's oil drop experiment would it have been possible to observe a droplet with a charge of 2.4×10^{-19} C? Explain.

2. The charge/mass ratio for a proton is 1,836 times smaller than the charge/mass ratio for an electron. What is the mass of a proton?

3. Solid selenium has a density of 4.81 g/cm^3. What is the average volume of a selenium atom?

4. If an FM radio station broadcasts at 104.5 MH$_z$, the frequency of the radiowaves is 104.5 MH$_z$. What is the wavelength of these waves? What is the energy per quanta and per mole?

5. What is the wavelength of light emitted when a hydrogen electron falls from the n = 5 level to the n = 2 level?

6. In the Lyman series, an emission line occurs for hydrogen at 97,492 cm^{-1} (cm^{-1} is a wavenumber, $\bar{\nu} = 1/\nu$). If the electron falls to n = 1, from what level did it fall?

7. What is the deBroglie wavelength of a car weighing 4,500 lbs and traveling at 40 mph?

The Electronic Structure of Atoms   Chapter 10

8. X-rays travel at the speed of light. What is the mass of an x-ray with a wavelength of 2.0 Å?

9. What orbital corresponds to the quantum numbers $n = 3$, $l = 1$, $m_l = 0$?

10. List all the possible sets of quantum numbers that a 2p electron could have.

11. Give one set of quantum numbers that the outer electron in Rb might have.

12. Which quantum shell would be the first (lowest energy) to contain a g subshell?

13. What is the total number of electrons that the $n = 4$ shell can hold?

14. Write the ground state electron configurations for the following atoms.
    a) Sr
    b) Cd
    c) C
    d) Cd
    e) V

15. Write the ground state electron configurations for the following atoms.
    a) Br^-
    b) Ti^{2+}
    c) Cu^{2+}
    d) S^{2-}
    e) Mn^{7+}

16. Predict the ground state electron configuration for Cu and Cu^{+1}. The true configurations are $[Ar]4s^1 3d^{10}$ and $[Ar]3d^{10}$. Why are these different than predicted?

17. What do Na^+, F^-, Ne, O^{2-} and Mg^{2+} have in common?

18. Name two species with the ground state configuration $[Ar]3d^6$.

19. How many unpaired electrons does Re^{3+} have?

20. Decide whether each of the following is paramagnetic or diamagnetic.
    a) B
    b) Al
    c) P
    d) Si
    e) Cu

21. Two atoms, A and B, combine to form a salt of formula A_2B. What are the likely electron configurations of A and B?

22. List the atoms, Na, Mg, Al, from low to high ionization energies.

23. Why does As have a larger I_1 value than Se?

24. Which of the following gas phase atoms should have the largest I_2 value? F, C, Li

25. List the following species from low to high electron affinity values.
    a) F, S, As, Sn
    b) Bi, Sb, As

CHAPTER 11
THE CHEMICAL BOND

11.1. Ionic Bonds

In Chapter 6 we discussed ionic and covalent bonds; we even calculated the percent ionic character of a bond from its measured dipole moment. The ionic bonds of most salts actually have some covalent character, and many covalent bonds have some polar character.

Solids with bonding that is predominantly ionic have structures quite different than covalent molecular structures. Ionic solids (and liquids) do not exist as discrete molecular units but rather have extended crystal lattice structures. Each cation is surrounded by as many anions as possible, usually 6 or 8, and each anion is surrounded by as many cations as possible, again 6 or 8. The ions are stacked, much like oranges in a crate with their edges touching. The ions can be treated as hard spheres with a measurable radius. The geometry of stacking is rather straightforward. The radius of each ion is constant enough that a table can be constructed of ionic radii. For instance, K^+ has radius of 1.33Å and F^- has a radius of 1.36Å. In solid KF the internuclear distance for adjacent K^+ and F^- ions is calculated to be $(1.33 + 1.36) = 2.69$ (Å). There is good agreement between actual internuclear distances and those calculated by totaling the ionic radii.

It is interesting to compare ionic radii with the original atomic radii. When F becomes F^- would you expect its size to increase, decrease, or remain the same? Using the theories developed in Chapter 10, we should expect F^- to be larger than F and indeed it is. Both species have 9 protons, but F has 9 electrons and F^- has 10 electrons. The tenth electron is well screened from the nucleus, and the excess negative charge results in a larger size.

$$\text{radius of F} = 0.71 \text{Å}$$

$$\text{radius of F}^- = 1.36 \text{Å}$$

The opposite trend is expected for cations. K, for instance, loses a 3s electrons in the third quantum shell and gets a net positive charge. The result is a smaller size.

$$\text{radius of K} = 2.35 \text{Å}$$

$$\text{radius of K}^+ = 1.33 \text{Å}$$

These changes in size are more dramatic in +2 and -2 ions.

Example 11.1. The species O^{2-}, F^-, Ne, Na^+ and Mg^{2+} are isoelectronic, yet their sizes vary greatly. List these species from smallest to largest.

All the ions and atoms have 10 electrons but each has a different number of protons and so a different effective nuclear charge.

Mg^{2+}	< Na^+	< Ne	< F^-	< O^{2-}
12P	11P	10P	9P	8P
$10e^-$	$10e^-$	$10e^-$	$10e^-$	$10e^-$

Ionic solids have such strong coulombic forces holding the crystal structure together that we expect the melting points and boiling points to be extremely high. This is certainly true. Why then do many salts readily dissolve in water? Hydration, dissolving in water, must more than compensate energetically for the ion-ion interactions that are lost. There are a number of different ways to view the ion-water interactions but they are generally of the Lewis acid-base type. When solid KF dissolves in water, the K^+ and F^- ions are dissociated. The K^+, a good Lewis acid, interacts with the lone electron pairs on a water molecule, a Lewis base. The F^- is a good Lewis base because of its four lone pairs of electrons which can interact with the δ^+ charge carried by the hydrogens of each water molecule. These constitute ion-dipole interactions which are quite strong.

Whether or not a salt dissolves in water can be found from the $\Delta G_{dissolving°}$, but this free energy is usually dominated by $\Delta H°_{dissolving}$. We can use a combination of thermochemical data to compute $\Delta H°_{hyd}$, the enthalpy of hydration when 1 mol of gas phase ions dissolve in water.

$$X^+_{(g)} + Y^-_{(g)} \rightarrow X^+_{(aq)} + Y^-_{(aq)}$$

$\Delta H°_{hyd}$ is negative, indicating the strength of the ion-water interactions. The charge and size of the ions are important factors in $\Delta H°_{hyd}$. For a series of +1 ions or -1 ions the size effect can be gauged. The smaller the ion is, the stronger its interaction with water, so $\Delta H°_{hyd}$ will be more exothermic. For a series of +1, +2, +3 ions the interaction increases greatly with charge, so $\Delta H°_{hyd}$ becomes very exothermic for a +3 species.

11.2. The Simplest Covalent Bonds

In ionic bonds the ions are held together by a strong coulombic attraction that is missing in covalent bonds. Covalently bonded atoms are generally similar enough in I and A values so that no electron transfer occurs, yet both atoms have some attraction for extra electrons. When the electronic properties of two atoms are identical or nearly the same, they form a *non-polar covalent bond*. A *polar covalent bond* forms when the atoms have unequally shared electrons resulting from dissimilar electronic properties.

Quantum mechanics can be applied to electrons shared by two nuclei to learn the probability of where these electrons are to be found. While we are limited to two electrons per bond in the Lewis dot model, quantum mechanics can describe bonding with one or three electrons as well. In Chapter 10 we saw that quantum mechanics gives us both a mathematical and physical description of atomic orbitals. When atoms are bonded the valence electrons and orbitals interact. We say that these electrons are in *molecular orbitals*. Molecular orbitals allow electron density to cover both atoms in a bond whereas atomic orbitals are localized on a single atom. Bonding electrons are at their lowest energy when positioned between two nuclei.

The simplest examples that we can consider are diatomic hydrogen and helium molecules. Let's begin with H_2, formed as two H atoms approach each other. Each atom has one electron in a 1s atomic orbital. As the atoms approach, the 1s orbitals overlap. Instead of viewing two distinct atomic orbitals we consider two molecular orbitals (MOs) as a *bonding molecular orbital* (σ). The bonding MO is a lower energy orbital than the antibonding MO. Both electrons reside here in the ground state. Any MO, just like any atomic orbital, can only hold two electrons. So the species H_2^- with three electrons would have a $\sigma^2\sigma^{*1}$ configuration. An H_2^+ with only one electron would have a σ^1 configuration. The bonding MO must necessarily be lower in energy than the two separate atomic orbitals. The antibonding MO is higher in energy. Figures 11.9 and 11.13 in the text show the shape of the σ and σ^* orbitals. The bonding molecular orbital has the greatest electron density between the two nuclei, and the antibonding molecular orbital has the greatest electron density away from the internuclear axis.

A dihelium species such as He$_2$ or He$_2^+$ uses only σ and σ^* orbitals from the overlap of 1s atomic orbitals. Of course the overall energy of a dihelium species is different than a dihydrogen species because of different nuclei, but the orbital interactions are similar.

Example 11.2. what is the configuration of electrons in the molecular orbitals of He$_2$ and He$_2^+$?

He$_2$	4 electrons	$\sigma^2 \sigma^{*2}$
He$_2^+$	3 electrons	$\sigma^2 \sigma^{*1}$

The He$_2$ species will not be stable because the number of electrons in the bonding MO equal the number of electrons in the antibonding MO.

Simple diatomic species such as H$_2^+$, He$_2^+$, and H$_2^-$ were not predicted or explained by Lewis dot structures because of the odd numbers of electrons, but are shown by quantum mechanics to be stable because each has more bonding electrons than antibonding electrons.

11.3. Atomic Orbitals and Chemical Bonds

The calculation method used to generate molecular orbitals for multielectron molecules is called the *linear combination of atomic orbitals* (LCAO). While solving the Shrödinger equation for the motion of electrons in the field of two fixed nuclei is an easy task, the results can be qualitatively understood by reference to wave mechanics. Again, let's consider two H atoms approaching each other, the 1s orbitals overlapping. Each 1s orbital is represented by a wavefunction, Π. When the two wavefunctions combine they can do so in a reinforcing or destructive way. The two new MOs can be represented by the addition or subtraction of the wavefunctions.

$$\sigma_2 = \Psi_a(1s) + \Psi_b(1s)$$
$$\sigma_s^* = \Psi_a(1s) - \Psi_b(1s)$$

Represented on an energy-level diagram, the separate atomic orbitals are at zero energy when separated. At the average bond distance, the MOs form σ at lower energy, σ^* at higher energy.

$$\overline{\sigma^*}$$

$$\overline{1s_a} \qquad \overline{1s_b}$$

$$\overline{\sigma}$$

We can fit the dihydrogen and dihelium species to a similar energy diagram remembering to use all previous rules for orbital-filling.

Example 11.3. Draw the ground state MO configurations for H_2 and H_2^-.

H_2 has 2 electrons, both in the σ orbitals

$$\overline{}\;\sigma^*$$
$$\underline{\uparrow}\quad\underline{\uparrow}$$
$$1s_a\quad\;1s_b$$
$$\underline{\uparrow\downarrow}\;\sigma$$

H_2^- has 3 electrons, 2 in σ, 1 in σ^*

$$\underline{\uparrow}\;\sigma^*$$
$$\underline{\uparrow}\quad\underline{\uparrow\downarrow}$$
$$1s_a\quad\;1s_b$$
$$\underline{\uparrow\downarrow}\;\sigma$$

In the Lewis dot and valence bond methods bonds may be single or double. In molecular orbital treatment these are the equivalent of a σ and π bond, respectively. σ orbitals result from atomic orbital overlap along the internuclear axis, π orbitals result from atomic orbital overlap above and below the internuclear axis as formed by two parallel p_x or two p_y orbitals. For every π molecular orbital formed there is also a π^* antibonding molecular orbital.

In the Lewis electron-dot method, lone-pair electrons were assigned to a specific atom. For instance, in the H_2O molecule, the oxygen was considered as having two lone pairs of electrons residing totally in oxygen atomic orbitals. Calculations show that these electrons, though non-bonding in nature, also reside in molecular orbitals, with density covering all atoms in the molecule. The *delocalized molecular orbital theory* considers all the electrons in any molecule to belong to all the atoms. The *localized molecular orbital theory* is a bit easier to use and localizes non-bonding electrons on one atom.

Example 11.4. Name the possible kinds of orbital overlaps that can lead to σ.

Any two atomic orbitals that overlap along the internuclear axis can form a σ orbital. This means an s orbital can overlap with *any* other orbital lobe, head on. Since the z axis is considered by convention to be the internuclear axis, $s+p_z$ forms a σ orbital. Of course $s + s$, as in the H_2 molecule, forms a σ orbital, but also an s combined with any d or f orbital that has a z component, that is, a lobe on the z axis.

Chapter 11 The Chemical Bond

Example 11.5. Name the possible kinds of orbital overlaps that can lead to π MOs.

The requirement for π orbitals is parallel atomic orbitals. So a p_x+p_x or p_y+p_y will lead to a π orbitals. Also, two parallel d orbitals, two parallel f orbitals, or a parallel p + d, p + f, or d + f can form a π orbital. The shapes of each MO depend on the wavefunctions of the atomic orbitals, so there is quite a variety.

While delocalized molecular orbital theory explains the energy levels of bonding and antibonding orbitals as well as the configuration of low-lying excited states, it fails to account for molecular geometry, and is too complicated for routine use in describing polyatomic molecules. The next section will explain a localized molecular orbital approach to the explanation of geometry.

11.4. Hybridization

The valence shell electron pair repulsion theory discussed in Chapter 6 empirically predicted that a CH_4 molecule should have tetrahedral geometry. Measurements of the CH_4 molecule show all C—H bonds to be identical in length and strength, and all bond angles to be the same. Carbon must contribute four atomic orbitals to the molecular orbitals and each hydrogen contributes one. Yet the four atomic orbitals that C uses for bonding are *not equivalent*. C has the electron configuration $1s^2 2s^2 2p^2$ and uses all four orbitals of its valence shell ($2s$, $2p_x$, $2p_y$, $2p_z$) in CH_4, yet how can four non-equivalent atomic orbitals result in four equivalent localized molecular orbitals?

Linus Pauling suggested that carbon's bonding orbitals were *hybridized*. We can think of these four hybrid orbitals as being the result of an s orbital and three p orbitals being mixed. Each hybrid orbital would have a 1/4 contribution from s and 3/4 contribution from p orbitals. Thus the term sp^3 hybrid describes the four bonding orbitals of C that overlap with the four 1s orbitals of the H atoms. And, as we know, the geometry expected for four equivalent orbitals whose electrons repel each other is tetrahedral. The shape of an sp^3 orbital is also a hybrid of 25% s and 75% p contributions.

The VSEPR and the hybridization models predict the same geometry, but hybridization specifies the orbitals used and their shapes.

The geometry of BF_3 can be described by sp^2 hybridization. Boron has an electron configuration of $1s^2 2s^2 2p^1$ and needs three equivalent orbitals for bonding. The 2s and two 2p orbitals are hybridized (the remaining p orbital is empty and un-hybridized). The result sp^2 set of orbitals have 33% s character and 67% p character. The shape the three sp^2 orbitals assume is trigonal planar and each one overlaps with a 2p of an F atom.

Formaldehyde, $H_2C=O$, is also a trigonal planar molecule. The carbon has hybridized sp^2 orbitals *in its σ framework*. The remaining unhybridized p orbital on C is situated parallel to a p orbital on O, forming a π molecular orbital. So it is the hybridized, σ orbitals that determine the geometry. The π orbitals have no density on the internuclear axis, and are positioned above and below a σ orbital. They do not contribute to the molecular geometry except as a correction to the ideal case. In the $H_2C=O$ molecule this means that the σ and π orbitals localized between the C and O occupy more space than only a σ orbital. The four electrons shared by C and O therefore repel the C—H electrons. As a consequence, the H—C—H bond angle is less than 120°.

The Chemical Bond   Chapter 11

An sp hybrid set, 50% s character and 50% p character, consists of two equivalent orbitals 180° apart. Both BeH_2 and CO_2 are examples where the central atom has sp hybridized orbitals. In CO_2 there are also two π orbitals formed.

Example 11.6. What set of hybrid orbitals should be assigned to N in NH_4^+?

All four N—H bonds are equivalent so the nitrogen atomic orbitals must be an sp^3 hybrid set.

Example 11.7. What set of hybrid orbitals should be assigned to C in CO_3^{2-}? Explain the resonance in terms of molecular orbital theory.

If necessary, use the VSEPR model to begin.

$$\begin{bmatrix} O \\ \diagdown \\ C = O \\ \diagup \\ O \end{bmatrix}^{2-}$$

We know there are also two other resonance forms. The σ framework contains three bonds so the C orbitals are sp^2 hybrids. The un-hybridized p orbital that remains is parallel to a p on oxygen and forms a π orbital. Since there are three oxygen molecules, with parallel p orbitals, the π orbital is delocalized over all four atoms. The term resonance in the VSEPR model translates to delocalized in MO theory.

Example 11.8. What hybridization is associated with the carbon atoms in acetylene, C_2H_2?

$$H-C\equiv C-H$$

The σ framework for each C is sp. Each C has two p orbitals that are not hybridized, which can form two π bonds. The molecule is linear since the sp orbitals are 180° apart.

Octet structures around a central atom are readily accounted for by sp, sp^2 and sp^3 hybrid orbitals, but many atoms ($z \geqslant 13$) have expanded octet structures. Since there is a total of only four s and p orbitals in a valence shell, an expanded octet requires additional orbitals of approximately equal energy. For the elements Al through Cl there are empty, low-lying 3d orbitals which can be hybridized with the 3s and 3p orbitals. An atom that can form five σ bonds has an sp^3d hybrid orbital set, and an atom that can form six σ bonds has an sp^3d^2 hybrid orbital set. The geometry is that predicted by VSEPR; trigonal bipyramidal for sp^3d and octahedral for sp^3d^2.

Example 11.9. What hybrid orbital set is associated with Br in the molecule BrF_3?

The Lewis dot structure gives Br five pairs of electrons, two non-bonding pairs and three bonding pairs.

Chapter 11 The Chemical Bond

$$\begin{array}{c} F \\ | \\ \ddot{\underset{..}{Br}}- F \\ | \\ F \end{array}$$

The hybridization is sp³d and the molecule is T-shaped. The non-bonding electron pairs occupy two of the trigonal planar positions

---

11.5. Results of Quantitative Bonding Calculations

This section shows that the Hartree-Fock method is successful in predicting bond energies and bond distances. Calculations of this type are computer based.

The Hückel method is a simpler technique for calculating bond energies that is especially effective for molecules with *conjugated double bonds*. Many carbon molecules have conjugation which means alternating single and double C—C bonds in a chain or ring. In fact, the Hückel method shows that treating these bonds merely as isolated single and double bonds does not show how stable they are. The delocalized molecular orbital model correctly predicts that the π electrons are delocalized over all C atoms with a subsequent lowering in energy. This delocalization is what we refer to as resonance in the valence shell model.

Problems

1. How does size vary in the ionic series I^-, Br^-, Cl^-, and F^-?

2. Predict the variations in size for H, H^+, and H^-.

3. List the following from smallest size to largest: K^+, Ar, Cl^-, Sc^{3+}, Ca^{2+}.

4. Which salt, MgF_2 or MgO, might be expected to have the greatest ΔH^o_{hyd}? What factors are most important?

5. Which ion do you expect to release the greatest amount of heat per mole when it is hydrated, Sc^{3+}, Ti^{3+}, V^{3+}?

6. The heat of hydration of a salt can be calculated by a Born-Haber cycle that takes advantage of ΔH being a state function. Using the data below calculate ΔH^o_{hyd} for $ZnBr_2$.

    $Zn^{2+}_{(g)} + 2Br^-_{(g)} \rightarrow ZnBr_{2(s)}$ lattice energy = $\Delta H^o_1 = -2615 \frac{kJ}{mol}$

    $Zn_{(s)} \rightarrow Zn_{(g)}$ heat of atomization = $\Delta H^o_2 = 130.7 \frac{kJ}{mol}$

    $Zn_{(g)} \rightarrow Zn^{2+}_{(g)}$ 2nd ionization energy = $\Delta H^o_3 = 1733.3 \frac{kJ}{mol}$

    $Br_{2(l)} \rightarrow Br_{2(g)}$ heat of vaporization = $\Delta H^o_4 = 30.9 \frac{kJ}{mol}$

    $\frac{1}{2} Br_{2(g)} \rightarrow Br_{(g)}$ bond dissociation energy = $\Delta H^o_5 = 96.4 \frac{kJ}{mol}$

The Chemical Bond   Chapter 11

$Br_{(g)} \rightarrow Br^-$ electron affinity = ΔH_6^o = $- 324.7 \dfrac{kJ}{mol}$

$ZnBr_{2(s)}$ heat of formation = ΔH_7^o = $- 328 \dfrac{kJ}{mol}$

7. Compare the molecular orbital descriptions of H_2, H_2^+, and H_2^-. Decide how the overall energy of H_2 changes when an electron is added or removed and to what these changes are due.

8. Why should He_2^+ be expected to exist but not He_2?

9. Do you expect H_2^{2-} to exist? Why?

10. How do the energies of two atomic valence orbitals compare with those of the bonding and antibonding molecular orbitals they form?

11. Draw the Lewis electron-dot structure of the molecule propene, CH_3CHCH_2.

    a) How many bonds does this molecule have? How many bonds are σ, how many π?
    b) What is the hybridization of each carbon?

12. For each of the following species, identify the orbital hybridization of the central atom.

    a) O_3 e) SO_2 i) HNO_2

    b) H_2O f) NO_3^- j) CN^-

    c) H_3O^+ g) SO_3

    d) BCl_3 h) PF_3

13. For each of the following species, identify the orbital hybridization of the central atom.

    a) PF_5 d) XeF_4

    b) ICl_4 e) SiF_6

    c) $BrCl_3$

14. There are many potentially harmful molecules produced in photochemical smog. One such molecule is peroxyacetylnitrate (PAN) with the formula CH_3COONO_2. Draw its structure and determine the orbital hybridization around each C and the N.

CHAPTER 12
MOLECULAR ORBITALS

12.1. Orbitals for Homonuclear Diatomic Molecules

The two simplest diatomics, H_2 and He_2, were discussed in Chapter 11. The molecular orbitals, σ and σ^* were formed by the linear combination of atomic 1s orbitals. If the two wave functions are reenforcing they form a bonding molecular orbital, lower in energy than the separate atomic orbital. If the wave functions interact destructively they form an anti-bonding molecular orbital, higher in energy than the separate atomic orbitals.

The situation gets more complicated when diatomic molecules are formed from elements with more than just a 1s orbital occupied. We will look at period 2 diatomics, Li_2, Be_2, B_2, etc. as we extend our molecular orbital discussion to cover the 2s and 2p orbitals of two atoms overlapping. Atoms such as Li or N have a filled 1s orbital that lies close to the nucleus. To simplify the molecular orbital description of molecules we shall consider inner core electrons, like those in the 1s, to be localized on one atom and essentially non-bonding in nature. Only the valence atomic orbitals shall be considered as contributing to the molecular orbitals.

When two 2s orbitals overlap it is analogous to the 1s situation in H_2 and He_2 but the σ_{2s} orbital formed is larger than the corresponding σ_{1s} and is of higher energy. Yet the σ_{2s} orbital is lower in energy than the separated 2s orbitals, and the σ_{2s}^* is higher in energy.

Each atom has three energetically equivalent 2p orbitals that can overlap with the 2p orbitals of a second atom. Unlike the spherically symmetric s orbitals, p orbitals can overlap in a number of different ways (see figures 12.3 and 12.4 of the text). If we assign the z coordinate to the internuclear axis, then the two $2p_z$ orbitals can meet in a head-on fashion with overlap of two lobes. This overlap along the internuclear axis constitutes a σ_p bonding orbital and σ_p^* anti-bonding orbital.

The interaction between two neighboring $2p_x$ or $2p_y$ orbitals is not an overlap along the internuclear axis, but above and below this axis. This leads to the formation of two ψ_p bonding orbitals and two ψ_p^* anti-bonding orbitals.

$$\pi 2p_x \simeq N[\psi_A(2p_x) + \psi_B(2p_x)]$$

$$\pi 2p_x^* \simeq N^*[\psi_A(2p_x) - \psi_B(2p_x)]$$

$$\pi 2p_y \simeq N[\psi_A(2p_y) + \psi_B(2p_y)]$$

$$\pi 2p_y^* \simeq N^*[\psi_A(2p_y) - \psi_B(2p_y)]$$

where N and N^* are normalization factors so that the electron is always accounted for in some space.

Molecular Orbitals   Chapter 12

Thus for period 2 diatomic molecules, a total of light atomic orbitals (the 2s and 2p orbitals from two different atoms) form light molecular orbitals, four bonding and four antibonding. But how do the σ_{2s}, σ_{2s}^*, σ_{2px}, σ_{2pz}^*, π_{2px}, π_{2px}^*, π_{2py} rank in relative energies? The answer is that the molecular orbital energy levels are different for every diatomic molecule but the period 2 homonuclear diatomic molecules fall into two general categories whose existence has been verified experimentally.

The molecules $Li_2, Be_2, B_2, C_2,$ and N_2 have molecular orbitals of the approximate ranking shown below. Note that while the energy increases from σ_{2s} to σ_{2p}^*, there is no scale listed because each molecule has a different overall energy.

$$-\sigma_{pz}^*$$

$$\pi_{px}^* - \quad - \pi_{py}^* \qquad \text{energy diagram}$$

$$-\sigma_{pz} \qquad \text{for } Li_2, Be_2, B_2,$$

$$\pi_{px} - \quad - \pi_{py} \qquad C_2, N_2$$

$$-\sigma_{2s}^*$$

$$-\sigma_{2s}$$

The molecules O_2 and F_2 have a slightly different pattern for their orbitals, as shown below.

$$-\sigma_{pz}^*$$

$$\pi_{px}^* - \quad - \pi_{py}^* \qquad \text{energy diagram}$$

$$\pi_{px} - \quad - \pi_{py} \qquad \text{for } O_2, F_2$$

$$-\sigma_{pz}$$

$$-\sigma_{2s}^*$$

$$-\sigma_{2s}$$

Chapter 12   Molecular Orbitals

The main difference between the two diagrams is the relative ranking of the σ_p and π_p orbitals. This is due to the interaction, in the case of Li_2, Be_2, B_2, C_2, and N_2, between the σ_{2s} and σ_{2s}^* electrons, and the σ_{2p} electrons. For O_2 and F_2, the σ_{2s} and σ_{2s}^* electrons do not interfere with and σ_{2p} electrons. As a possible explanation we can visualize the 2s orbitals of the lighter elements to interact so strongly that any other electrons along the internuclear axis are repelled or raised in energy. Thus the σ_{2p} for the lighter elements is raised above the energy of the π_p orbitals.

Example 12.1. Indicate the molecular orbital configuration for the O_2 molecule. How is this fundamentally different from the Lewis electron dot configuration?

$$\begin{array}{c}
-\sigma_{pz}^* \\
\\
\pi_{px}^* \;\uparrow \quad\quad \uparrow\; \pi_{py}^* \\
\\
\pi_{px} \;\uparrow\downarrow \quad \uparrow\downarrow\; \pi_{py} \\
\underline{\uparrow\downarrow}\,\sigma_{pz} \\
\underline{\uparrow\downarrow}\,\sigma_{2s}^* \\
\underline{\uparrow\downarrow}\,\sigma_{2s}
\end{array}$$

O_2 has 12 valence electrons total

The molecular orbital diagram indicates that O_2 should have two unpaired electrons and so be paramagnetic. The electron-dot model implies that all the electrons are paired in $O = O$. Experimentally it is quite a simple matter to show that O_2 *is* paramagnetic.

The bond order, or number of bonds between two atoms, can be found by the following formula.

$$\text{bond order} = \frac{\text{\# electrons in bonding orbitals} - \text{\# electrons in anti-bonding orbitals}}{2}$$

For the O_2 example above this would give $\frac{8-4}{2} = 2$ bonds, in agreement with the Lewis structure.

Higher bond orders generally indicate stronger and shorter bonds. As we shall see, 1/2 bond order increments are easily diagrammed.

Molecular Orbitals — Chapter 12

Example 12.2. Draw a molecular orbital energy diagram for C_2^+. What is the bond order?

$$—\sigma^*_{pz}$$

$$\pi^*_{px} \quad — \quad — \quad \pi^*_{py} \qquad C_2^+ \text{ has 7 valence electrons total}$$

$$—\sigma_{pz}$$

$$\pi_{px} \; \underline{\uparrow\downarrow} \quad \underline{\uparrow} \; p_{py}$$

$$\underline{\uparrow\downarrow}\,\sigma^*_{2s}$$

$$\underline{\uparrow\downarrow}\,\sigma_{2s}$$

bond order $= \dfrac{5-2}{2} = 1\tfrac{1}{2}$ bonds.

There is no simple Lewis electron-dot equivalent of 1 1/2 bonds for C_2^+.

Example 12.3. Which of the following should have the greatest dissociation energy, C_2^+, C_2, C_2^-?

	C_2^+	C_2	C_2^-
	7 valence electrons	8 valence electrons	9 valence electrons
	bond order $= 1\tfrac{1}{2}$	bond order $= 2$	bond order $= 2\tfrac{1}{2}$

Chapter 12 — Molecular Orbitals

C_2^- should have the shortest and strongest bond, therefore it will have the greater dissociation energy.

12.2. Heteronuclear Diatomic Molecules

The previous examples of diatomic molecules were homonuclear, therefore the atomic orbitals of the separate atoms were at identical energy levels, and electrons in the resultant molecular orbitals were equally shared by both atoms. In heteronuclear diatomic molecules the atomic orbitals, while close together in energy, are not equal and the electrons are not shared equally.

When the two atoms have nearly the same atomic number as in C and N or N and O, the MO diagram is similar to that of homonuclear diatomics. When the atoms have a large difference in atomic number, the bonding electrons will be more closely associated with one atom than the other leading to a polar bond, as in HF, or to the extreme case of an ionic bond in LiF.

Example 12.4. Give the electron configuration of CO and the bond order.

Since CO is isoelectronic with N_2, and C and O have similar atomic numbers, the MO description is merely a perturbation of the N_2 diagram. Since O has a higher nuclear charge than C, its 2s and 2p orbitals are at a lower energy than the carbon orbitals. Once again the core electrons in the 1s orbitals are considered to be non-bonding.

Molecular Orbitals — Chapter 12

Note that the bonding molecular orbitals are nearer in energy to the oxygen atomic orbitals while the anti-bonding molecular orbitals are nearer to the carbon atomic orbitals. This indicates that the bonding electrons spend more time under the influence of the O nucleus than the C nucleus. This leads to a dipole moment, with partial negative charge in the O and partial positive charge on the C.

$$\text{bond order} = \frac{8-2}{2} = 3 \text{ bonds}$$

Example 12.5. LiF is a good example of an ionic bond, but its electron configuration can still be represented by a molecular orbital diagram. Draw a reasonable MO diagram for this formula.

The bonding electrons come from the interaction of lithium's 2s orbital with a fluorine 2p orbital. Since F has a much higher atomic number, the 2p orbitals are quite a bit lower in energy than the 2s of Li.

$$-\sigma^*_{s+p}$$

2s ↑

lithium

↑↓ ↑↓   non-bonding   ↑↓ ↑↓ ↑ 2p

↑↓ s_{s+p} fluorine

Two of the fluorine 2p orbitals are essentially non-bonding so the energy level does not change. The σ_{s+p} is very close to the 2p energy level of the F and not energetically near the Li 2s orbital. The Li valence electron is transferred to F.

12.3. Triatomic Molecules

The big question when dealing with triatomic molecules is can we predict and explain why some are linear and some are bent? In many cases the VSEPR model makes the correct shape prediction but does not necessarily account for variations in bond energies. Molecular orbital theory provides a more satisfying, albeit more complex, treatment of triatomic molecules. The empirical results will be presented here. Students are referred to the text for the theoretical explanations.

Chapter 12   Molecular Orbitals

The geometry of triatomic hydrides and their ions depends on the number of valence electrons. The important example of CH_2 with 6 valence electrons is found to have characteristics of both the linear and bent MO diagrams below.

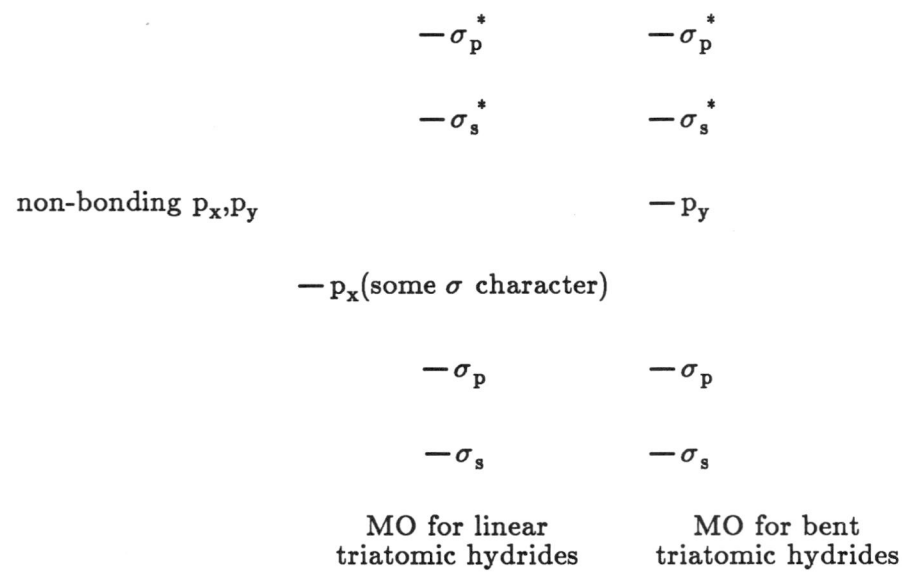

CH$_2$ has, in its ground state, two unpaired electrons as in the linear MO diagram but it is also slightly bent. Triatomic hydrides with 4 or fewer valence electrons are found to be linear in the ground state; those molecules with five or more valence electrons are bent because the p_x orbital lowers in energy as with bending. This bent configuration would be expected to become less favored if the σ_s^* and σ_p^* orbitals are occupied because these orbitals are lower in energy whenever the atoms are linear (see figure 12.13 in the text). With this in mind let's look at an example.

Example 12.6. Compare the configurations and expected geometries of BH_2^+, BH_2, and BH_2^-.

BH_2^+ has 4 valence electrons. The configuration is $\sigma_s^2\sigma_p^2$. It may not seem that either bent or linear would be preferred energetically, but the orbitals are slightly lower in energy in the linear configuration. BH_2 has 5 valence electrons. The configuration should be $\sigma_s^2\sigma_p^2 p_x^1$ and the molecule is bent. BH_2^- has 6 valence electrons. The configuration should be $\sigma_s^2\sigma_p^2 p_x^2$ and the molecule is bent.

Example 12.7. Predict the geometry of CH_2^+, CH_2^-, NH_2^-, and H_2S.
They are all predicted to be bent.

Molecular Orbitals   Chapter 12

The same question of linear or bent geometry must be asked of triatomic nonhydrides. Because the number of valence electrons may be as high as 24, the MO diagrams are messier. Yet the molecules generally follow one of the energy schemes below.

$$
\begin{array}{ll}
 & -\sigma^*p_z \\
-\sigma_p^* & -\sigma_s^* \\
-\sigma_s^* & -\pi_y^* \\
\pi_x^* - - \pi_y^* & \\
 & -- \text{ nonbonding p with } \pi \text{ character} \\
-- \text{ nonbonding p orbitals with } \pi \text{ character} & -\pi_y \\
\pi_{px} - - \pi_{py} & - \text{ nonbonding p with } \sigma \text{ character} \\
-\sigma_p & -\sigma_{pz} \\
-\sigma_s & -\sigma_s \\
-- \text{ nonbonding 2s orbitals} & -- \text{ nonbonding 2s orbitals} \\
\text{MO for linear triatomic molecules} & \text{MO for bent triatomic molecules}
\end{array}
$$

In general, molecules with 16 valence electrons or less are linear. If more than 16 valence electrons are present, the bent configuration is energetically favored so that an anti-bonding orbital need not be occupied until at least 19 electrons are present.

Example 12.8. An interesting series to compare are NO_2, NO_2^+, and NO_2^-. The O-N-O bond angles are 135°, 115°, and 180°, respectively. Discuss the bond order and configuration of each.

-207-

NO_2^+ has 16 valence electrons and so should be linear. All bonding and non-bonding orbitals are filled, and all anti-bonding orbitals are empty. The species is diamagnetic and the total bond order = 4. The Lewis electron-dot structure agrees with this.

$$\ddot{\underset{..}{O}}=N=\ddot{\underset{..}{O}}$$

NO_2 has 17 valence electrons. The lower energy configuration is bent. There will be one unpaired electron in a non-bonding p orbital, but still no occupied anti-bonding orbitals. The bond order is 3, or $1\frac{1}{2}$ per O.

NO_2^- has 18 valence electrons and a bent geometry like NO_2, but with all electrons paired. The bond order is still 3 since the extra electron occupies a non-bonding p orbital.

Problems

1. In VSEPR theory the interaction between two N atoms in N_2 is called a triple bond. Describe a triple bond in molecular orbital terminology.

2. Consider the following series:

$$O_2 \; , \; O_2^+ \; , \; O_2^- \; , \; O_2^{2-}$$

   a) What is the MO configuration of each?
   b) Which species has the longest bond length?
   c) Which species has the shortest bond length?
   d) Which species are paramagnetic?

3. What is the bond order of Br_2^-?

4. What is the highest energy occupied molecular orbital (HOMO) and unoccupied molecular orbital (LUMO) in F_2^+?

5. Compare the bond lengths and strengths of each of the following pairs.

   a) O_2 and N_2.

   b) Cl_2 and I_2.

6. How many unpaired electrons are present in C_2^-? What orbitals do these electrons occupy?

Molecular Orbitals   Chapter 12

7. As a result of adding two electrons to N_2 to form N_2^{2-}, the bond distance and bond strength change. In what direction?

8. If B_2 is reduced to B_2^-, what orbital does the added electron occupy?

9. Draw the MO configuration of NO.

   a) What is the bond order?
   b) Is NO paramagnetic or diamagnetic?
   c) Why should NO^+ be very easy to make?

10. What is the bond order of ClO?

11. Predict whether each of the following should have linear or bent geometry.

    a) O_3

    b) I_3^-

    c) N_3^-

12. Predict whether each of the following should have linear or bent geometry.

    a) NNO

    b) SCl_2

    c) ClO_2

CHAPTER 13
PERIODIC PROPERTIES

13.1. The Periodic Table

The long form of the periodic chart orders the elements by atomic number. The behavior of these elements is cyclic in nature, or periodic, so those elements with similar behaviors are placed in a vertical row or *group*. The horizontal row, or *period,* shows a progression of certain characteristics that we shall investigate. We now know that the periodic behavior of the elements is due to the number of valence electrons and the strength with which they are held. Mendeleev, Meyer, and the 19th century scientists that helped to develop the modern periodic table knew nothing of protons, electrons, and quantum theory. They based the organization of the table on atomic weights and the known descriptive chemistry of each element, though only 60 or so were known at the time.

Groups IA-VIIIA are called the *representative elements.* Some individual groups also have common names such as *alkali metals* (Group IA), *alkaline-earth metals* (Group IIA), *chalcogens* (Group VIA), *halogens* (Group VIIA) and the *inert or noble gases* (Group VIIIA). The B groups, IIIB-IIB from left to right include the *transition elements* and *post-transition elements.* The two long periods located below the main portion of the table are the lanthanide and actinide series or *inner-transition elements.*

13.2. Periodic Properties

There are several properties of the elements that can be easily correlated with electron configuration and atomic size. Looking at the period trends and group trends for these properties can put a wealth of chemical information at our fingertips, provided we have a periodic table in front of us.

The most basic division of the elements separates them into metals, semimetals, and nonmetals. Metal atoms self-bond with 8-12 neighbor atoms. The few valence electrons tend to be delocalized throughout the whole lattice. Because of the extended delocalization of valence electrons, metals are good conductors of electricity.

Nonmetals self-bond in very small molecules like H_2, N_2, or P_4. The bonds are covalent and the valence electrons are localized on each molecule. This means electrical conductivity is poor because electrons are not easily passed from one molecule to another.

The semimetals and some allotropic forms of nonmetals (like C) exist as extended covalent solids. Although the self-bonding is extensive as in metals, it is also localized and covalent as in nonmetals. Thus semimetals have small but measurable electrical conductivities.

Although we have discussed the trends in ionization energy in Chapter 10 it is useful to mention these trends again in terms of metallic character. *Generally* I increases across a period (left to right) and decreases down a group. Metallic characteristics are associated with a low ionization energy. As the I values decrease, the metallic properties increase. Nowhere is this more apparent than in the carbon group, IVA. Carbon is a nonmetal; Si and Ge are semimetals; and Sn and Pb are metals. This increase in metallic behavior parallels the decrease in ionization energy.

Periodic Properties   Chapter 13

By looking at both the ionization energy and electron affinity of an atom, we can get a good idea of the atom's attraction for its valence electrons as well as its attraction for extra electrons. A property that combines these characteristics is *electronegativity*.

Electronegativity is a measure of an atom's attraction for electrons in a bond. It is given a dimensionless value, the lowest of which is 0.8 and the highest 4.0, on the Pauling scale. The period and group trends of electronegativity (χ) are the same as those of ionization energy and electron affinity. The most electronegative elements are found in the upper right-hand portion of the table while the least electronegative elements are in the lower left-hand portion of the table. When the electronegativity difference between two bonded elements is large, the bond will be predominantly ionic; if the $\Delta\chi$ value is small the bond will be predominantly covalent.

Example 13.1. Using only a periodic table, rank the following atoms according to their electronegativities: N, Cl, Na, Ca, H.

The predicted ranking is Na$\simeq$Ca<H<N<Cl. Sodium and Ca should be approximately equal since they are diagonally situated.

Example 13.2. Which bond is more polar, Li-Br or P-Cl?

Again, the periodic table is all you need to answer this. Phosphorus and chlorine are much closer together on the table than lithium and bromine and consequently should have more similar electronegativity values. So a P-Cl bond is less polar, less ionic, than a Li-Br bond.

There are many similarities to be found in the oxidation states of a group. As would be expected, elements with high electronegativities tend to have negative oxidation states while low electronegativity elements tend to form positive oxidation states. Metals usually have positive oxidation states loosely related to the group number. Main group metals in IA and IIA only have available +1 and +2 states, respectively (other than zero). Group IIIA metals have +3 and +1 states. Many transition metals have a *maximum* oxidation state equal to their group number. For example, Mn has a +7, +4, +3, and +2 state that can readily form. The electron configurations of Mn^{+7} ([Ar]) and Mn^{+2} ([Ar]3d^5) are predictably stable, but Mn^{+4} and Mn^{+3} are not expected based on our knowledge to this point in the course. After group VIIB with 7 valence electrons, it becomes more difficult to achieve the high oxidation states especially in the 3d transition metals. The +2 and +3 states are most common for Fe, Co, and Ni, but higher states are found for the 4d and 5d group members.

A different trend is observed for the nonmetals where oxidation states may be positive or negative. Again, the group number indicates the number of valence electrons and also the maximum positive oxidation state. But (8 - group number) gives the maximum negative oxidation state because it equals the number of extra electrons needed to fill the valence s and p orbitals (that is, to complete the octet). Let's use N to illustrate this point. Nitrogen has five valence electrons and needs three more for an octet. It's common that oxidation states range from +5, as in HNO$_3$, to -3, as in NH$_3$, in increments of two. But the trend for the rest of the group of positive states becomes more important since there is a decrease in electronegativity.

While filling or half-filling a shell may be a useful predictor of oxidation states, we must keep in mind that these states are a formality and that they have no basis in reality in many cases. The N^{+5} ion does not exist, but it is a convenient way for a chemist to label an N atom that shares five electrons with a more electronegative element.

Chapter 13 — Periodic Properties

The atomic size or radius correlates inversely with electronegativity (and ionization energy and electron affinity). Atomic size decreases from left to right in a period since the more tightly held valence electrons are naturally constrained to a smaller volume. This steady decrease in size is shown in table 13.6 of the text where period 4 metals shrink from a radius of 2.35 Å for K, to 1.39 Å for Ge. The group trend also correlates inversely with χ, I, and A values. Proceeding from top to bottom in a given group, the addition of a new quantum shell leads to larger size. The anomalies that occur are interesting. Their existence is actually due to the same trend which they don't seem to follow. For instance, Zr and Hf have nearly identical radii, yet Hf has 32 more electrons and the additional 5p, 6s, and 4f orbitals associated with them. It is because of the 4f orbitals that Hf is so small. The lanthanide series of elements shows a steady decrease in size, as it should. The decrease is so large that element 72 even with 32 more electrons is the same size as element 40. This phenomenon is called the lanthanide contraction.

Example 13.3. Predict the possible oxidation states of Br.

Since Br has seven valence electrons, it can share up to seven electrons with a more electronegative element or gain one electron from a less electronegative element. So the range is +7 to -1, likely in increments of two.

Example 13.4. Which atom should have the largest atomic radius: Ne, Cl, or Ar?

A chlorine atom should be the largest, a neon atom the smallest.

13.3. Chemical Properties of the Oxides

Binary oxides form for nearly all elements. Studying these oxide compounds can provide additional insight into periodic chemical behavior. The metal oxides such as K_2O and CaO are ionic solids. When these oxides dissolve in water, the oxide ion, O^{2-}, is not stable and rapidly becomes OH^-, yielding a basic solution.

$$K_2O + H_2O \rightarrow 2K^+ + 2OH^-$$

The nonmetal oxides such as SO_2 and BrO_3 are covalent bonded molecular compounds that yield acidic aqueous solutions.

$$SO_2 + H_2O \rightarrow H_2SO_3$$

Metals near the semimetal region form oxides that are insoluble in water but dissolve in either an acid or a base. These oxides, such as Al_2O_3, are *amphoteric*.

$$Al_2O_3 + 6H^+ \rightarrow 2Al^{3+} + 3H_2O$$

$$Al_2O_3 + 2OH^- + 3H_2O \rightarrow 2Al(OH)_4^-$$

Periodic Properties   Chapter 13

There are elements that can form more than one oxide because there are several common oxidation states. The trend in these cases is clear: the higher the oxidation state of the element bonded to oxygen, the more acidic the oxide.

Example 13.5. Phosphorus forms many oxides. Two of the more important compounds are P_4O_6 and P_4O_{10}, both of which dissolve in water to yield acidic solutions. Which oxide is more acidic? Write balanced equations for their reaction with water.

$$P_4O_6 + 6H_2O \rightarrow 4H_3PO_3 \quad \text{phosphorus acid}$$

$$P_4O_{10} + 6H_2O \rightarrow 4H_3PO_4 \quad \text{phosphoric acid}$$

The oxidation state of P in P_4O_{10} is higher than in P_4O_6, so P_4O_{10} should be more acidic. This is true, and phosphoric acid is a stronger acid than phosphorus acid.

13.4. The Properties of Hydrides

Binary hydrogen compounds are called hydrides, but the hydrogen is not necessarily negatively charged, as the name implies. In any given period, the metal hydrides do indeed contain H^- ions, but the nonmetal hydrides contain a covalently bonded H^+. When a metal hydride dissolves in water, the strong reducing agent, H^-, reacts with H_2O to form H_2. The solution will be basic.

$$NaH + H_2O \rightarrow Na^+ + OH^- + H_2$$

Hydrides of the nonmetals react with water to form acidic solutions.

$$HCl + H_2O \rightarrow H_3O + Cl^-$$

Following the trend in hydride behavior across period 2, LiH and BeH_2 are basic, B_2H_6 (the dimer of BH_3) and CH_4 are neither acidic nor basic, NH_3 is a weak base, H_2O (admittedly a weak acid) is a stronger acid than NH_3, and HF is a moderately strong acid. We see the general progression from basic to acidic. For the nonmetal groups VIA and VIIA there is also an increase in acidity moving down the row.

Transition metal hydrides are unusual in several respects. Hydrogen gas combined with transition metals is universally used for adding hydrogen to a carbon compound with double bonds as shown below, yet little is known about how the H_2 and the transition metal interact. It appears as though the H_2 molecules fill in spaces in the metal lattice without forming compounds of specific proportions, because the H_2 can easily be released again.

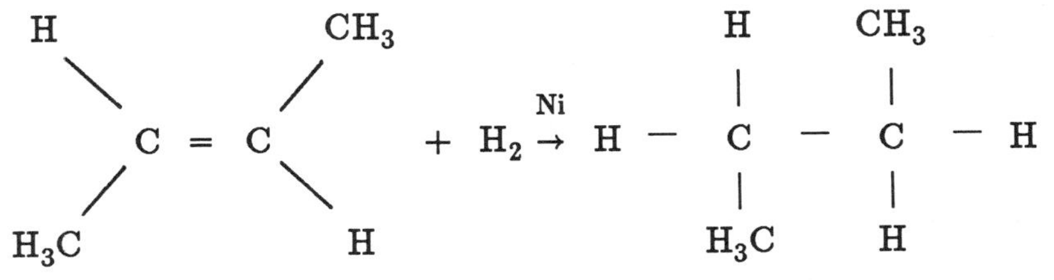

$$\underset{H_3C}{\overset{H}{}}C=C\underset{H}{\overset{CH_3}{}} + H_2 \xrightarrow{Ni} H-\underset{H_3C}{\overset{H}{\underset{|}{C}}}-\underset{H}{\overset{CH_3}{\underset{|}{C}}}-H$$

Chapter 13 Periodic Properties

The hydride of oxygen is water. Because it is ubiquitous here on earth we take it for granted, but it has many unusual characteristics crucial to life. Besides its obvious abundance, water is unusual because it exists on the surface of the earth in all three phases - solid, liquid, and gas.

As a comparison, let's look at the group trend in melting and boiling points of hydrides.

	Group IVA			Group VIA	
	m.p.	b.p.		m.p.	b.p.
CH_4	-182° C	-164° C	H_2O	0° C	100° C
SiH_4	-185° C	-112° C	H_2S	-86° C	-61° C
GeH_4	-165° C	-88° C	H_2Se	-60° C	-41° C
SnH_4	-150° C	-52° C	HeTe	-49° C	-2° C

The trend in group IV hydrides is predictable. Descending the group, the molecules become larger and the van der Waals forces increase so both the melting points and boiling points show an increase. If H_2O is ignored, the group VI hydrides show the same trend. Why is water different? The H_2O molecules can hydrogen bond. While each H-bond is rather weak (about 21 kJ/mol) there are many of them. The only elements capable of forming H-bonds are O, N, F and possibly Cl -- all small, electronegative atoms.

Each H_2O molecule can form four hydrogen bonds. Each lone pair of electrons on oxygen, which carries a δ- charge, H-bonds to a $\delta+$ charged H from another water molecule. Large scale hydrogen bonding in solid water leads to a hexagonal arrangement of oxygen atoms and an observable "snowflake" pattern. Liquid water has localized hydrogen bonding but lacks the extended network of the solid state. For this reason $H_2O_{(l)}$ is denser than $H_2O_{(s)}$ and ice floats on rivers and lakes.*

Example 13.6. HF, liquid and solid, also exhibits hydrogen bonding. Below are the melting points of the other group VIIA hydrides. By analogy with the group VIA hydrides, what do you predict for the melting point of HF?

HCl   m.p. = $-115°$ C ,   HI   m.p. = $-51°$ C

HBr   m.p. = $-88°$ C

---

*Just as ionic solids can have different lattice arrangements, so can water. Several of these different forms of $H_2O_{(s)}$ can occur at high pressures. For a fictional account of a scientist who develops a new ice crystal form, stable at room temperature, read Kurt Vonnegut Jr.'s book "Cat's Cradle."

Periodic Properties   Chapter 13

If HF had the equivalent number and strength of hydrogen bonds as H_2O, its melting point should be -35 to -25° C. *If* HF had no hydrogen bonds then its melting point would be lower than -115° C. So the conclusion that can be drawn from the real melting point of -83° C is that HF does have some hydrogen bonds. In fact an F—H intermolecular bond is stronger than that of oxygen, but fewer hydrogen bonds are possible per molecule than in H_2O.

Problems

1. Using only a periodic table name the halogen with the greatest electron affinity.

2. Which alkaline-earth metal has the largest atomic volume?

3. List the following elements according to size, smallest atomic radius to largest: Cs, W, Mo, Au, Bi.

4. Which elements possess two unpaired 4p electrons?

5. List the following elements from high to low electronegativities, using only a periodic table: As, Sn, S, F, Tl.

6. Label each of the following bonds as ionic, polar covalent, or nonpolar covalent. Then arrange the bonds in order of increasing ionic character.

    B—Cl,   Mg—Cl,   Na—Cl,   C—Cl,   O—Cl

7. Name the group of elements that has two unpaired valence electrons and forms a -2 ion in combination with metals.

8. In many texts electronegativity values of the noble gases are not listed. Why is this so? If the values were listed, would you expect them to be high or low?

9. An unknown element, x, formed an oxide with the formula x_2O_3 and a chloride with the formula xCl_3. What are the possible groups of which x could be a member?

10. What are the possible oxidation states of Cr? Which ones might be especially stable because of the electron configuration?

11. What are the possible oxidation states of Cl? With which elements will it have a positive oxidation state?

12. Write the formulas for binary oxides of all period 3 elements. There will be more than one oxide of P, S, and Cl.

13. Which of the following hydrides are covalent?

    KH, MgH_2, SrH_2, AsH_3

14. Lithium aluminum hydride, $LiAlH_4$ is commonly used for a reducing agent in chemical labs, but care must be exercised because $LiAlH_4$ reacts violently with H_2O. Balance the

Chapter 13   Periodic Properties

reaction below and decide what has been oxidized and what has been reduced.

$$H_2O + LiAlH_4 \rightarrow H_2 + Al(OH)_3 + LiOH$$

15. List the following oxides in order of increasing acidity in water: MgO, Al_2O_3, SO_3, P_4O_{10}.

16. Complete and balance the following equations.

   a) $H_2C=CH_2 +$ _____ $\rightarrow H_3C-CH_3$

   b) $ZnO + HCl \rightarrow$

   c) $NaH + H_2O \rightarrow$

   d) $N_2O_5 + H_2O \rightarrow$

   e) $ZnO + NaOH \rightarrow$

17. What anhydride dissolves in water to form $HClO_3$?

18. What boiling point trend do you expect for NH_3, PH_3, AsH_3, SbH_3?

19. Predict the trend in ΔH_{vap}^0 values for H_2O, H_2S, H_2Se, H_2Te.

20. Why can't CH_4 form intermolecular hydrogen bonds?

CHAPTER 14
THE REPRESENTATIVE ELEMENTS: GROUPS I-IV

14.1. The Alkali Metals

The alkali metals are group IA of the periodic table, and include Li, Na, K, Rb, Cs, and Fr. Though H is listed as a group IA member, its physical and chemical properties make it different enough to not be included here as a metal. The element francium, while naturally occurring, exists only as short-lived radioactive isotopes present in very small quantities in the Earth's crust. Thus we will limit our discussion of alkali metals to Li, Na, K, Rb, and Cs.

Most properties follow a smooth, predictable trend from low atomic number (Li) to high (Cs). The size increases steadily and the ionization energy decreases steadily because of it. The +1 ions form easily but the I_2 values are prohibitively high, so no other oxidation states occur. The ions are considerably smaller than the corresponding atoms. The trend in size remains the same, with Li^+ being smaller than Na^+ which is smaller than K^+, and so on.

All alkali metals naturally occur in the +1 oxidation state. The elemental forms are highly reactive with water. All the alkali metals produce hydrogen gas and basic solutions in water with varying degrees of vigor: Lithium and Na generate some mild heat and fizzing while K will flame and Rb and Cs react explosively.

Example 14.1. Write a balanced equation for the reaction of elemental potassium and water.

$$2K_{(s)} + 2H_2O_{(l)} \rightarrow 2K^+_{(aq)} + 2OH^-_{(aq)} + H_{2(g)}$$

All the alkali metals are strong reducing agents. The trend in $\epsilon°$ values for the reduction potentials of Li through Cs does not follow an easily identifiable pattern. Lithium is the strongest reducing agent, then Cs, followed by K and Rb, and lastly Na. Compare the data for ionization energies and $\epsilon°$ values below.

	Li	Na	K	Rb	Cs
$I_1(M_{(g)} \rightarrow M^+_{(g)} + e^-)$ kJ/mol	520	496	419	403	376
$\epsilon°\,(M^-_{(aq)} + e^- \rightarrow M_{(s)})$ V	-3.04	-2.71	-2.94	-2.94	-3.03

In the gas phase it is easier to remove an electron from Na than Li, yet Li is a stronger reducing agent in water. This makes sense only if we realize that there is more involved in the aqueous system that just the removal of an electron from the metal atom. We can break the aqueous reduction into three steps, one of which is ionization.

Chapter 14 — The Representative Elements: Groups I-IV

$$M_{(s)} \rightarrow M_{(g)} \qquad\qquad \Delta H_f^0 \text{ of } M_{(g)}$$

$$M_{(g)} \rightarrow M_{(g)}^+ + e^- \qquad\qquad I_1$$

$$M_{(g)}^+ + H_2O \rightarrow M_{(aq)}^+ \qquad\qquad \Delta H_{hyd}^0$$

The energies involved in all three steps help determine the reduction potential. Lithium ion is very small and so ΔH_{hyd}^0 is a large negative number for Li^+ which can more than compensate for the endothermic ΔH_f^0 and I_1. Cesium is nearly as strong a reducing agent as Li even though its +1 ion releases much less heat on hydration. This is because ΔH_f^0 of $M_{(g)}$ and I_1 are much smaller for the large Cs atom than for Li.

The alkali metals are found on Earth mainly as salts. Most of the alkali salts are soluble so sea water is a rich source of these ions. The free metals can be made by electrolysis of a molten salt.

$$2NaCl_{(l)} \rightarrow 2Na_{(s)} + Cl_{2(g)}$$

Or one active metal can be used to reduce (replace) another at high temperatures.

$$Ca_{(s)} + 2RbCl_{(s)} \rightarrow CaCl_{2(s)} + 12Rb_{(g)}$$

The alkali metals easily form stable halides and oxygen compounds (either oxides, peroxides, or superoxides).

$$2K_{(s)} + Cl_{2(g)} \rightarrow 2KCl_{(s)}$$

$$Li_{(s)} + O_{2(g)} \rightarrow 2Li_2O_{(s)}$$

$$Rb_{(s)} + O_{2(g)} \rightarrow RbO_2$$

All the oxides dissolve in water to give strongly basic solutions.

$$Na_2O_{(s)} + H_2O_{(l)} \rightarrow 2Na_{(aq)}^+ + 2OH_{(aq)}^-$$

The peroxides and superoxides are used in self-contained breathing apparatus to generate O_2.

$$2Na_2O_{2(s)} + 2CO_{2(g)} \rightarrow 2Na_2CO_{3(s)} + O_{2(g)}$$

Sodium hydroxide is among the top ten chemicals produced and consumed each year in the United States. It is made by the electrolysis of a NaCl solution.

$$2NaCl_{(aq)} + 2H_2O_{(l)} \rightarrow 2NaOH_{(aq)} + H_{2(g)} + Cl_{2(g)}$$

14.2. The Alkaline-Earth Metals

The alkaline-earth metals are group IIA of the periodic table: Be, Mg, Ca, Sr, Ba, and Ra. As the atomic weight increases, atomic size increases and as a direct result, the ionization energies decrease and the melting and boiling points decrease. These are the same general trends observed for group IA; but it is interesting to compare the two groups. The group IIA elements have a smaller atomic size than the adjacent group IA elements. We have previously ascribed this to the increase in effective nuclear charge for the alkaline-earth metals. The melting and boiling points (as well as $\Delta H°$ values for these phase changes) are higher for group IIA than IA. This is indicative of stronger metallic bonding which in turn reflects the smaller size due to higher effective nuclear charge.

It takes more energy to remove the first electron from the alkaline-earth metals than from the alkali metals but it is still relatively easy. The alkaline-earth metals are never found in nature in the elemental form. They are found in the +2 oxidation state. Losing two electrons, while endothermic, is not excessive in its energy requirements. The necessary energy input is more than compensated for by the higher lattice energies of the salts formed or the more negative heats of hydration in water. The resulting environment makes the +2 state more stable than +1 or +3.

A qualitative view of how the +2 ions can be stabilized, relative to the +1 ions can be seen from the interactions with water. When a cation is dissolved in water there are strong ion-dipole interactions between the cation and the δ- of H_2O. Most positive ions align six water molecules in an octahedral fashion, oxygen atoms toward the ion. Some small ions like Be^{2+} or even H^+ interact with four waters in this first coordination sphere. An aqueous metal ion, usually written as $M^{+n}_{(aq)}$, is better described by the formula $M(H_2O)_6^{+n}$ or $M(H_2O)_4^{+n}$. In any case, a +2 ion will interact more strongly with the water molecules than a +1 ion, and draw them closer because of the colombic attraction.

The alkaline-earth metals are found as ions in sea water, in fresh water that is "hard," and in insoluble deposits of carbonate, silicate, and phosphate rocks. The group IIA salts are in general not as soluble as those of group IA salts. The hydrides and oxides of alkaline-earth metals are basic in water with the exception of Be which forms bonds that are appreciably covalent in character. For example, $BeCl_2$ does not conduct electricity well because the compound is molecular in nature, not ionic. Thus BeO in water is amphoteric.

$$Be^{2+}_{(aq)} + H_2O \rightarrow BeOH^+ + H_3O^+$$

$$Be^{2+}_{(aq)} + 4OH^- \rightarrow Be(OH)_4^{2-} + 4H_2O$$

Even Mg^{2+} which is not quite as small as Be^{2+} forms some covalent bonds with carbon, though the bonds with oxygen and the halogens are ionic.

The reduction potentials of the M^{2+} ions are quite large. In general, the alkaline-earth ions are not as strong reducing agents as the alkali ions and there is no clear trend within the group.

Chapter 14 The Representative Elements: Groups I-IV

Example 14.2. Helium atoms have the same valence electron configuration as the alkaline-earth metals, ns^2. Why is He not a member of this group?

Helium has much larger I_1 and I_2 values than all group IIA elements because its outer electrons are its only electrons. He does not easily form ions and so is atomic in nature with no metallic characteristics.

Example 14.3. The two barium salts $BaCO_3$ and $BaSO_4$ are insoluble in base or pure water, but $BaCO_3$ will dissolve in acid solution. Write the chemical equation describing the process.

$$BaCO_{3(s)} + 2H^+_{(aq)} \rightarrow Ba^{2+}_{(aq)} + CO_{2(g)} + H_2O_{(l)}$$

14.3. The Elements of Group IIIA

While all the group IA and IIA members are metals, group IIIA contains metals and semimetals, thus a wider variety of properties are expected and found. Boron is the only true semimetal of the group, and metallic character increases from top to bottom. Since the valence configuration is ns^2np^1 for all these elements, the +3 oxidation state is predictably important but In and Tl chemistry involves the +1 state also. Boron, while having the +3 oxidation state in common with the other group members, does not exist as B^{+3} cations. The oxidation state is a mere formality since B bonds covalently. Thus the BCl_3 molecule is trigonal planar with bond angles of 120° and the boron orbitals are sp^2 hybrids. It is the empty, non-hybridized p orbital on B that makes many boron compounds electron pair acceptors or Lewis acids.

$$BCl_3 + :N(CH_3)_3 \rightarrow (CH_3)_3N-BCl_3$$

Trisubstituted aluminum compounds show similar behavior.

Example 14.4. Lithium aluminum hydride, $LiAlH_4$, can be made by reacting lithium hydride with aluminum chloride. Write the balanced equation for this process and identify the Lewis acid and base.

$$4LiH + AlCl_3 \rightarrow LiAlH_4 + 3LiCl$$

Hydride, H^-, is the Lewis acid and the aluminum chloride, $AlCl_3$, is the initial Lewis base.

The atomic radius increases from B to Tl and the heat of atomization (ΔH_f^0 to $M_{(g)}$) decreases as expected. The melting point, boiling point and ionization energy trends are not as consistent though there is a *general* decrease from the top of the group to the bottom. Aluminum is the strongest reducing agent of the group since B^{3+} does not exist, but group IIIA elements are weaker reducing agents than the IA and IIA metals. We don't think of Al as being easily oxidized since the apparently stable metal can be found everywhere in our homes (foil, cars, pots and pans, screen doors, etc.). What we view as stability or inertness is actually due to a thin film of Al_2O_3 that rapidly forms on a fresh aluminum surface. The oxide coating binds so strongly to the surface that it excludes additional oxidation.

The Representative Elements: Groups I-IV　　　　　　　　　　　　　　　　　　　　　Chapter 14

Aluminum has many uses that take advantage of its light weight. The external fuel tanks of the space shuttle contain liquid H_2 and O_2, and are made of aluminum. The solid-rocket boosters (SRB's) of the shuttle burn Al as fuel. The powdered Al and the oxidizing agent, NH_4ClO_4, are embedded in an epoxy which is ignited on lift-off. The tremendous heat generated expands the gas products providing the lift. The huge white cloud seen when the SRB's ignite is primarily due to the Al_2O_3 that is produced.

One of the unusual features of group IIIA chemistry and of boron in particular, is the 3-center bridge bond. The simplest boron hydride compound, BH_3, exists as a dimer, B_2H_6, where each B has 2 normal single bonds to H, and has two bridge bonds to H atoms shared with the other boron.

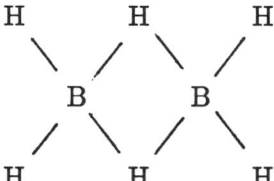

The boron atoms can be considered as having sp^3 hybrids. The two bridging H atoms form only one bond each, but it covers three atoms with only two electrons.

The other group members show analogous behavior. $AlCl_3$ is a solid monomer, but in the gas phase Al_2Cl_6 forms with two Cl bridging atoms.

14.4. The Elements of Group IVA

The elements of group IVA, C, Si, Ge, Sn, and Pb, encompass nonmetals (C), semimetals (Si, Ge), and metals (Sn, Pb). Consequently, there is great variety in the behavior of the elements and their compounds. The trends in ionization energies and heats of atomization show a steady decrease descending the group from C to Pb. This is easily correlated with the steady increase of atomic radius from C to Pb. Elemental carbon occurs in nature but all other group members occur in compound form, although the elements, once made, are relatively stable under normal conditions of temperature, pressure, and environment.

The two allotropic forms of carbon are diamond and graphite, vastly different in character. Diamond is hard, translucent, and an insulator, while graphite is soft, shiny, gray, and a fairly good conductor of electricity. The differences are due to the molecular structures. Each carbon atom in diamond is tetrahedrally bonded to four other carbon atoms to form a three-dimensional network. The carbon atoms in graphite are sp^2 hybridized and are bonded in a plane to three other C atoms. The graphite structure is therefore planar or sheet-like. Each sheet is weakly attracted to the next by van der Waals forces, resulting in the soft structure. The delocalized π electrons account for the electrical conductivity of graphite.

Example 14.5. The ΔH_f^0 of $C_{(graphite)}$ is 0 and the ΔH_f^0 of $C_{(diamond)}$ is 2.90 kJ/mol. Which form is more stable at room temperature? What is the equilibrium constant for the conversion of diamond to graphite?

$$C_{(d)} \rightarrow C_{(gr)}$$

Graphite is the more stable allotrope at 25° C and 1 atm, although diamond is more stable at very high pressures. Therefore diamond will spontaneously convert to graphite but the process is slow and unobservable.

$$K = e^{-\Delta G^\circ / RT} = e^{2.90/ (8.314 \times 10^{-3})(298)}$$

$$K = 3.22$$

Crystalline silicon and germanium have a diamond-type lattice. Both elements are valuable in the semiconductor industry. Although pure Si and Ge are not good conductors of electricity, addition of small amounts of boron or phosphorous enhances the conductivity. The development of semiconductor devices has revolutionized modern technology.

Because C—C bonds are relatively strong (348 kJ/mol) there are millions of compounds containing long carbon chains. This ability to form long chains is called *catenation* and provides the huge variety of organic molecules on which life is based. Silicon, germanium, and the other group IVA members do not catenate. Si—Si bonds, for example, are weak (177 kJ/mol) by comparison with C—C bonds and Si—O bonds (369 kJ/mol). This accounts for the instability of compounds like disilane, Si_2H_6, in air.

Silicon is abundant in the Earth's crust as SiO_2 (sand and quartz) and silicates such as feldspar and talc. Glass is made from SiO_2. Addition of other components like lime (CaO), soda (Na_2O), or lead (PbO) give glass different characteristics.

Germanium is the least abundant member of group IVA. Lead and tin occur as the simple oxides and sulfides, PbS and SnO_2. The oxides of carbon, silicon and germanium (CO_2, SiO_2, GeO_2) are weakly acidic while the oxides of tin and lead (SnO, SnO_2, PbO) are amphoteric. PbO_2 is a strong oxidizing agent in water.

Example 14.6. Write balanced equations for PbO dissolving in acid and base.

$$PbO_{(s)} + H_2O \rightarrow Pb(OH)_{2(s)}$$

$$Pb(OH)_2 + 2H^+ \rightarrow Pb^{2+} + 2H_2O$$

$$Pb(OH)_2 + 2OH^- \rightarrow Pb(OH)_4^{2-}$$

Tin and lead have many everyday applications. For example, steel cans are coated with tin for protection, car batteries are dependent on lead chemistry, lead used to be a common additive in paint and gasoline before its toxic effects were well documented, and tin in its B allotropic form has been used to make organ pipes.

The B form is metallic in nature and resists corrosion. An α form also exists, and is nonmetallic and brittle. The B form of tin slowly converts to the α form when temperatures

The Representative Elements: Groups I-IV — Chapter 14

are below 13° C. In unheated European cathedrals this tin conversion resulted in the disintegration of many organ pipes.

Example 14.7. In basic solution, the Sn^{2+} ion exists as $Sn(OH)_4^{2-}$, which disproportionates to tin metal and the tin (IV) hexahydroxide complex ion. Write a balanced disproportionation reaction.

$$2Sn(OH)_4^{2-} \rightarrow Sn + Sn(OH)_6^{2-} + 2OH^-$$

Problems

1. Explain why there is no direct correlation between the I_1 values of the alkali metals and the $\epsilon °$ values.

2. Write balanced equations for the reactions that occur when the following chemicals are combined.

    a) metallic potassium and water
    b) sodium hydride and water
    c) sodium oxide and water
    d) lithium peroxide and carbon dioxide

3. Sodium metal can be prepared by electrolysis from molten NaCl but not from aqueous NaCl. Why?

4. Write balanced equations for the reactions that occur when the following chemicals are combined.

    a) calcium hydride and water
    b) strontium oxide and water
    c) $Mg(OH)_2$ (milk of magnesia) and HCl (stomach acid)

5. Limestone, $CaCO_3$, will dissolve in ground waters which contain $CO_{2(aq)}$. Write a balanced equation that describes this process.

6. What is the main reason that group IA elements are stronger reducing agents than the corresponding group IIA elements?

7. When aqueous NaOH is added to Al^{3+}, a white precipitate forms. Continued addition of NaOH causes the precipitate to redissolve. What reactions are occurring?

8. In the following reaction, what is the Lewis acid?

$$2LiH + B_2H_6 \rightarrow 2LiBH_4$$

Chapter 14 The Representative Elements: Groups I-IV

9. Boric acid can be written as H_3BO_3 or $B(OH)_3$. Boric acid can act as a Lewis acid in water or OH^- solutions. Write a balanced equation for the reaction of boric acid with water or OH^-. Which formula fits the character of boric acid better?

10. Aluminum is nearly twice as abundant as iron in the earth's crust, yet Al is more expensive to produce. Can you present a reason why this might be the case?

11. Write balanced equations for the following reactions.

   a) calcium carbide and water produce acetylene (C_2H_2)
   b) tin dissolves in hydrochloric acid

12. Do you expect the chlorides of the group IVA elements to be ionic or covalent?

13. Tin (II) hydroxide is amphoteric. Write equations for the reaction of $Sn(OH)_2$ with acid and base.

14. We calculated that graphite is more stable than diamond at 25° C and 1 atm. At very high pressures the diamond form is favored. What does this indicate about the densities of diamond and graphite?

15. Lead storage batteries depend on the thermodynamic stability of Pb^{2+} relative to Pb and Pb^{4+}. Lead and lead (IV) oxide react in sulfuric acid solution to produce lead (II) sulfate. Write this equation.

CHAPTER 15
THE NONMETALLIC ELEMENTS

15.1. The Elements of Group VA

The group VA elements provide excellent examples of trends in chemical and physical behavior. Nitrogen and phosphorus are nonmetals, arsenic and antimony are semimetals, and bismuth is a metal. As expected, the atomic radius increases from N to Bi while ionization energies and electronegativity values decrease. All group members have five valence electrons (ns^2np^3) which means that the possible oxidation states range from +5, when combined with atoms of higher electronegativity, to -3, when combined with atoms of lower electronegativity. The +3 state is particularly common where only the p electrons are shared or transferred. Nitrogen exhibits all these possible oxidation states while other group members exhibit +3 most often, and +5 and -3 with less frequency.

Not all the group VA elements form hydrides, but there is a clear trend among those that exist. The hydrides are less basic and less stable with increasing atomic weight. The oxides of N and P are acidic, those of As and Sb are amphoteric and Bi_2O_3 is basic.

Example 15.1. Dinitrogen pentoxide dissolves in water, antimony trioxide (Sb_2O_3) is amphoteric because it dissolves in acid or base, and bismuth (III) oxide is basic, dissolving only in acid. Write balanced equations describing this solubility in H_2O, HCl or NaOH.

$$N_2O_{5(s)} + H_2O_{(l)} \rightarrow 2\ NHO_{3(aq)}$$

$$Sb_2O_{3(s)} + 6HCl_{(aq)} \rightarrow 2SbCl_{3(s)} + 3H_2O_{(l)}$$

$$Sb_2O_{3(s)} + 6NaOH_{(aq)} \rightarrow 2Na_3SbO_{3(aq)} + 3H_2O_{(l)}$$

$$Bi_2O_{3(s)} + 6HCl_{(aq)} \rightarrow 2BiCl_{3(s)} + 3H_2O$$

Only nitrogen of all group members occurs naturally in the elemental form, N_2. Phosphorus is found in phosphate rocks, and arsenic, antimony, and bismuth occur as oxides or sulfides of small abundance. Why should N_2 be so stable while the other elements are not? The answer lies in the strong triple bond which is quite unreactive (fortunately so, or a N_2/O_2 environment could lead to dilute nitric acid oceans).

Example 15.2. Although P_2 does exist in the gas phase, elemental phosphorus usually occurs as P_4 molecules or polymers. P_2 would be expected to have a triple bond like N_2. Can you think of an explanation as to why the phosphorus triple bond should be weaker than that of nitrogen?

Chapter 15 — The Nonmetallic Elements

$$P \equiv P \qquad 481 \text{ kJ mol}^{-1} \text{ bond energy}$$

$$N \equiv N \qquad 945 \text{ kJ mol}^{-1} \text{ bond energy}$$

One plausible explanation is that due to the greater size of P atoms there is less overlap of the p orbitals which results in weaker bonds.

There are many important compounds containing group VA elements, some of which shall be mentioned here because of our everyday encounters with them. Your text has a more thorough survey of these compounds.

Many tons of NH_3 are produced each year throughout the world. A large portion of the NH_3 is used for fertilizer. Phosphorus too is a major component of fertilizers. Any gardener will recognize that a fertilizer has numbers such as 10-15-5 or 5-8-0 associated with it. These numbers refer respectively to the percent by mass of N (as ammonium salts), P (as dihydrogen phosphate salts) and potash (as potassium salts). Liquid ammonia, much like water, can dissolve many salts because it is polar and exhibits intermolecular hydrogen bonding. Solid NH_3 is even thought to be a major component of Saturn's rings.

Nitrogen oxides are both useful and problem-causing compounds. The oxide N_2O is relatively inert and can be used as an anesthetic (laughing gas) or as a propellant for whipped cream (because it dissolves in fats), or even to increase performance in the engine of a race car. On the other hand, NO and NO_2 are present in smog and can lead to tissue damage in living organisms.

The -3 oxidation state is quite important for N and P (other than NH_3 and PH_3) although a N^{-3} or P^{-3} ion may only exist with a group I or IIA metal. The bonding in nitrides and phosphides is covalent with semimetals and other nonmetals. This -3 state is less important for As, Sb, and Bi with the exception of gallium arsenide, GaAs, a compound with semiconductor properties that is receiving a great deal of attention in that industry.

Another group of interesting nitrogen compounds are the hydrazines. Hydrazine, N_2H_4, is used in the space shuttle auxiliary power unit (APU) as fuel. Methylhydrazine, $CH_3N_2H_3$, is the fuel and dinitrogen tetroxide the oxidant in the orbital maneuvering engines that bring the shuttle out of orbit. The reaction (shown below) generates a large quantity of heat as well as the stable $N \equiv N$.

$$CH_3N_2H_3 + N_2O_4 \rightarrow N_2 + H_2O + CH_3OH$$

Example 15.3. Draw the Lewis electron-dot structures of methylhydrazine and dinitrogen tetroxide.

methylhydrazine
```
          H
          |      ..    ..
    H  —  C  —  N  —  N  —  H
          |     |     |
          H     H     H
```

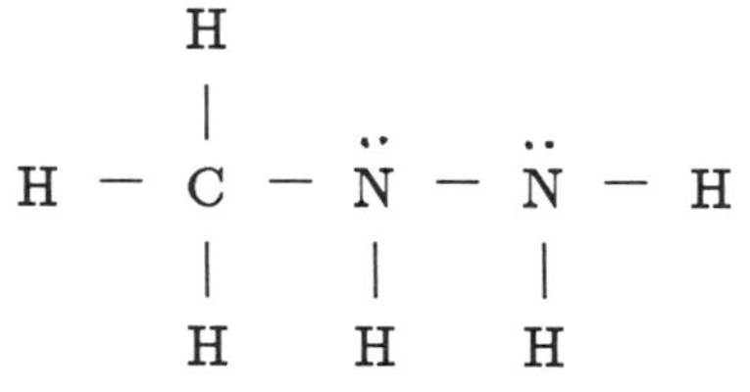

The Nonmetallic Elements   Chapter 15

dinitrogen tetroxide

(structure of N₂O₄ with two O atoms double-bonded to each N, and N—N single bond)

15.2. The Elements of Group VIA

The elements of group VIA: oxygen, sulfur, selenium, tellurium, and polonium are also known as the *chalcogens*. These elements, like those of group VA, show an increase in metallic character with increasing atomic weight. Oxygen and sulfur are nonmetals, selenium and tellurium are semimetals, and polonium is a metal. Polonium is a rare element, first isolated by Marie Curie and named for her homeland, Poland. All isotopes of polonium are radioactive and are encountered as part of a nuclear decay scheme.

The remaining four group members show predictable trends in atomic size (increasing with increasing atomic number), ionization energy, and electronegativity (decreasing with increasing atomic number). The elements all form hydrides, H_2Te and H_2Se being more acidic in water and better reducing agents than H_2S and H_2O. As the atomic size increases there is also a decreased tendency to form multiple bonds. Oxygen forms many double, even triple bonds. S, Se, and Te form single bonds, frequently with an expanded octet. These larger atoms also catenate well.

In elemental form, only oxygen exists as a gas. Its allotropic forms are O_2 and O_3. Sulfur, Se, and Te are solids at room temperature and room pressure in all allotropic forms. But only oxygen and sulfur occur in nature elementally. A lot of O_2 exists on Earth, but it is reactive. Even more oxygen exists in water, sand, and silicates. The supply of O_2 is constantly being replenished by photosynthesis.

$$CO_2 + H_2O \xrightarrow{h\nu} C,H,O \text{ compound} + O_2$$

The necessary free energy for this "uphill" reaction comes from sunlight.

Example 15.4. How much free energy is required to produce glucose, $C_6H_{12}O_6$, from CO_2 and H_2O?

	$6CO_{2(g)}$	+	$6H_2O_{(l)}$	→	$C_6H_{12}O_{6(s)}$	+	$9O_{2(g)}$
ΔG_f^0 (kJ/mol)	-394		-237		-919		-

$\Delta G_{rxn}^0 = [-919 + 9(0)] - [6(-394) + 6(-237)]$

$\Delta G_{rxn}^0 = +2867$ kJ

Chapter 15 — The Nonmetallic Elements

Oxygen usually is found in the -2 oxidation state in compounds. The only exception occurs with O—F compounds. Since fluorine is the most electronegative element, oxygen must be assigned a positive oxidation state when bonded to fluorine. Other negative oxidation states are also encountered in peroxides and superoxides.

Example 15.5. Assign oxidation states for each O atom in OF_2, H_2O_2, and KO_2. Draw the Lewis electron-dot structures of these compounds.

OF_2 +2 −1   F—O—F

H_2O_2 +1 −1 per atom   H—O—O—H

KO_2 +1 −1 −2 per atom   There is no good Lewis structure for the superoxide ion. Molecular orbital theory predicts a bond order of $\frac{3}{2}$.

Sulfur (selenium and tellurium also) have several positive oxidation states in addition to -2 because of the many compounds formed with oxygen and the halogens. Probably the single most important sulfur-oxygen compound is sulfuric acid, H_2SO_4. Its synthesis from $S_{(s)}$ is detailed in the text. Sulfuric acid is important in the manufacture of fertilizers, dyes, paper, steel, and just about everything else we use daily. More sulfuric acid is produced than any other chemical in the world. The sulfur for acid production comes from a variety of sources, but mainly elemental S is used. It is found near volcanos and hot springs where it is produced from H_2S.

Example 15.6. In the heat of volcanic activity, H_2S reacts with O_2 to form SO_2 which can in turn react with more H_2S to form S. Write equations showing these processes.

$$2H_2S + 3O_2 \rightarrow 2SO_2 + 2H_2O$$

$$SO_2 + 2H_2S \rightarrow 3S + 2H_2O$$

Most coal and petroleum reserves contain some sulfur which is converted to SO_2 (and some small amounts of SO_3) when burned. Sulfur dioxide and sulfur trioxide, along with nitrogen oxides produced primarily in car exhaust, constitute the pollutants that form *acid rain*. In some heavily industrialized areas of Europe and North America the pH of the rain and lakes is so low it kills many plants and animals.

However, the food and wine industry relies on SO_2 as a beneficial chemical because it acts as a fungicide on grapes and in wine. It also preserves fresh fruits and vegetables from spoiling or discoloration. Recently the use of SO_2 and sulfites, SO_3^{2-} salts, has come under attack because some people develop an allergic reaction to food treated with these chemicals.

Example 15.7. Power plants that burn coal and oil usually have "scrubbers" in their smoke stacks that remove SO_2 and SO_3 before it reaches the environment. The gaseous pollutants are reacted with lime, CaO, in the stacks. What are the products?

$$SO_{2(g)} + CaO_{(s)} \rightarrow CaSO_{3(s)}$$

$$SO_{3(g)} + CaO_{(s)} \rightarrow CaSO_{4(s)}$$

Selenium and tellurium behave similarly to sulfur. The elements occur naturally as selenides and tellurides of Cu, Pb, or Au in the sulfide ores. While Se is a necessary minor human nutrient, its biggest industrial use is to coat the drums in copy machines.

15.3. The Elements of Group VIIA

The elements of group VIIA, fluorine, chlorine, bromine, iodine, and astatine are called the *halogens*, meaning salt-formers. All isotopes of At are short-lived and radioactive so little is known of their chemistry. The other four halogens are nonmetals, though even in this group there is an increase in metallic characteristics as the atomic weight increases. Iodine in the solid and liquid state has a small but perceptible electrical conductivity.

In elemental form, all the halogens are diatomic; F_2 and Cl_2 are gases, Br_2 is a liquid, and I_2 is a solid at 25 °C. Each element has a high electronegativity value but that of fluorine is the highest. Consequently, F exists only in -1 oxidation state with other elements. Cl, Br, and I although readily forming the -1 halide ion, can also exist in the +1, +3, +5 and +7 states with oxygen, fluorine, and any other halogen of higher electronegativity. When found in nature, F, Cl, and Br are anions, while I occurs both as iodide and as iodate, IO_3^-, salts.

All the halogens are good oxidizing agents, F_2 being the strongest and I_2 the weakest. F_2 is such a powerful oxidizing agent that it cannot be prepared from F^- by usual electrochemical means. Electrolysis must be used. The F—F bond is relatively weak (158 kJ/mol) while X—F bonds are quite strong. The H—F bond for instance has a bond-dissociation energy of 568 kJ/mol. This is the main reason that F_2 is such a strong oxidant. We can also see why HF is a weak acid, the H—F bond is not easily broken. The other hydrogen halides -- HCl, HBr, and HI are strong acids in water. Hydrofluoric acid, though weak, can etch or dissolve glass. Frosted glass and light bulbs are produced using HF.

Example 15.8. Write a balanced equation to show the effect of HF solution on glass (mostly SiO_2).

$$SiO_{2(s)} + 6HF_{(aq)} \rightarrow 2H^+_{(aq)} + SiF_6^{2-}{}_{(aq)} + H_2O$$

Because of intermolecular H-bonding, HF has a higher boiling point than HCl.

Except for fluorine, each halogen forms a series of oxyanions. Comparing the acid strength of an analogous series, the trend is clear.

	HOCl	HOBr	HOI
$pK_a=$	7.5	8.7	11

The key factor seems to be the electronegativity of the central halogen. The OCl^- is more stable in water than OBr^- and OI^-.

How do we explain the variation of acid strength for a single halogen oxyacid series?

	HOCl (HClO)	HO_2Cl ($HClO_2$)	HO_3Cl ($HClO_3$)	HO_4Cl ($HClO_4$)
$pK_a=$	7.5	2.0	<0	<0

As the oxidation state of the Cl increases and additional oxygens are added, the H is held less tightly. Although the hydrogen is attached to only one oxygen, there is an inductive effect. The Cl and other O atoms induce a stronger pull on the electrons shared by the H and O, and the H^+ is more easily released.

There are many interesting uses of halogen compounds. Stannous fluoride (SnF_2) and sodium monofluorophosphate (Na_2FPO_4) known as the trademark MFP, are fluoride additives in toothpastes. The F^- ion replaces a small portion of the Ca^{2+} in tooth enamel and at the same time increases its resistance to acid. An entire group of compounds called *fluorocarbons* are quite inert. Teflon is a fluorocarbon polymer, and freons, used as refrigerants, contain both fluorine and chlorine.

$$\text{Teflon} \quad \left(\begin{array}{cc} F & F \\ | & | \\ -C - C - \\ | & | \\ F & F \end{array} \right)_n \quad \text{a polymer of tetrafluoroethene}$$

$$\text{Freon 12} \quad \begin{array}{c} F \\ | \\ F - C - Cl \\ | \\ Cl \end{array} \quad \text{dichlorodifluoromethane}$$

PVC tubing is a *polyvinylchloride* product.

The inertness of some halocarbons also leads to problems. Compounds such as PCB's (polychlorinated biphenyls), DDT, and chlordane exist in the environment even after their purpose is served and in the long run may cause more harm than good.

Large amounts of Cl_2 and ClO^- are used to bleach paper and pulp products each year. One of the biggest uses of bromine is as AgBr which is used to coat black and white film. Iodine is a required nutrient in small amounts and is found in iodized salt as KI.

Example 15.9. Until recently, the I^- in table salt was stabilized by the addition of sodium thiosulfate, $Na_2S_2O_3$, which keeps the I^- from being oxidized to I_2 according to the following equation.

$$2I_2 + S_2O_3^{2-} + 6OH^- \rightleftarrows 2SO_3^{2-} + 3H_2O + 4I^-$$

Under what pH conditions is $S_2O_3^{2-}$ most effective in preventing I^- oxidation?

Le Chatelier's principle predicts that the I^- will be most stable in the presence of $S_2O_3^{2-}$ at high pH values.

The *interhalogens* are compounds formed from two or more halogens. Although the compounds are not of commercial importance, the molecular structures are of interest. VSEPR theory accurately predicts most interhalogen geometries.

Example 15.10. Draw the Lewis electron-dot structure of ClF_3 and predict the shape of the molecule.

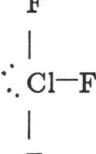

This molecule involves dsp^3 hybride orbitals and should be T-shaped. The lone electron pairs are approximately 120° apart, and the F—C—F bond angles are approximately 90°.

15.4. The Noble Gas Compounds

Helium, neon, argon, krypton, xenon, and radon have been called the rare gases, the inert gases, and now the noble gases. They are neither as rare nor as inert as scientists originally thought.

Helium is found in natural gas deposits. Ne, Ar, Kr, and Xe are all present in our atmosphere. Argon accounts for 1% of the air we breathe. Radon is radioactive and so is not generally included in a discussion of the noble gases.

Chapter 15 The Nonmetallic Elements

All the group VIIIA elements are monoatomic, colorless, odorless gases. The ionization energy values decrease with increasing atomic weight and size. The boiling points increase with size due to London forces.

The practical uses of the noble gases are limited to providing inert environments and to discharge tubes, the common neon lights with which we are familiar. Not all neon lights contain Ne however. Each noble gas has a distinct color when voltage is passed through it. Helium discharge tubes are pink, neon tubes are orange/yellow and argon tubes are blue/violet.

The group VIIIA elements Xe and Kr do display some chemistry. Since 1962 chemists have known that Xe will form compounds, but only with the most electronegative elements O and F. Then Kr was found to combine with F also. Although the Xe—O, Xe—F, Kr—F bonds are not strong, there is nothing unique about them. The octet is not as sacred or untouchable as chemists had thought. It is reasonable that Kr and Xe would be the most reactive of the noble gases because of their lower ionization energies. Xenon forms compounds in which it has +2, +4, +6, or +8 oxidation states, while Kr forms only a +2 state.

Problems

1. Can you think of a reason why N—N bonds are weaker than C—C bonds?

2. Household products that contain ammonia or hypochlorite ions always include a warning not to mix the two solutions. Poisonous hydrazine gas will form (N_2H_4). Write a balanced equation for this reaction.

3. Nitrous acid solutions can be formed by N_2O_3 and water or by NO and NO_2 with water (N_2O_3 dissociates to NO and NO_2 in the gas phase). Write balanced equations for the formation of HNO_2.

4. Draw Lewis electron-dot structures of the following compounds.

    a) N_2F_4 c) $POCl_3$

    b) $ClNO_2$

5. Use molecular orbital theory to predict the geometry of ClNO.

6. P_4O_{10} will dissolve in excess water to yield phosphoric acid. Write the equation.

7. Both Cu and Ag metals will *not* dissolve in HCl solution but will dissolve in HNO_3 solution. Why?

8. Pure NO_2 is reddish-brown in color and is paramagnetic, but under certain conditions becomes colorless and diamagnetic. What is happening?

9. Baking powder contains sodium bicarbonate and sodium dihydrogenphosphate. When heated the two reactants give off CO_2 gas which causes baked goods to rise. Write the balanced equation for this reaction.

10. Small amounts of pure O_2 can be generated in a laboratory by dropping water on sodium peroxide. Write the equation.

The Nonmetallic Elements   Chapter 15

11. Draw the structure of $S_2O_3^{2-}$.

12. Using molecular orbital theory, compare the bond order in O_2^+, O_2, O_2^-, O_2^{2-}.

13. Silver tarnish, Ag_2S, can be removed by replacing the Ag^+ with a more reactive metal, like aluminum. Placing tarnished silver in a warm salt solution in an aluminum pot often will remove the tarnish. Write the redox reaction that is occurring.

14. If all the bond angles in an S_8 molecule are equal, what are they?

15. What advantage is there to putting SF_6 gas into a tennis ball rather than air?

16. The acid HX where X = halogen, can be formed by reacting solid NaX with concentrated sulfuric acid. Give the balanced reaction.

17. What are the names of the following acids?

    a) HBrO
    b) $HBrO_2$
    c) HOI
    d) HIO_3

18. When IF gas is heated, the compound disproportionates. Write a balanced equation for this process.

19. Predict the geometry of the following species.

    a) I_3^-
    b) ClF_4^+
    c) IF_5

20. Find the equilibrium constant at 25 °C for the reaction below.

$$I_{2(aq)} + I^-_{(aq)} \rightleftharpoons I_3^-{}_{(aq)}$$

    ΔG° (kJ/mol)   16.4   -51.6   -51.4

21. Use the standard reduction potential diagram given for chlorine in the text to calculate $\epsilon °$ for the Cl_2 disproportionation in acid.

$$Cl_{2(g)} + H_2O + H^+_{(aq)} \rightarrow H^+_{(aq)} + Cl^-_{(aq)} + HOCl_{(aq)}$$

22. Predict the geometry of XeF_2, XeF_4, and XeF_6.

23. Prior to 1962 when Bartlett prepared $Xe^+PtF_6^-$, why had no one (except Linus Pauling) expected noble gases to form compounds?

CHAPTER 16
THE TRANSITION METALS

16.1. General Properties of the Elements

The transition metals include all the d-block and f-block elements. The 3d, 4d, and 5d transition metals are located in the central portion of the periodic table. The 4f and 5f metals (the lanthanide and actinide series) are located below the main portion of the periodic table. In many respects these 58 elements are similar, yet differences, as well as period and group trends, exist. Some of the elements, like iron and copper are abundant on earth while others such as iridium and hafnium are rare. Nearly all the lanthanide and actinide elements are rare or non-existent in nature. Therefore, our discussion will concentrate on the 3d transition elements, the most abundant and familiar of these metals.

The transition metals generally have high melting points, boiling points, ΔH^0_{vap}, ΔH^0_{fus}, electrical conductivity, and hardness values. All are solids except mercury, and all are metallic in appearance. A variety of oxidation states are possible for the transition metals. Most result in colored compounds, both in the pure state or in solution. Transition metal chemistry is colorful chemistry. We will see why in section 16.3.

The groups headed by Sc, Ti, V, Cr, Mn (formerly the IIIB-VIIB groups) have a maximum oxidation state equal to the number of s+d electrons in the valence shell. Thus chromium has oxidation states from 0 to +6. Examples of some Cr compounds are shown below.

Compound	Cr Oxidation State	Cr Configuration
Cr	0	$3d^5 4s^1$
$CrCl_2$	+2	$3d^4$
$[Cr(H_2O)_6]Cl_3$	+3	$3d^3$
K_2CrF_6	+4	$3d^2$
$CrOF_3$	+5	$3d^1$
H_2CrO_4	+6	$3d^0$

As expected, the highest possible oxidation state of the elements from Sc to Mn becomes increasingly harder to obtain. So MnO_4^-, with Mn^{7+} is a stronger oxidizing agent than V_2O_5, with V^{5+}.

All the 3d transition metals, except Sc and Ti, can form +2 ions in water. Recall that these ions are best represented as hydrated species with six H_2O molecules coordinated octahedrally to the metal ion. The size of the +2 ions, whether in a crystalline salt or as the

The Transition Metals

$M(H_2O)_6^{2+}$ species, shows a small decrease across the period as expected. As we mentioned in an earlier chapter, the size of group members shows little variation due to the lanthanide contraction.

16.2. Transition Metal Complexes

A transition metal complex or *coordination compound* consists of a central metal atom or ion that is surrounded by a group of molecules or ions called *ligands*. The ligands are in the first *coordination sphere* of the metal center. Both parts of the complex, the metal atom or ion and the ligands, are stable species alone, but are more stable when associated. A metal complex may have no net charge or may be anionic or cationic in nature.

Example 16.1. Determine the oxidation state of the central metal atom in $Fe(H_2O)_6^{3+}$ and $Fe(CN)_6^{4-}$.

$Fe(H_2O)_6^{3+}$ contains the Fe^{3+} species since the six water ligands are neutral

$Fe(CN)_6^{4-}$ contains the Fe^{2+} species since the six cyanide ions have a -1 charge (CN^-)

The interaction between the ligands and the metal is that of a Lewis acid and base. The metal center is usually electron deficient while the ligands are electron donors such as H_2O, NH_3, or CN^-. The electron-dot structures of the ligands show one or more lone pairs of electrons that can be donated or shared with a metal center. Whether this interaction is better described as ionic or covalent is the topic of the next section in this chapter.

If any part of a neutral compound is a metal complex, it is put in brackets.

Example 16.2. Determine the oxidation state of Cr in $K_3[Cr(CN)_6]$.

This compound consists of $3K^+$ ions and a $Cr(CN)_6^{3-}$ complex ion. The oxidation state of Cr must be +3 since cyanide has a -1 charge.

The *coordination number* of a complex refers to how many ligands are in the first coordination sphere. The most common coordination numbers are 2, 4, and 6. Notice in the examples below that there are two possible geometries for a coordination number of 4.

Chapter 16 — The Transition Metals

Complex	Coordination Number (C.N.)	Geometry
$Ag(NH_3)_2^+$	2	linear
MnO_4^-	4	tetrahedral
$Pt(NH_3)_4^{2+}$	4	square planar
$Co(H_2O)_6^{3+}$	6	octahedral

Isomers are compounds with the same molecular formula but different arrangement of atoms. Isomers can be different in physical and chemical properties. The general types of isomers are described below, with examples shown.

Structural Isomers - the atoms are attached differently

Ionization Isomers - yield different ions in solution
$[Co(NH_3)_5Cl]SO_4 \rightarrow [Co(NH_3)_5Cl]^{2+} + SO_4^{2-}$
$[Co(NH_3)_5SO_4]Cl \rightarrow [Co(NH_3)_5So_4]^{1+} + Cl^-$

Linkage Isomers - result from ligands that have optional points of attachment
$Fe(SCN)_4^-$ vs $Fe(NCS)]_4^-$

Stereoisomers - all bonds are the same but there is a different spatial arrangement of atoms

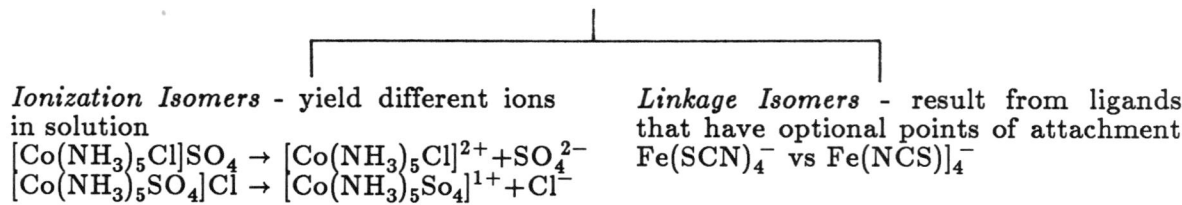

Geometric Isomers - can occur in square planar or octahedral complexes where ligands can be either next to or diagonal to each other. Cis-$Pt(NH_3)_2Cl_2$ has the two NH_3 molecules 90° apart and the two Cl^- ions 90° apart. Trans-$Pt(NH_3)_2Cl_2$ has the two NH_3 atoms 180° apart as well as the two Cl^- ions 180° apart.

Optical Isomers - results from having non-superimposable mirror images. Only tetrahedral complexes with 4 different ligands or octahedral complexes with cis ligands show this type of isomerism. See figure 16.7 in the text.

The rules of nomenclature are given in the text, but several other guidelines may be helpful. When naming a neutral salt, name the cation first and then the anion, even though one or both ions may be a metal complex. Brackets are generally used to separate the anion and cation species. When more than one kind of ligand is to be named, list them in alphabetical order, without dropping any letters.

The Transition Metals   Chapter 16

Example 16.3. Name $[Cr(H_2O)_6]Cl_3$.

The cation $Cr(H_2O)_6^{3+}$ is named first as one word, followed by the anion. There is no need to say trichloride because it is understood that a neutral salt must have three chlorides to balance this cation.

hexaaquachromium (III) chloride

Example 16.4. Name $[V(en)_2Cl_2]Br_2$. The abbreviation en means ethylenediamine, $NH_2CH_2CH_2NH_2$, a bidentate ligand.

The cation $V(en)_2Cl_2^{2+}$ contains a vanadium metal center with a 4+ oxidation state. The prefix di- can be used for the chloro ligand but not with the en ligand, where bis- is appropriate.

dichlorobis(ethylenediamine) vanadium (IV) bromide

16.3. Bonding Theories for Transition Metal Complexes

Any good bonding theory for transition metal complexes must account for experimental evidence such as the occurrence of colors in the visible range of light that vary from metal center to metal center, and ligand to ligand. Also, some metal centers such as Co^{3+}, a $3d^6$ configuration, is sometimes paramagnetic and sometimes diamagnetic, depending on the ligand. Another fact to be accounted for is that some coordination numbers of 4 lead to tetrahedral geometry while others have square planar geometry.

The two extremes of bonding can be considered. The strictly covalent model, where the metal center shares the lone pair electrons of the ligands is called *valence bond theory* (VBT). The metal center can be thought to contribute empty hybrid orbitals to overlap with filled ligand orbitals. The valence bond model does little more than account for geometry.

Complex Geometry	Metal Hybrid Orbitals
linear	sp
tetrahedral	sp^3
octahedral	d^2sp^3
square planar	dsp^2(s, p_x, p_y, $d_{x^2-y^2}$)

Chapter 16 — The Transition Metals

The other extreme, completely ionic bonding, is much better at explaining the experimental facts. This model is called *crystal field theory* (CFT). Crystal field theory treats the complex as a M^{n+} ion surrounded by 6 (in the octahedral case) or 4 (in the tetrahedral or square planar cases) negative point charges. The 3d orbitals of a free, gaseous metal ion are of equal energy or degenerate. But in the presence of a negative electric field of ligands, this degeneracy disappears. Let's look individually at the three most common crystal fields.

The octahedral complex can be viewed as six ligands approaching the metal ion along the x, y, and z axes. Recall the geometry of the d orbitals; the $d_{x^2-y^2}$ and d_{z^2} orbitals are centered on the cartesian axes while the d_{xy}, d_{xz}, d_{yz} orbitals are centered between the axes. Thus the negatively charged ligands on the axes will repel any electrons in the $d_{x^2-y^2}$ and d_{z^2} orbitals, and the d_{xy}, d_{yz}, d_{xz} orbitals will be of lower energy. The previously five-fold degenerate orbitals now separate into a three-fold generate set called the t_{2g} orbitals and a two-fold degenerate set called the e_g orbitals. These designations are from group theory which need not concern us here. On an energy diagram we see the results of the d orbitals splitting in an octahedral crystal field.

$$\underline{\quad x^2-y^2 \quad} \underline{\quad z^2 \quad} \; e_g$$

d orbitals in a spherically symmetric field

$$\underline{\quad xy \quad} \underline{\quad yz \quad} \underline{\quad xz \quad} \; t_{2g}$$

d orbitals of a gaseous M^{n+} ion (no field)

d orbitals in an octahedral field

The energy difference between the t_{2g} and e_g orbitals is called Δ_0, the crystal field splitting energy. This corresponds to the amount of energy needed to promote an electron from the t_{2g} set to the e_g set. This is generally energy available from visible light, so most transition metal complexes are colored. The *energy average* of the d orbitals in an octahedral field is the same as in a spherical field so the e_g orbitals are $\frac{3}{5}\Delta_0$ higher in energy and the t_{2g} orbitals are $\frac{2}{5}\Delta_0$ lower in energy than the spherical field.

The Transition Metals — Chapter 16

```
                       octahedral field

                            ___
                            e_g           3/5 Δ_0

    ___ ___ ___ ___ ___    .......
    spherical field                       2/5 Δ_0

                          ___ ___ ___
                            t_{2g}
```

The value of Δ_0 changes with the ligands present and with the metal center. As you expected, the smaller and more highly charged that the metal ion is, the stronger the interaction will be with the ligands, and so Δ_0 is larger. But the prediction that negatively charged ligands should result in larger Δ_0 values than neutral ligands is not consistent with experimental evidence. The *spectrochemical series* below shows the empirical results. Some polar molecules such as NH_3 and H_2O are stronger field ligands (that is, result in greater d orbital splitting and large Δ_0 values) than some negative ligands like hydroxide and the halides. This inability of crystal field theory to account for the spectrochemical series is obviously a shortcoming, but CFT does account for colors and for magnetic properties, as we'll see after this next example.

Example 16.5. Predict the relative Δ_0 values for the following complexes:

$$Co(NH_3)_6^{2+}, \quad Co(NH_3)_6^{3+}, \quad Ni(NH_3)_6^{2+}$$

Let's first compare $Co(NH_3)_6^{2+}$ and $Co(NH_3)_6^{3+}$. The difference between them will be due to the charge on the cobalt ion. Co^{3+} should be smaller and have a greater charge to size ratio, resulting in a stronger interaction with the NH_3 ligands. $Co(NH_3)_6^{3+}$ is expected to have a larger Δ_0 than $Co(NH_3)_6^{2+}$. Now compare the Co^{2+} and Ni^{2+} complexes. They have the same charge but Ni^{2+} should be a slightly smaller ion and so should have a slightly larger Δ_0.

Predicted Δ_0 values: $Co(NH_3)_6^{2+} < Ni(NH_3)_6^{2+} < Co(NH_3)_6^{3+}$

actual values: $Co(NH_3)_6^{2+}$, 121 kJ/mol

$Ni(NH_3)_6^{2+}$, 130 kJ/mol

$Co(NH_3)_6^{3+}$, 272 kJ/mol

Chapter 16	The Transition Metals

Now let us see how changing ligands can change the electronic configuration and magnetic properties of a metal ion. A Co^{3+} ion has a $3d^6$ electron configuration. In the presence of a weak field ligand such as F^-, Co^{3+} has 4 unpaired electrons. This corresponds to a $t_{2g}^4 e_g^2$ configuration.

$$CoF_6^{3-} \qquad \Delta_0 \text{ is small}$$

But the same Co^{3+} center in the presence of a stronger field ligand such as NH_3, is diamagnetic with a t_{2g}^6 configuration.

$$Co(NH_3)_6^{3+} \qquad \Delta_0 \text{ is large}$$

We say that the CoF_6^{3-} complex is *high spin* and the $Co(NH_3)_6^{3+}$ complex is *low spin*. For configurations d^1, d^2, d^3, d^8, d^9, d^{10} there is only one possible way to fill the t_{2g} and e_g orbitals, regardless of the ligand, but d^4 through d^7 species can have both high and low spin complexes, depending on the ligand.

Example 16.6. Why should we expect the d^3 metal centers V^{2+}, Cr^{3+}, and Mn^{4+} to be stable in an octahedral field?

A d^3 metal center in an octahedral field should have a t_{2g}^3 configuration, regardless of the ligand. Since the t_{2g} orbitals are lower in energy than the degenerate d orbitals in a spherical field, there is a net stabilization associated with the d^3 octahedral complexes. This stabilization energy is equal to $3 \times \frac{2}{5} \Delta_0$, or $\frac{6}{5} \Delta_0$.

Example 16.7. The violet $Ti(H_2O)_6^{3+}$ complex absorbs visible light at 493 nm, which corresponds to promoting a t_{2g} electron to the e_g orbitals. What is Δ_0 in kJ/mol?

$$\Delta_0 = E = h\nu = \frac{hc}{\lambda}$$

The Transition Metals — Chapter 16

$$\Delta_0 = \frac{(6.626 \times 10^{-34} \text{J·s})(3.0 \times 10^8 \text{m·s}^{-1})(6.02 \times 10^{23} \text{ mol}^{-1})}{493 \times 10^{-9} \text{ m}}$$

$$\Delta_0 = 2.43 \times 10^5 \text{ J/mol} = 243 \text{ kJ/mol}$$

The d orbitals also split in a tetrahedral field but in the opposite manner. Whereas the octahedral ligands were considered to be on the x, y, z axes, the tetrahedral ligands approach the metal center between axes. Figure 16.5 in the text illustrates this geometry. Since the t_{2g} orbitals (d_{xy}, d_{xz}, d_{yz}) are located between the axes, any electrons there would be repelled or raised in energy, while electrons in the e_g orbitals along the axes are lowered in energy.

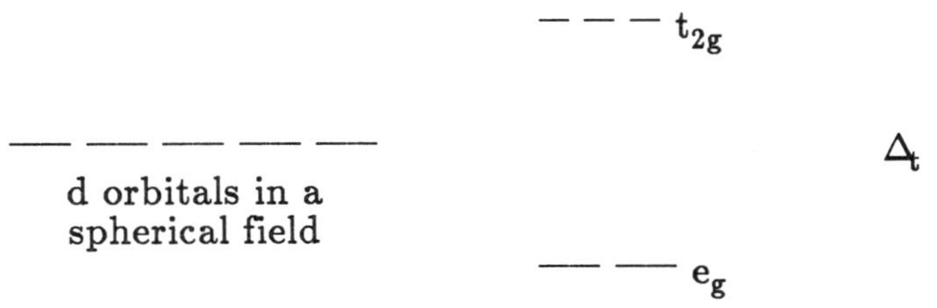

The value, Δ_t, is the splitting energy in the tetrahedral crystal field. The tetrahedral geometry is less common than the octahedral and Δ_t values are smaller than Δ_0 values. This is understandable because 4 ligands should not be capable of splitting the d orbitals to the same extent as 6 ligands. However, tetrahedral complexes are found for a small metal center with large ligands, like $ZnBr_4^{2-}$.

Square planar complexes are most easily envisioned as an octahedral complex in which the z axis ligands have been removed, leaving the four x and y plane ligands. This lowers the energy of the orbitals with a z component which is offset by raising the energy of the orbitals with only x and y components, and these ligands move closer to the metal ion.

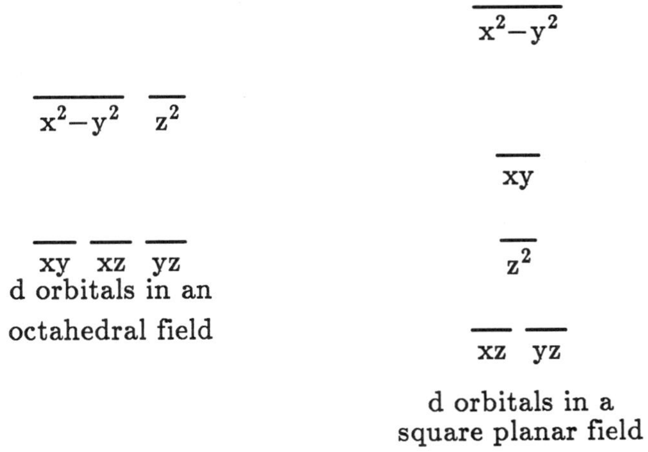

Some complexes are intermediate between octahedral and square planar geometries. The z axis ligands are farther from the metal center than the xy ligands but are not completely removed. This is a tetragonal distortion. While d^8 metal centers are usually square planar, many d^9 complexes show tetragonal distortion. There are several Δ values, or energy transitions, possible in a square planar complex.

Example 16.8. Scandium (III) complexes are octahedral but colorless. Why should $Sc(H_2O)_6^{3+}$ be colorless?

The Sc^{3+} ion has a d^0 configuration. The d orbitals may be split in the octahedral field, but since there are no d electrons to occupy the orbitals, there are no electrons to promote from t_{2g} to e_g, thus no measurable Δ_0 value.

Crystal field theory is not a perfect model because it treats the $M^{n+}-L$ bonds as strictly ionic, but many bonds do have some covalent character. This can be accommodated in ligand field theory (LFT) which applies molecular orbital theory to transition metal complexes. The combination of 3d, 4s, and 4p orbitals from the metal with six filled orbitals from the ligands (in the octahedral case) gives 15 molecular orbitals (see figure 16.18 in the text). The six pairs of electrons from the ligands occupy six σ orbitals. The d electrons from the metal center occupy non-bonding and anti-bonding orbitals.

The spectrochemical series of ligands can be explained in this way. Ligands with no π bonds, but extra lone electron pairs besides those used in the σ M—L bond (like Cl^-, Br^-, OH^-) can feed this electron density back toward the occupied t_{2g} orbitals of the metal, making them less stable. This L→M interaction is destabilizing overall, so Δ_0 is small. Ligands that do have π bonds and empty π^* orbitals can accept electron density from the t_{2g} orbitals of the metal. The t_{2g} and π^* orbitals are positioned for such a delocalization. This M→L interaction, sometimes called π *back-bonding*, stabilizes the complex and Δ_0 is large, as with CN^-, CO, or NO_2^-.

The carbonyl complexes are particularly good examples of this. The CO ligands stabilize the zero oxidation state of many transition metals. The π back-bonding is strong enough to notice a concomitant weakening of the CO. But the overall effect is stabilizing. This delocalization can be simplistically viewed as:

$$M-C\equiv O \quad \rightleftarrows \quad M=C=O$$

Before leaving this discussion of bonding in the transition metal complexes, some mention should be made of the 2nd and 3rd row transition metals and how they compare with their 1st row group members. In general, the 4d and 5d orbitals interact more strongly with ligands than 3d orbitals, leading to greater splitting of the t_{2g} and e_g sets. The Δ_0 values are large enough for these metal centers to be in low spin configurations under most circumstances.

16.4 The Lanthanides

Since the 4f orbitals do not interact with ligands as strongly as the d orbitals, the lanthanide series of elements, La through Lu, show only some behaviors in common with the transition metals. All the lanthanides display a +3 oxidation state and are similar to one another in their chemical reactions.

The Transition Metals   Chapter 16

16.5. Transition Metal Chemistry

Below is a summary of the common and not-so-common oxidation states of the 3d transition metals and some of the most common industrial uses.

Oxidation States

Metal	Common	Not-so-common	Uses
Sc	+3		none
Ti	+3,+4	+2	lightweight steel alloys, TiO_2 - white paint pigment
V	+2,+3,+4,+5		corrosion resistant steel alloy
Cr	+2,+3,+6	+5	stainless steel, chrome plating, CrO_2 - coating for recording tape
Mn	+2,+4,+7	+3,+6	hardener for steel, $MnCl_2$, MnO_2 - dry cell and alkaline batteries
Fe	+2,+3		steel
Co	+2,+3		alloys, ^{60}Co - radiation therapy
Ni	+2	+3,+4	catalysts, nickel plating
Cu	+1,+2		coins, brass, bronze, wiring $CuCl_2, Cu(NO_3)_2$ - wood preservatives
Zn	+2		galvanizing steels, brass, bronze, electrodes in alkaline and dry cell batteries

It is worth noting that the hexaaqua M^{2+} and M^{3+} complexes are all stronger acids than H_2O. Some pKa values are given below.

Species	pKa
H_2O	14.0
$Ca(H_2O)^{2+}$	12.6
$Mn(H_2O)^{2+}$	10.6
$Zn(H_2O)_6^{2+}$	8.8
$Sc(H_2O)_6^{3+}$	5.5
$Fe(H_2O)_6^{3+}$	2.2

We can see the trend from Ca^{2+} (d^0) to Zn^{2+} (d^{10}) is an increase in acidity. The general trend is not surprising. A water molecule in the first coordination sphere of a M^{n+} ion should be more easily ionized than in pure water. We should also expect the +3 ions to be more ionizing than +2 ions, both because of the greater charge and the smaller size. However, the trends across a period are not smooth and the extra stabilization that occurs with the splitting of the d orbitals must be considered as part of a more rigorous explanation.

16.6. Copper, Silver, and Gold

These special elements are known as the *coinage metals* because they easily can be cast into different shapes and are not easily corroded. They are very good conductors of heat and electricity but because of their cost, silver and gold are used only for specialty wiring.

Element	Oxidation States	Uses
Cu	+1, +2	coins, brass, bronze, wiring
Ag	+1	jewelry, flatware, mirror backing, AgBr, AgI - photography
Au	+1, +3	jewelry, dentistry, electronic wiring, coins

An interesting use of silver is in lenses that are photosensitive. The glass contains a dispersion of AgCl or AgBr which is white. When light hits the glass, some of the AgCl is converted to Ag+Cl. The silver is black and so the lenses darken. In the absence of light, the Ag and Cl atoms recombine and the glass lightens to the original state.

The biggest use of cyanide salts is in the purification of gold. Gold can be dissolved by forming the stable dicyanoaurate (I) complex which can then be reduced to gold metal.

$$2Au_{(s)} + \tfrac{1}{2}O_{2(g)} + 4CN^-_{(aq)} \rightarrow 2Au(CN)_{2\,(aq)}^- + 4OH^-_{(aq)}$$

$$2Au(CN)_{2\,(aq)}^- + Zn_{(s)} \rightarrow Zn(CN)_{4\,(aq)}^{2-} + 2Au$$

Copper metal is the most reactive of the coinage metals and forms a green coating of $CuCO_3$, $CuSO_4$, and $Cu(OH)_2$ after long exposure to the atmosphere. Silver will tarnish when exposed to SO_2 but is otherwise quite inert.

The Transition Metals   Chapter 16

16.7. Zinc, Cadmium, and Mercury

These elements are post-transition metals because the d shells are completely filled in both the zero oxidation state and the common +2 state. Only Hg has another oxidation state, +1 in Hg_2^{2+}, a dimeric ion. Mercury is also unusual because it is the only liquid transition metal.

Metal	Oxidation States	Uses
Zn	+2	galvanizing steel, brass, bronze, electrodes in batteries
Cd	+2	electrodes in rechargeable (NiCad) batteries
Hg	+1,+2	previously used for dental amalgams, thermometer liquid

Problems

1. What are the electron configurations of all common manganese oxidation states, Mn, Mn^{2+}, Mn^{4+}, Mn^{7+}?

2. What is the electron configuration of Cu?

3. A compound with the empirical formula $CrCl_3 \cdot 4H_2O$ dissolves in water to yield only two ions. What are they?

4. Name the following compounds and ions.

    a) $Pt(en)Cl_4$

    b) $K_2[Ni(CN)_4]$

    c) $[Co(NH_3)_2(CN)_4]^-$

    d) FeF_6^{3-}

    e) $Fe(CO)_5$

    f) $[Fe(OH)_2Cl_2I_2]^{3-}$

5. Write the formula for each of the following compounds.

    a) tetraamminecopper (II) ion
    b) dibromotetraammineruthenium (III) nitrate
    c) aquachlorobis(ethylenediamine)rhodium (III) chloride
    d) potassium diamminedioxalatocobaltate (III)
    e) calcium hexacyanoferrate (II)

6. Draw all the octahedral structures in problem #4 and indicate which will have geometric isomers.

Chapter 16 The Transition Metals

7. For each of the following compounds decide if there are optical isomers.

   a) $PtCl_3Br$ (square planar)

   b) $[Zn(OH)BrClI]^{2-}$ (tetrahedral)

   c) $Co(NH_3)_3Cl_3$

8. Invisible ink is a solution of the pale pink compound hexaaquacobalt (II) chloride which when heated converts to the blue compound cobalt (II) tetrachlorocobaltate (II) and water. Write the balanced reaction.

9. How many unpaired electrons are there in CoF_6^{3-}?

10. Give the d electron configuration for each of the following complexes.

    a) $CoCl_4^{2-}$

    b) $Fe(CN)_6^{2-}$

    c) $IrCl_6^{2-}$

11. Which 3d transition metals form d^3 ions?

12. In each of the following pairs, choose the complex that should absorb light of the shortest wavelength.

    a) $Co(H_2O)_6^{3+}$ or $Co(NH_3)_6^{3+}$

    b) $Fe(H_2O)_6^{2+}$ or $Fe(H_2O)_6^{3+}$

    c) $CoClub4^{2-}$ or $CoCl_6^{4-}$

13. The Δ_0 value for $Cr(NH_3)_6^{3+}$ is 258 kJ/mol. What wavelength of light is absorbed in this electron transition?

14. Why are Zn^{2+} complexes colorless?

15. Which two 3d transition metals do not easily form +2 ions in water?

16. Solutions of $Co(H_2O)_6^{3+}$ must always be freshly prepared. Why?

17. In the presence of O_2, CO_2, and water vapor, copper metal will oxidize to a mixture of copper (II) hydroxide and copper (II) carbonate. Write the balanced equation.

18. Vitamin B_{12} contains a cobalt metal center. At one stage of its physiological function, vitamin B_{12} is thought to contain a Co^{+1} ion. What coordination geometry might a Co^{+1} prefer?

The Transition Metals   Chapter 16

19. Without referring to any reduction potentials, choose which one of the following pairs should be the stronger oxidizing agent.

   a) TiO_2 or MnO_4^{2-}

   b) V_2O_5 or $K_2Cr_2O_7$

   c) Co^{3+} or Ni^{3+}

CHAPTER 17
ORGANIC CHEMISTRY

17.1. The Alkanes

The alkanes are a class of hydrocarbons (compounds containing only carbon and hydrogen) that have only single bonds. That means that all the C atoms have four bonds and tetrahedral geometry. A hydrocarbon with only single bonds is said to be *saturated*. Alkanes have the general formula C_nH_{2n+2}. Carbon can form infinitely long chains by bonding with other carbon atoms. This can lead to long straight chains of C atoms or branched chains. The simplest alkane is methane, CH_4. Next comes ethane, C_2H_6, propane, C_3H_8, butane, C_4H_{10}, and pentane C_5H_{12}, etc. The common prefixes meth-, eth-, prop-, and but- mean 1,2,3,4 carbons, respectively. Carbon chains of 5 or more use the standard Greek prefixes. The -ane ending indicates an alkane.

A great variety of alkanes are found as part of the Earth's petroleum products (see the list below).

Petroleum Products

Common Names	Alkanes
Natural Gas (methane)	C_1
LPG (propane and butane)	C_3, C_4
Gasoline	C_5–C_{10}
Kerosene	C_{10}–C_{18}
Diesel Fuel	C_{12}–C_{20}
Lubricating Oils, Paraffins, Asphalt	C_{20} or more

To name a hydrocarbon you must draw a skeleton structure and identify the longest chain of C atoms. This is the parent molecule. The branches are named and the number of the carbon atom to which they are attached is given.

Example 17.1. Name the following compound.

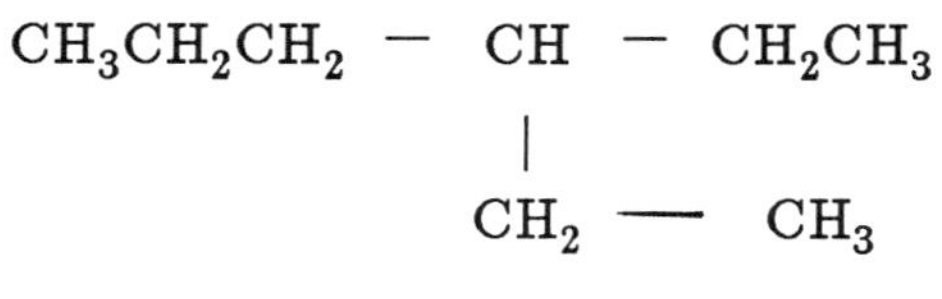

$$CH_3CH_2CH_2 - CH - CH_2CH_3$$
$$|$$
$$CH_2 - CH_3$$

Organic Chemistry   Chapter 17

The longest chain, no matter how you count, is 6 carbons so the parent molecule is hexane. There is an ethyl group (2 carbons) attached to carbon #3 (this is always done to give the *lowest* carbon number possible).

$$\text{name:} \quad \text{3-ethylhexane}$$

Example 17.2. Name the following.

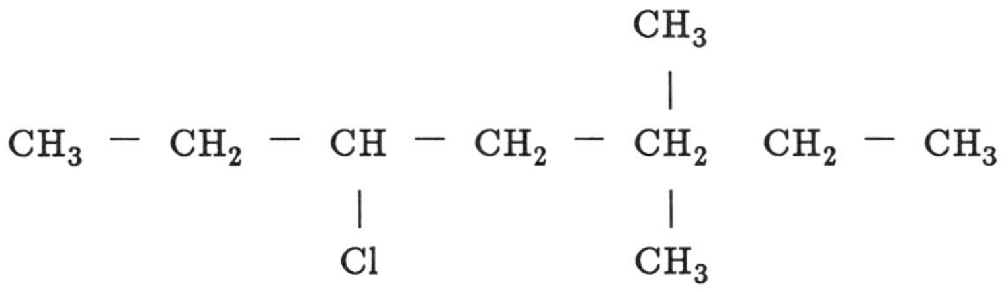

$$\text{CH}_3-\text{CH}_2-\underset{\underset{\text{Cl}}{|}}{\text{CH}}-\text{CH}_2-\underset{\underset{\text{CH}_3}{|}}{\overset{\overset{\text{CH}_3}{|}}{\text{CH}_2}}\quad\text{CH}_2-\text{CH}_3$$

The longest carbon chain is 7, a heptane. There are 2 methyl groups and 1 chloro group. These could be numbered 3-chloro-5,5-dimethylheptane *or* 3,3-dimethyl-5-chloroheptane. The second name is preferred because it uses the smallest numbers possible.

In general, as the number of carbons (and the molecular weight) increases in a series of hydrocarbons, so do the melting points and boiling points. This is to be expected. Both carbon and hydrogen have similar electronegativity values so the C—H bond is virtually non-polar. As a class, hydrocarbons are non-polar and unreactive. The melting and boiling points of these compounds are therefore a measure of London forces, which increase with size and molecular weight. A branched hydrocarbon has a lower melting and boiling point than the analogous straight chain compound. The lower surface area of the branched alkane results in weaker intermolecular forces and so the molecules are more easily separated.

As mentioned in Chapter 16, compounds with tetrahedral geometry may exhibit optical isomerism if all four ligands are different. Any organic molecule which has one or more carbon centers bonded to four different atoms or groups will have an optical isomer or an *enantiomer* (a non-superimposable mirror image).

Example 17.3. How many optically active sites (chiral centers) does the following molecule have?

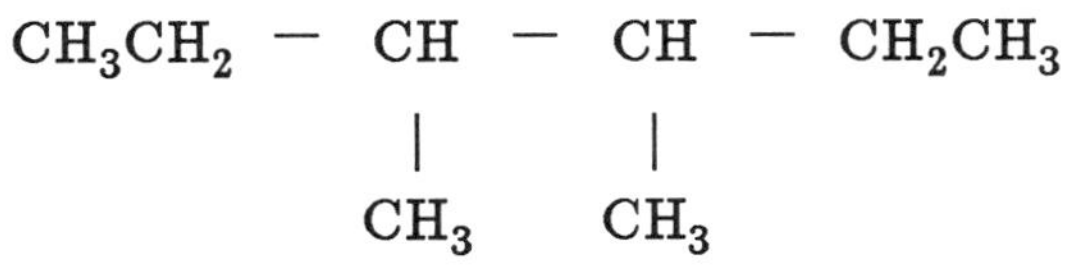

$$\text{CH}_3\text{CH}_2-\underset{\underset{\text{CH}_3}{|}}{\text{CH}}-\underset{\underset{\text{CH}_3}{|}}{\text{CH}}-\text{CH}_2\text{CH}_3$$

There are two chiral centers, indicated by the * in the expanded structure below.

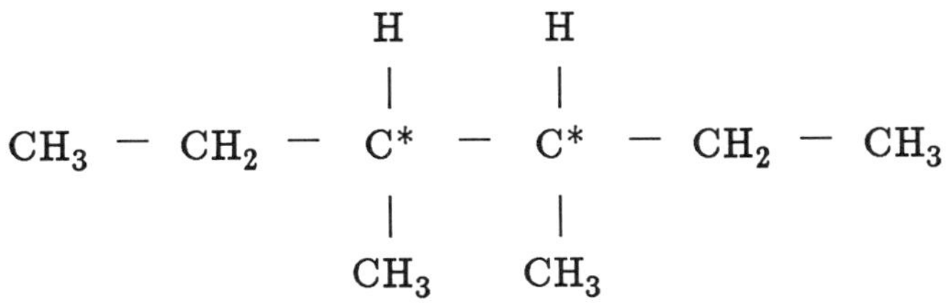

$$\text{CH}_3-\text{CH}_2-\underset{\underset{\text{CH}_3}{|}}{\overset{\overset{\text{H}}{|}}{\text{C*}}}-\underset{\underset{\text{CH}_3}{|}}{\overset{\overset{\text{H}}{|}}{\text{C*}}}-\text{CH}_2-\text{CH}_3$$

Can you name the compound? It is 3,4-dimethylhexane.

17.2. The Cycloalkanes

It is possible for an alkane structure to consist of a ring of carbon atoms. These compounds are called cycloalkanes and have the empirical formulas C_nH_{2n}. In the small rings cyclopropane (C_3) and cyclobutane (C_4) there is ring strain because the C—C—C bond angles are considerably smaller than the ideal tetrahedral 109°. There is little ring strain in the larger cycloalkanes because the carbon atoms do not remain in one plane but twist out of the plane in order to achieve bond angles nearer 109°. In cyclohexane this leads to conformers called chair, boat or twist forms that allow for minimum CH_2 repulsions and normal tetrahedral bond angles.

17.3. Unsaturated Hydrocarbons

Unsaturated hydrocarbons contain C=C and C≡C bonds, called *alkenes* and *alkynes*, respectively. Alkenes are named with an -ene ending and, when necessary, a number indicating placement of the double bond. Alkynes are named with a -yne ending.

Example 17.4. Name $CH_3CH=CHCH(CH_3)_2$.

The longest chain is 5 carbons, and the lowest numbering system gives a name of 1-methyl-3-pentene. This means the double bond is between carbons 3 and 4.

Example 17.5. Draw the structure of 1,3-butadiene.
$$H_2C=CH-CH=CH_2$$

Double bonded carbons have sp^2 hybridized orbitals which are planar. There is a chance of cis/trans geometric isomers if the groups attached to the carbons are different. For instance, propene does not exhibit geometric isomerism, but 1,2-dichloroethene does.

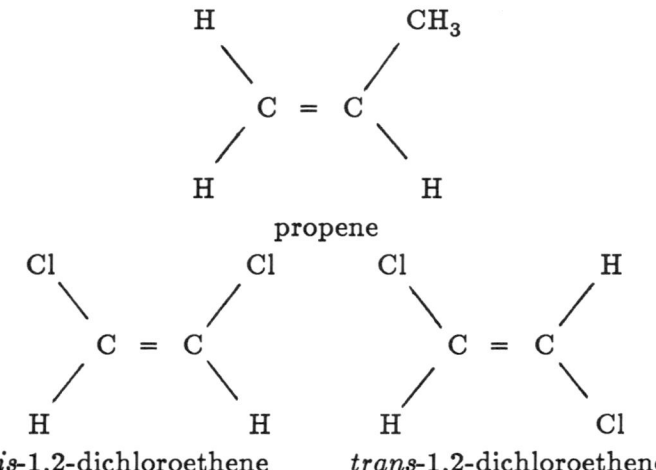

-250-

Alkynes have no geometric isomers because the sp hybrid orbitals dictate a linear geometry about the triple bond.

Many compounds contain multiple sites of unsaturation. These double and triple bonds may be isolated in the molecule so that they do not interact, but frequently they occur close together, on alternate carbon atoms. This is called *conjugation*. Conjugated alkenes can be written using resonance structures but are better understood in terms of delocalized π electrons. Benzene, the name used for cyclohexatriene, can be written as the resonance structures below.

$$\begin{array}{ccc}
\text{CH} & \text{CH} & \text{CH} \\
\diagup\!\!\diagup \quad \diagdown & \diagup \quad \diagdown\!\!\diagdown & \diagup \quad | \quad \diagdown \\
\text{HC} \quad\quad \text{CH} & \text{HC} \quad\quad \text{CH} & \text{HC} \quad\quad \text{CH} \\
| \quad\quad \| & \| \quad\quad | & \| \quad\quad \| \\
\text{HC} \quad\quad \text{CH} & \text{HC} \quad\quad \text{CH} & \text{HC} \quad\quad \text{CH} \\
\diagdown\!\!\diagdown \quad \diagup & \diagdown \quad \diagup\!\!\diagup & \diagdown \quad | \quad \diagup \\
\text{CH} & \text{CH} & \text{CH}
\end{array}$$

Benzene is actually more stable than any one resonance structure predicts in terms of bond strengths. All the C atoms are in a plane, restricted by the double bonds. Each C atom has a p orbital perpendicular to the plane. These six p orbitals form molecular orbitals that allow the π electrons to be delocalized over the entire ring. Compounds like benzene are called *aromatic*.

17.4. Functional Groups

A great deal of interesting chemistry results when hydrocarbons have other elements incorporated in the structure. These other elements are mainly oxygen and nitrogen. Sulfur, phosphorus and the halogens also change the physical and chemical properties of the organic chemical properties of the organic compounds that contain them. Some common functional groups found in organic molecules are surveyed below.

Alcohols contain an OH group, or hydroxyl group, attached to a C atom. Because O has lone pairs of electrons and is the second most electronegative element, alcohols have a measurable dipole moment, unlike most hydrocarbons. Alcohols are usually water soluble because of their polarity. The names of alcohols use the hydrocarbon root with an -ol ending.

$$\text{CH}_3\text{CH}_2\text{CH}_2\text{CH}_2\text{OH} \quad\quad \text{CH}_3\text{CH}_2\text{CHOHCH}_3$$
$$\text{1-butanol} \quad\quad\quad\quad \text{2-butanol}$$

Alcohols can be classified as primary, secondary, and tertiary depending on the carbon that is bonded to the hydroxyl group. If the hydroxyl-bonded carbon is bonded to only one other carbon, the alcohol is primary, as in 1-butanol. A secondary alcohol is one in which the hydroxyl group is bonded to a carbon attached to two other carbon atoms, as in 2-butanol. A tertiary alcohol has a carbon bonded to an OH as well as three other carbon atoms, as shown in the following molecule.

Chapter 17   Organic Chemistry

$$\begin{array}{c} CH_3 \\ | \\ CH_3 - CH - CH_3 \\ | \\ OH \end{array}$$

2-methyl-2-propanol
(common name: tert-butyl alcohol)

Ethers contain an oxygen atom attached to two hydrocarbon moieties. Ethers usually have a smaller dipole moment than alcohols so don't dissolve in water as easily, but are nonetheless Lewis bases because of the two lone pairs of electrons on the O atom. The systematic names of ethers include the -oxy ending for the smaller hydrocarbon portion.

$CH_3CH_2OCH_2CH_3$ $HOCH_2CH_2CH_2OCH_2CH_3$

ethoxyethane   3-ethoxypropanol
(common name
diethyl ether)

Aldehydes and *ketones* both have the general formula

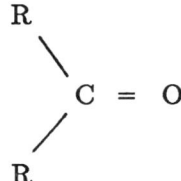

where both R and R′ are hydrocarbon groups for ketones, but only one of which is a hydrocarbon group for aldehydes (the other R=H). The $>C=O$ functional group is called a *carbonyl group*. Aldehydes are good reducing agents because the terminal carbonyl group is reactive. Ketones are less reactive and are not easily oxidized. The names of aldehydes end in -al while ketones end in -one.

Example 17.6. Name the following compound.

$$(CH_3)_2CHCH_2-\underset{\underset{O}{\|}}{CH}$$

4-methyl-butanol is the correct name. The aldehyde is understood to be at carbon #1 so the name 4-methyl-1-butanol is redundant.

Organic Chemistry Chapter 17

Example 17.7. The common chemical acetone (nail polish remover) has the IUPAC name 2-propanone. Draw the structure of acetone.

$$CH_3-\underset{\underset{O}{\|}}{C}-CH_3$$

Carboxylic acids contain a COOH group with the structure shown below.

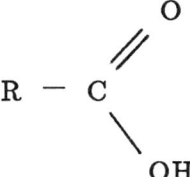

Most small carboxylic acids show appreciable acidity and dissolve in water because of intermolecular hydrogen bonding. The names of carboxylic acids end in -oic acid.

$$CH_3-\underset{\underset{CH_3}{|}}{\overset{\overset{CH_3}{|}}{C}}-COOH$$

2,2-dimethylpropanoic acid

The salt names end in -ate.

$$CH_3-\underset{\underset{CH_3}{|}}{\overset{\overset{CH_3}{|}}{C}}-COO^-\ Na^+$$

sodium 2,2-dimethylpropanoate

Esters are compounds formed from the reaction of a carboxylic acid and an alcohol. The resulting compound has the following formula.

$$R-\underset{\underset{O}{\|}}{C}-OR'$$

The acid portion ends in an -ate while the alcohol portion ends in -yl when the ester is named. A few common esters and their names are shown below.

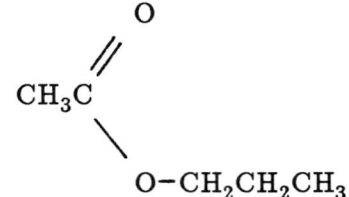

ethyl butanoate
(common name: ethyl butyrate)

$$CH_3C\begin{matrix} \nearrow O \\ \searrow O-CH_2CH_2CH_3 \end{matrix}$$

butyl ethanoate
(common name: butyl acetate)

Both the esters shown have odors that are reminiscent of pineapples. Many esters have fruity aromas.

The *amines* are the final functional group in this survey. As with alcohols, there are primary, secondary, and tertiary amines. A primary amine has the formula RNH_2 and is amino-substituted ammonia. Secondary amines have the formula

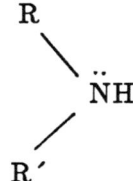

and tertiary amines

$$\begin{matrix} R \searrow \\ \ddot{N}-R' \\ R' \nearrow \end{matrix}$$

A fourth substitution also can be made on the N atom, usually resulting in quaternary ammonium salt.

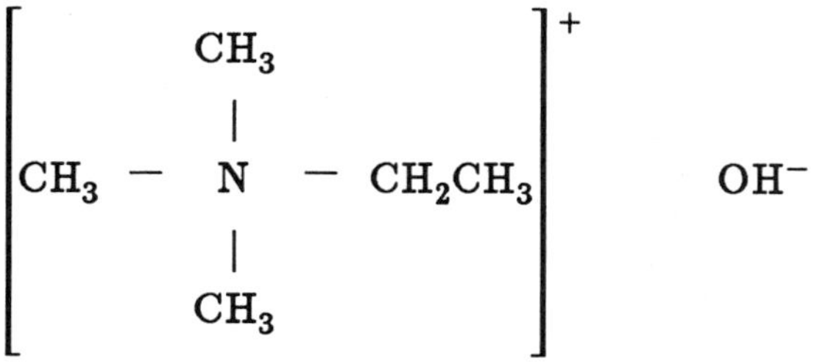

Organic Chemistry   Chapter 17

$$\left[CH_3 - \underset{\underset{CH_3}{|}}{\overset{\overset{CH_3}{|}}{N}} - CH_2CH_3 \right]^+ \quad OH^-$$

trimethylethylammonium hydroxide

Most amines are named for the attached hydrocarbon groups, but oxygen functional groups take precedence in the nomenclature.

$$CH_3CH_2NH_2$$

ethylamine

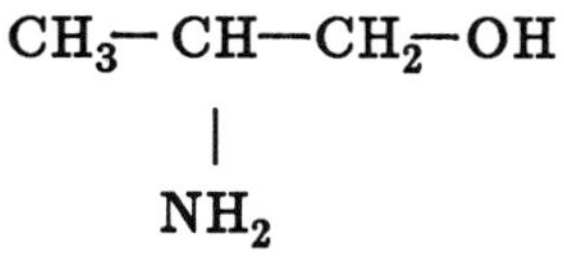

$$CH_3-CH-CH_2-OH$$
$$|$$
$$NH_2$$

2-aminopropanol

17.5. Reactivity of Functional Groups

The three most important and common reaction types associated with the organic molecules we have discussed are displacement or substitution reactions, addition reactions (or its reverse, elimination reactions), and oxidation-reduction reactions. Some examples of these reactions are shown in the next few pages. The mechanisms are not indicated as they vary with the reactant.

Addition/Elimination Reactions

alkenes, alkynes $\quad \diagup\!\!\!\!\!\diagdown C=C \diagdown\!\!\!\!\!\diagup + H_2 \xrightarrow{Ni} -\underset{|}{\overset{|}{C}}-\underset{|}{\overset{|}{C}}-$
                                                            HH

$-C\equiv C- + Br_2 \rightarrow \quad \diagup\!\!\!\!\!\diagdown C=C \diagdown\!\!\!\!\!\diagup \quad \rightarrow -\underset{|}{\overset{|}{C}}-\underset{|}{\overset{|}{C}}-$
$ Br Br Br Br$

with Br, Br above the right product.

$\diagup\!\!\!\!\!\diagdown C=C \diagdown\!\!\!\!\!\diagup + HBr \rightarrow -\underset{|}{\overset{|}{C}}-\underset{|}{\overset{|}{C}}-$
$ H Br$

-255-

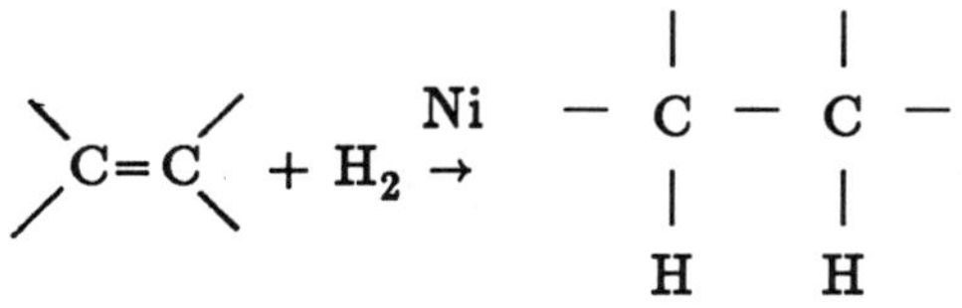

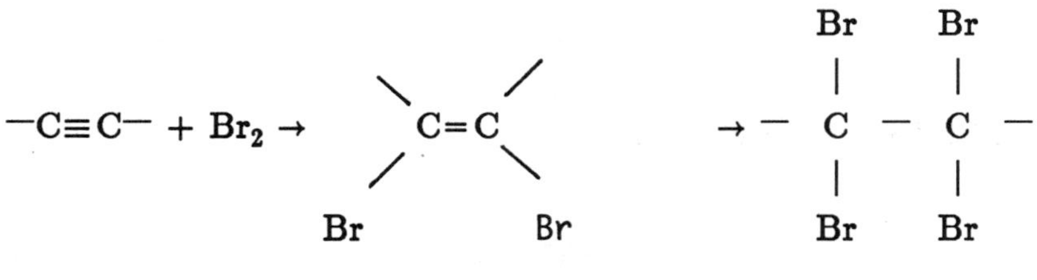

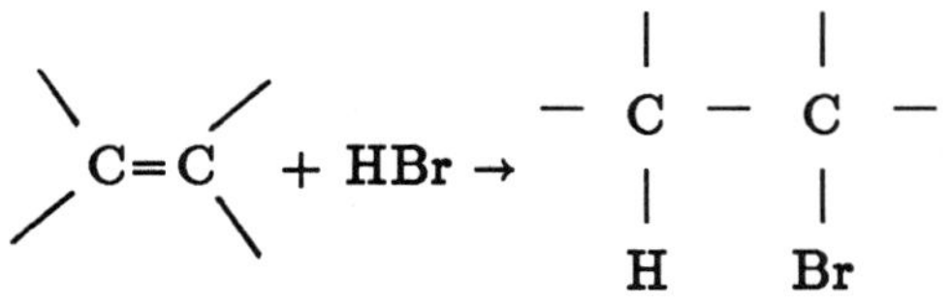

Chapter 17   Organic Chemistry

$$\ce{>C=C< + H2O ->[H^+]} \begin{array}{c} | \quad | \\ -C-C- \\ | \quad | \\ H \quad OH \end{array}$$

alcohols $\quad \begin{array}{c} | \quad | \\ -C-C- \\ | \quad | \\ H \quad OH \end{array} \xrightarrow{H_2SO_4} \ce{>C=C<} + H_2O$

aldehydes, ketones $\quad \ce{>C=O} + \underset{\text{Grignard Reagent}}{RMgX} \rightarrow \begin{array}{c} | \\ -C-R \\ | \\ OMgX \end{array}$

$$\begin{array}{c} | \\ -C-R \\ | \\ OH \end{array}$$

Displacement Reactions

alkanes $\quad \begin{array}{c} | \\ -C-H \\ | \end{array} + Br_2 \xrightarrow{h\nu} \begin{array}{c} | \\ -C-Br^+ \\ | \end{array} + HBr$

order of reactivity: $Cl_2 > Br_2$

alcohols $\quad R-OH + HBr \rightarrow RBr + H_2O$

order of reactivity: $HI > HBr > HCl$
and tertiary $-OH >$ secondary $>$ primary

Oxidation-Reduction Reactions

alkenes, alkynes $\quad \ce{>C=C<} \xrightarrow{KMnO_4} \begin{array}{c} | \quad | \\ -C-C- \\ | \quad | \\ OH \quad OH \end{array}$

Organic Chemistry					Chapter 17

$$\text{alcohols} \quad -\underset{H}{\overset{|}{C}}- OH \quad \xrightarrow{Cr_2O_7^{2-}} \quad \diagdown_{\diagup}C=O \quad \text{ketone}$$

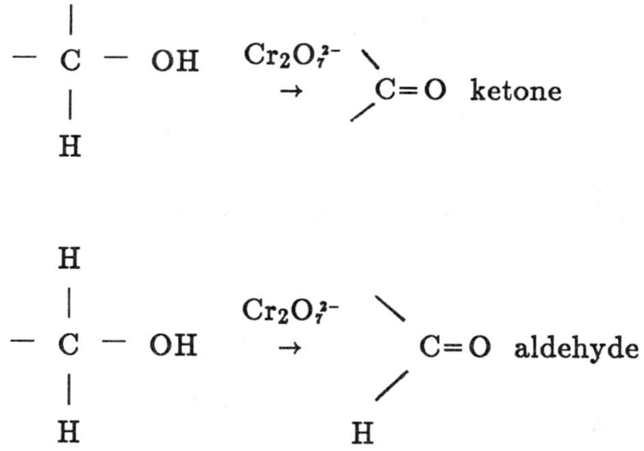

$$-\underset{H}{\overset{H}{\underset{|}{\overset{|}{C}}}}- OH \quad \xrightarrow{Cr_2O_7^{2-}} \quad \underset{H}{\diagdown_{\diagup}}C=O \quad \text{aldehyde}$$

$$\text{aldehydes} \quad \underset{H}{\diagdown_{\diagup}}C=O \quad \xrightarrow{Cr_2O_7^{2-}} \quad \underset{HO}{\diagdown_{\diagup}}C=O$$

$$\diagdown_{\diagup}C=O \quad \xrightarrow{LiAlH_4} \quad -\underset{H}{\overset{|}{\underset{|}{C}}}- OH$$

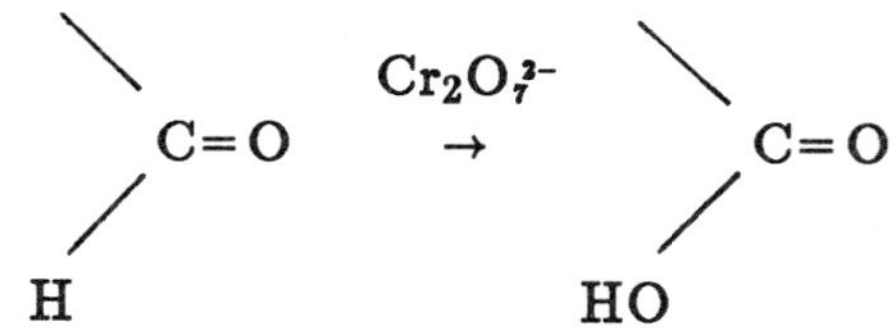

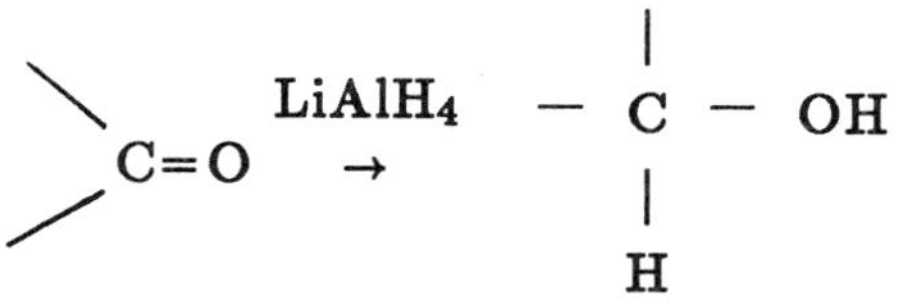

Problems

1. Draw a structure for each of the following compounds.

    a) 2,2,4-trimethylpentane (also known as iso-octane)
    b) 2-bromobutane
    c) 3,4-dimethyl-4-ethylheptane
    d) 2,2,3,3-tetramethylpentane

2. How many different structures does the formula C_4H_{10} have? Name them.

3. A compound with the formula C_5H_{10} could be an alkene or a cycloalkane. Draw all possible structures and name them.

4. Name the following compounds.

    a) HC≡CH

    b) $HOCH_2CH=CHCH_2CH_2CH=C(CH_3)_2$

    c) $CH_2=\underset{\underset{CH_3}{|}}{C}-CH_2CH_2CH_3$

-257-

Chapter 17   Organic Chemistry

5. What isomers (geometric or optical) exist for the compounds in problem #4?

6. Four different alcohols have the molecular formula C_4H_9OH. Name them. Do any of these alcohols have a chiral center?

7. Give the correct name for di-isopropyl ether, $(CH_3)_2CHOCH(CH_3)_2$.

8. Name the following.

   a) $CH_3CH=CHCHO$

   b) $CH_3\underset{CH_3}{CH}CH_2\underset{O}{\overset{\|}{C}}CH_3$

   c) $CH_3CH_2\underset{OH}{CH}CH_2CHO$

   d) $CH_3COOCH_2CH_3$

   e) Cl_3CCOOH

   f) $HCOO^-NH_4^+$

   g) $Ca(CH_3CH_2COO)_2$

9. Complete the following equations.

   a)   2-pentene + Cl_2 $\xrightarrow{h\nu}$

   b) $H-C\equiv C-H + 2H_2$ $\xrightarrow{Ni}$

   c)   1-butene + H_2O $\xrightarrow{H^+}$

   d)   propene + HCl →

10. An oxidation-reduction reaction that all hydrocarbons undergo is combustion. Write a balanced reaction for the combustion of methanol and identify the changes in oxidation states.

11. Predict the products of the following reactions.

   a) CH_3CH_2OH $\xrightarrow{H_2SO_4}$

   b) $(CH_3)_3COH$ $\xrightarrow{H_2SO_4}$

   c) $(CH_2CH_3)_3N + HCl →$

   d) CH_3CH_2OH $\xrightarrow{K_2Cr_2O_7}$

   e) $CH_3CH_2\underset{CH_3}{CH}CHO$ $\xrightarrow{NaBH_4}$

-258-

Organic Chemistry Chapter 17

    f) $CH_3OH + CH_2=CHCH_2COOH \rightarrow$

    g) $CH_3CH_2CH_2CH_2MgBr + (CH_3)_2C=O \rightarrow$

12. The reaction of potassium permanganate and ethene yields manganese (IV) oxide and 1,2-ethanediol.

$$H_2C=CH_2 + KMnO_4 \rightarrow \underset{\underset{OH}{|}\underset{OH}{|}}{H_2C-CH_2} + MnO_2$$

Balance this redox equation in acid solution.

CHAPTER 18
BIOCHEMISTRY

18.1. The Cell

As you know, living things are composed of cells, and cells are simply organized collections of molecules with specific functions. A large portion of each cell is water but organic molecules account for most of the non-aqueous matter.

A great deal of the chemistry we have learned can be applied to the understanding of biological processes. Frequently the biochemical molecules are large but their functions can be rationalized in terms of structure, thermochemistry, and kinetics.

18.2 Biochemical Energetics and ATP

Metabolism is the sum of all chemical activities in an organism. There are two general categories of metabolism. *Catabolism* is the breaking down of large molecules into smaller molecules, usually accompanied by an overall release of energy. An example would be the conversion of a sugar to CO_2 and H_2O via combustion. *Anabolism* is the reverse of catabolism. It is the construction of large molecules from smaller units, which usually consumes energy. Green plants build sugar molecules from CO_2 and H_2O during photosynthesis. Both anabolism and catabolism occur in the same organism. The energy derived from catabolism is used to drive anabolism. This amounts to coupling a spontaneous reaction ($\Delta G = -$) with a non-spontaneous reaction ($\Delta G = +$).

In most organisms this is accomplished by using the reaction below.

$$ATP + H_2O \rightarrow ADP + HPO_4^{2-} \qquad \Delta G° \simeq -31 \text{ kJ/ mol @ pH} = 7$$

Metabolic processes occur in many steps, some of which can put energy into ATP production or release energy by ATP destruction. For example, the combustion of 1 mole of glucose, below, releases an enormous amount of energy, but it does so through an 18-step process.

$$C_6H_{12}O_6 + 6O_2 \rightarrow 6H_2O + 6CO_2 \qquad \Delta G° = -2870 \text{ kJ/ mol}$$

A total of 38 ATP molecules are produced from ADP molecules per glucose molecule oxidized. The ATP is essentially storing this energy for later release. Do not, however, think that this energy is stored in the new phosphate bond that forms in ATP. *Energy is not stored in bonds.* Bonds form due to a lowering of energy. What is true is that the third phosphate group in ATP is weakly bonded, so that when this weak bond is broken and the phosphate is free to interact with other molecules there is a net release of energy, *not* from the bond-breaking process (which must be endothermic) but from the formation of more stable bonds.

18.3. Lipids

There are many different kinds of lipids. They vary greatly in structure but all have one important property in common. Lipids are non-polar due to long hydrocarbon sections. They do not dissolve in water and so are called *hydrophobic*. The class of lipids we shall discuss here are the fats and oils. These lipids are the esters formed from glycerol and long chain carboxylic acids.

Biochemistry   Chapter 18

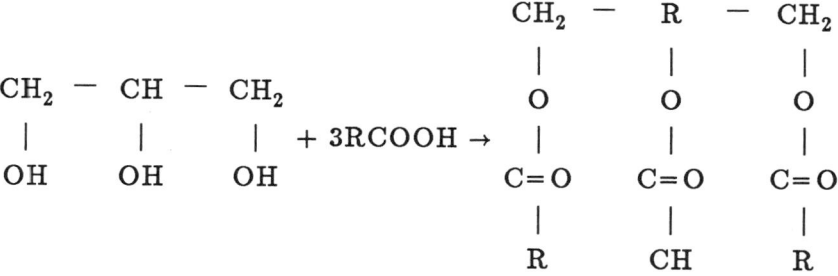

$$\underset{\text{glycerol}}{\begin{array}{c}CH_2 - CH - CH_2\\ |\quad\quad |\quad\quad |\\ OH\quad OH\quad OH\end{array}} + \underset{\text{acids}}{3RCOOH} \rightarrow \underset{\text{triglyceride ester}}{\begin{array}{c}CH_2 - R - CH_2\\ |\quad\quad |\quad\quad |\\ O\quad\quad O\quad\quad O\\ |\quad\quad |\quad\quad |\\ C=O\quad C=O\quad C=O\\ |\quad\quad |\quad\quad |\\ R\quad\quad CH\quad R\end{array}}$$

A triglyceride ester may have three identical or three different fatty acid moieties. If the acid groups are saturated then the triglyceride will be a solid at room temperature. This is a fat. If the fatty acids are unsaturated (have double bonds) the triglyceride will be liquid at room temperature. This is an oil. Oils can be converted to fats by hydrogenation.

$$\begin{array}{c}CH_2-O-C-(CH_2)_7CH=CH(CH_2)_7CH_3\\ |\\ CH-O-C-(CH_2)_7CH=CH(CH_2)_7CH_3 + 3H_2\\ |\\ CH_2-O-C-(CH_2)_7CH=CH(CH_2)_7CH_3\end{array} \xrightarrow{Ni} \begin{array}{c}CH_2-O-C-(CH_2)_{16}CH_3\\ |\\ CH-O-C-(CH_2)_{16}CH_3\\ |\\ CH_2-O-C-(CH_2)_{16}CH_3\end{array}$$

The ester linkages are broken in an organism by enzymes. The fatty acids are then metabolized step-wise, much like glucose, but fatty acids contain more carbons and so as a consequence, even more ATP molecules are produced from the combustion of fats and oils than from sugars.

18.4. Carbohydrates

Carbohydrates have the general formula $C_nH_{2n}O_n$ or $(CH_2O)_n$. The compounds are either polyhydroxy aldehydes (aldoses) or polyhydroxy ketones (ketoses). The simplest carbohydrates are the *monosaccharides* like fructose, glucose, or ribose. The open chain form of glucose is shown below. Each asterik (*) indicates a chiral center.

$$\underset{\text{glucose}}{\begin{array}{cccccc}1 & 2 & 3 & 4 & 5 & 6\\ & * & * & * & *\\ HC & - CH & - CH & - CH & - CH & - CH_2\\ \| & | & | & | & | & |\\ O & OH & OH & OH & OH & OH\end{array}}$$

There are 2^4 or 8 pairs of enantiomers possible, only three structures of which are abundant in nature. A ring structure of glucose and other monosaccharides is slightly more stable under conditions found in the body. The ring form has the same formula, but the aldehyde carbon

Chapter 18   Biochemistry

from the open chain form now has an -OH group attached and is chiral. The two enantiomers are called α and β-glucose.

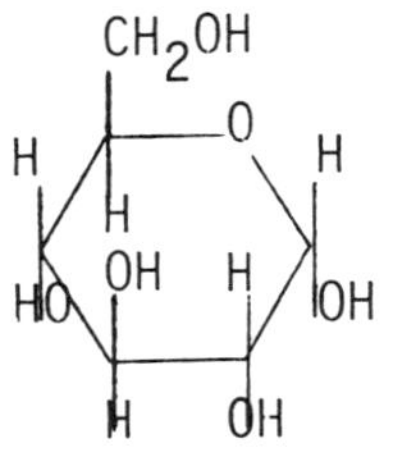

α−glucose   β−glucose

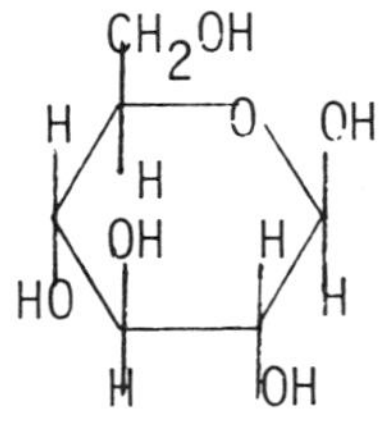

Disaccharides are two monosaccharides joined in a 1,4 linkage (carbon #1 to carbon #4 of the next molecule, via an O atom) by the loss of H_2O. This type of reaction is called a *condensation* and is common in living organisms.

Example 18.1. Maltose is the disaccharide of two α-glucose molecules. Draw the structure of maltose in ring form.

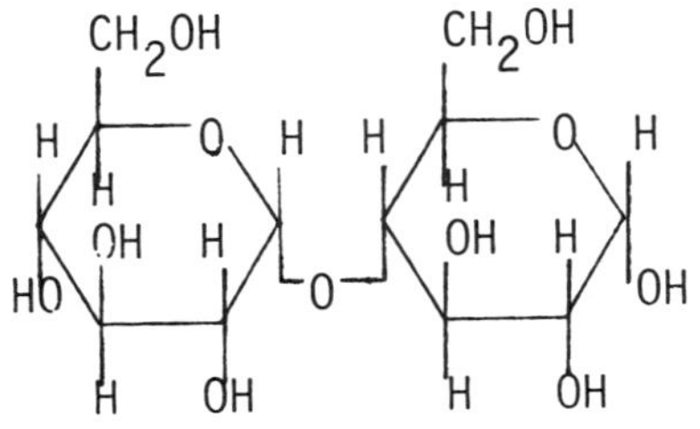

Mono and disaccharides are readily metabolized (or oxidized) in most organisms. Large quantities are not stored as the simple sugars, but rather as *polysaccharides*. Starch is one of these polysaccharides formed by long chains of α-glucose molecules. Polysaccharides in plants also serve as structural material. Cellulose is a polysaccharide of β-glucose.

Example 18.2. Assuming an average of 3500 glucose units per cellulose "molecule," what is the molecular weight of cellulose?

Each glucose unit, $C_6H_{12}O_6$, has a molecular weight of 180 g/mol. But the linkages required 1 mol H_2O to be lost per linkage. Only the end glucose does not lose a water of condensation.

$$MW = 3500(8180) - 3499(18) = 567,000 \text{ g/mol}$$

Biochemistry　　Chapter 18

In the metabolism of carbohydrates, pyruvate, a ketoacid anion, plays an important role. Each glucose molecule is converted to two pyruvates, each of which enters the Krebs cycle to become CO_2 and H_2O, generating ATP molecules along the way.

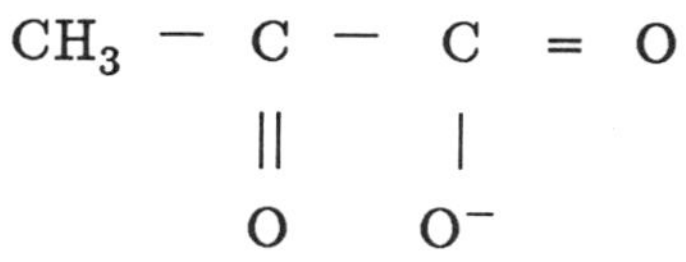

$$CH_3 - \underset{\underset{O}{\|}}{C} - \underset{\underset{O^-}{|}}{C} = O$$

pyruvate ion

Example 18.3. The K_a of pyruvic acid is 3.20×10^{-3}. What is the ratio of pyruvic acid to pyruvate ion at a pH of 7?

$$K_a = 3.20 \text{ time } 10^{-3} = \frac{[H^+][\text{pyruvate}]}{[\text{pyruvic acid}]}$$

$$\text{at } pH = 7, \quad [H^+] = 10^{-7}$$

$$\frac{[\text{pyruvate}]}{[\text{pyruvic acid}]} = 32{,}000$$

So the anion form dominates under most physiological conditions.

18.5. Proteins

All proteins are long chains of amino acids. There are twenty acids and a seemingly endless variety of proteins that exist. An *amino acid* is a carboxylic acid with an amine group on the α carbon and the carbon atom adjacent to the COOH. The general form of an amino acid is shown below.

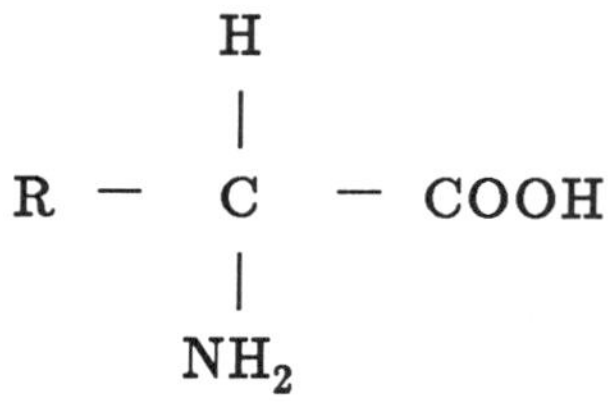

$$R - \underset{\underset{NH_2}{|}}{\overset{\overset{H}{|}}{C}} - COOH$$

R is a hydrogen atom in the simplest amino acid glycine, but R can be a variety of alkyl groups, some substituted with hydroxy, carbonyl, or amine groups. Except for glycine, all amino acids have a chiral center at the α carbon. However, all living organisms contain the set of amino acids designated as the L enantiomers.

Chapter 18   Biochemistry

Because each amino acid has a base site $(-NH_2)$ and an acid site $(-COOH)$ we expect pH to be important in determining the exact acid form. Near a pH of 7, the acid group is mostly deprotonated while the amine group is protonated. This leads to both an anion and cation in the same molecule, called a *zwitterion*.

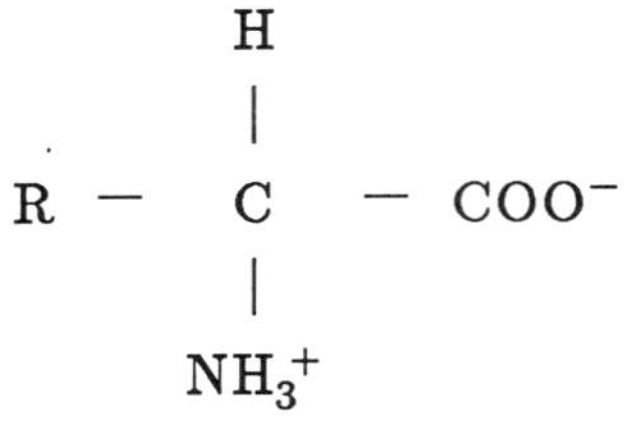

$$R - \underset{\underset{NH_3^+}{|}}{\overset{\overset{H}{|}}{C}} - COO^-$$

In addition, the R group may contain an acidic or basic functional group with its own K_a or K_b value.

Amino acids are bonded together to form peptide chains. The peptide bond forms by a condensation reaction. The amine group of one amino acid bonds to the acid group of another amino acid with a subsequent loss of a water molecule.

Example 18.4. Write the chemical equation for serine and valine, the two amino acids shown below, forming a dipeptide.

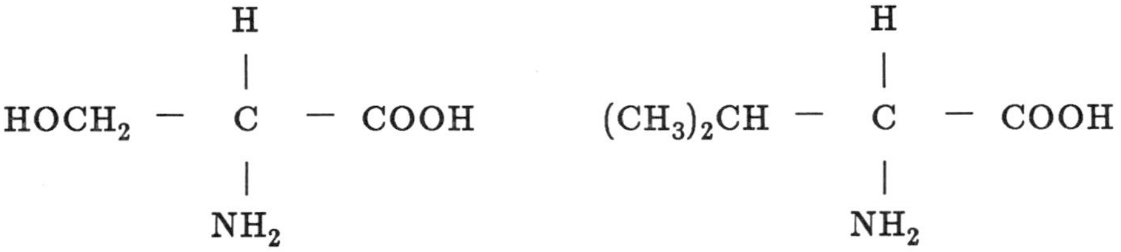

$$HOCH_2 - \underset{\underset{NH_2}{|}}{\overset{\overset{H}{|}}{C}} - COOH \qquad (CH_3)_2CH - \underset{\underset{NH_2}{|}}{\overset{\overset{H}{|}}{C}} - COOH$$

$$\text{serine} \qquad\qquad\qquad\qquad \text{valine}$$

There are *two* ways in which the peptide bond can form. The $-NH_2$ of serine plus the $-COOH$ of valine, *or* the $-NH_2$ of valine plus the $-COOH$ of serine. This results in two different dipeptides, ser-val or val-ser. By convention a peptide chain is written, left-to-right, to account for the acid portion of the left amino acid reacting with the amine portion of the right amino acid.

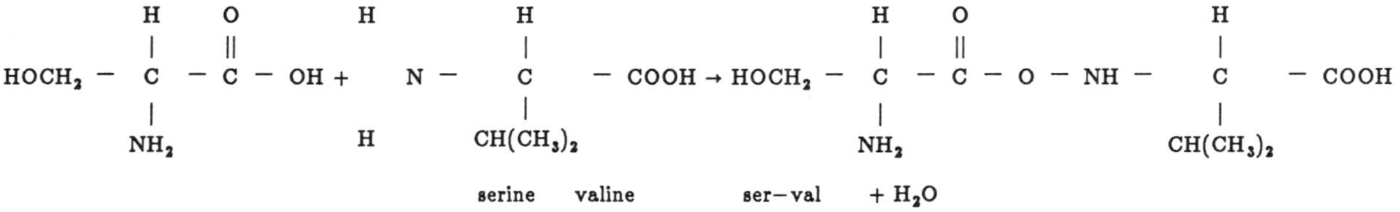

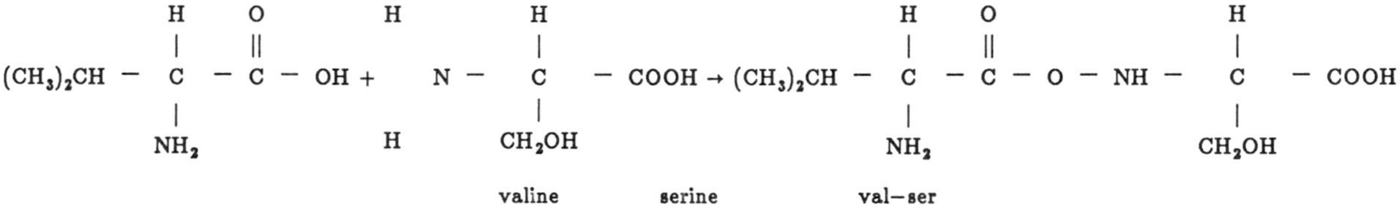

$$HOCH_2-\underset{\underset{NH_2}{|}}{\overset{\overset{H}{|}}{C}}-\overset{\overset{O}{\|}}{C}-OH + \underset{\underset{H}{|}}{\overset{\overset{H}{|}}{N}}-\underset{\underset{CH(CH_3)_2}{|}}{\overset{\overset{H}{|}}{C}}-COOH \rightarrow HOCH_2-\underset{\underset{NH_2}{|}}{\overset{\overset{H}{|}}{C}}-\overset{\overset{O}{\|}}{C}-O-NH-\underset{\underset{CH(CH_3)_2}{|}}{\overset{\overset{H}{|}}{C}}-COOH + H_2O$$

$$\text{serine} \qquad\qquad \text{valine} \qquad\qquad \text{ser-val}$$

$$(CH_3)_2CH-\underset{\underset{NH_2}{|}}{\overset{\overset{H}{|}}{C}}-\overset{\overset{O}{\|}}{C}-OH + \underset{\underset{H}{|}}{\overset{\overset{H}{|}}{N}}-\underset{\underset{CH_2OH}{|}}{\overset{\overset{H}{|}}{C}}-COOH \rightarrow (CH_3)_2CH-\underset{\underset{NH_2}{|}}{\overset{\overset{H}{|}}{C}}-\overset{\overset{O}{\|}}{C}-O-NH-\underset{\underset{CH_2OH}{|}}{\overset{\overset{H}{|}}{C}}-COOH$$

$$\text{valine} \qquad\qquad \text{serine} \qquad\qquad \text{val-ser}$$

The order of the amino acids in a protein constitutes the primary structure, but the subsequent shape that a protein takes (the secondary, tertiary, and quaternary structures) depends mainly on three types of interactions. The most common interaction type is hydrogen bonding. An amino acid chain can fold in such a way that one peptide linkage can H-bond with

$$\left(\begin{array}{c} - C - O - NH - \\ \parallel \\ O \end{array} \right)$$

another peptide linkage. While hydrogen bonds are weak individually, there can be so many that the resulting structure is very resistant to change. A second type of weak interaction also leads to folding of protein chains. Some amino acids have polar R groups attached to the α carbon and some have non-polar groups. Since polar sections of the molecule will be attracted to each other and to water, the chain may fold in such a way to allow these polar groups to be nearby. At the same time, non-polar groups do not have an affinity for water and so may aggregate in a hydrophobic portion of the molecule if the chain can fold in such a way to allow this. The third type of interaction between portions of one chain or between separate chains is due to covalent bonding of sulfur atoms. The amino acid cysteine contains an —SH group. When two cysteine residues are aligned properly an —S—S— bond can form to hold the protein in a specific configuration.

18.6 Nucleic Acids

Nucleic acids are yet another class of biological polymers. The two major types of nucleic acid are DNA (deoxyribonucleic acid) and RNA (ribonucleic acid). DNA molecules store and pass on all the hereditary information of an organism. DNA is found in the chromosomes of each cell. RNA puts the genetic information to work in the cell and is found throughout it.

The individual parts of these polymers are 1) a sugar, the five carbon ribose in RNA and dioxyribose (ribose minus an OH) in DNA, 2) a phosphate linked to the #5 carbon of the sugar, and 3) an organic base molecule linked to the #1 carbon of the sugar. The phosphate-sugar-base monomer is called a nucleotide. Nucleotides are used elsewhere in the cell too. But nucleic acids have large numbers of nucleotides in a chain. There are 5 organic bases used in most DNA and RNA. In DNA, adenine, guanine, thymine, and cytosine are found while in RNA, thymine is replaced by the related uracil.

RNA molecules are usually a single strand while DNA molecules are double-stranded in a helix where the two strands are held in place by intermolecular hydrogen bonding. The geometry of the double helix of DNA is such that guanine will H-bond only to cytosine while thymine will H-bond only to adenine. Each of the complementary base pairs forms 2 or 3 H-bonds of approximately the same distance.

18.7. Biological Functions of the Nucleic Acids

The two main functions of the nucleic acids are 1) to replicate so that the same genetic information is present in every cell of an organism and 2) to use the genetic information of DNA to make all the proteins in the organism.

The replication begins by the double-stranded DNA molecule unwinding to two separate complementary strands, each one of which acquires a new complementary strand, stepwise. The result is two complete DNA molecules.

Chapter 18   Biochemistry

When the second function, protein synthesis, occurs both DNA and RNA are important. A DNA molecule unwinds and a complementary messenger RNA (m-RNA) strand forms. This m-RNA molecule becomes attached to a ribosome in the cell where transfer RNA (t-RNA) builds a protein by forming a strand of bases complementary to the m-RNA (therefore the same as the original DNA). Each 3-nucleotide or 3-base piece of t-RNA brings with it an amino acid. As the t-RNA attaches to the m-RNA, the amino acids form a growing peptide chain and eventually a protein.

Problems

1. The reaction of ATP and water to yield ADP and HPO_4^{2-} releases 31 kJ/mol under intercellular conditions. What is the value of K_{eq} at 37°C?

2. Which of the 20 amino acids contains the highest percentage of nitrogen?

3. How many chiral centers are in the amino acid threonine?

4. In the livers of animals, pyruvate can be animated to make the amino acid alanine. Write the reaction, identifying the carbon that acquires the amine group.

5. Photosynthesis is the reverse of respiration.

$$6CO_{2(g)} + 6H_2O_{(l)} \rightarrow C_6H_{12}O_{6(g)} + 6O_{2(g)}$$

   a) What is $\Delta G°$ for this reaction at 298 K?
   b) What is the number of liters of O_2 that can be produced by a tree that converts 75 g of CO_2 each day? (assume T=298 K)

6. Draw the dipeptides leu-ala and ala-leu.

7. Why are fats more efficient for energy storage than carbohydrates?

8. The popular flavor enhancer, MSG, is monosodium glutamate, the sodium salt of glutamic acid. Draw the structure of MSG.

9. Which amino acids contain polar side chains?

10. If a DNA molecule is analyzed, what ratio of adenine to guanine is expected?

CHAPTER 19
THE NUCLEUS

19.1. The Nature of the Nucleus

Each nucleus contains protons and neutrons, whose numbers are given in the following shorthand notation.

$$^A_Z X$$

Z is the atomic number or number of protons in the nucleus of element X. A is the total number of nuclear particles or nucleons. Many elements have several isotopes which vary only in the number of neutrons. Some isotopes are stable. Some are not and are radioactive.

Although nearly all of an atom's mass is contained in the nucleus, it occupies only a very small part of the overall volume. The smallest nucleus belongs to the ^1H atom with a radius of 1.3×10^{-13} cm. The nuclear radii of other atoms can be calculated with the equation below.

$$r = (1.3 \times 10^{-13} \text{ cm}) A^{\frac{1}{3}}$$

This assumes that all nuclei are spherical in shape which is not true. Some nuclei are slightly elongated. The densities of nuclei are nearly constant at approximately 2×10^{14} g/cm^3, which roughly corresponds to 200 million tons of weight per mL of volume.

Example 19.1. How much larger is the nucleus of an average gold atom than the nucleus of a hydrogen atom?

$$r_H = 1.3 \times 10^{-13} \text{ cm}$$

$$r_{Au} = 1.3 \times 10^{-13} (197)^{\frac{1}{3}} = 7.6 \times 10^{-13} \text{ cm}$$

The radius of an Au atom is only 5.8 times that of an H atom but since $V \alpha r^3$, the volume of the Au nucleus will be 197 times larger than the H atom nucleus.

Example 9.2. Calculate the density of the ^{56}Fe nucleus which has a mass of 55.92068 g/mol.

$$r_{Fe} = (1.3 \times 10^{-13})(56)^{\frac{1}{3}} = 5.0 \times 10^{-13} \text{ cm}$$

$$V = \tfrac{4}{3}\pi r^3 = 5.2 \times 10^{-37} \text{ cm}^3$$

Chapter 19　　　　　　　　　　　　　　　　　　　　　　　　　　　　　　The Nucleus

$$\text{mass per nucleus} = \frac{55.92068}{6.02 \times 10^{233}} = 9.29 \times 10^{-23} \text{ g}$$

$$\text{density} = \frac{9.29 \times 10^{-23} \text{ g}}{5.2 \times 10^{-37} \text{ cm}^3} = 1.8 \times 10^{14} \text{ g/cm}^3$$

It is convenient when dealing with individual atoms and nuclei to use the atomic mass unit rather than grams.

$$6.022 \times 10^{23} \text{ amu} = 1.000 \text{ g}$$

$$1 \text{ amu} = 1.661 \times 10^{-24} \text{ g}$$

It also will be helpful to recall the mass of a proton, electron, and neutron.

proton	1.0072765 amu
neutron	1.0086650 amu
electron	0.00054866 amu
^{1}H (proton + electron)	1.0078250 amu

Example 19.3. Predict the mass of a $_9^{19}$F atom in amu.

The *predicted* mass can be calculated by adding the mass of 9 protons, 9 electrons and 10 neutrons or by adding 9 H atoms and 10 neutrons.

$$9(^1\text{H}) + 10\text{n} = 9(1.0078250) + 10(1.0086650)$$

$$= 19.157075 \text{ amu}$$

The mass of a ^{19}F atom is not what we predicted but is a somewhat smaller value, 18.9984033 amu. What has happened to the remaining 0.1586717 amu? The missing mass has been converted to energy that was released when the nucleus was formed. This is called the *binding energy* and represents how much lower in energy a nucleus is than its separate nucleons, or how much energy would be required to separate the nucleons completely. The interconversion of mass and energy is common in nuclear reactions. Einstein's famous equation relates mass and energy via the speed of light.

$$E = mc^2$$

The Nucleus   Chapter 19

Example 19.4. What is the binding energy of a ^{19}F atom?

The mass lost, 0.1586717 amu, has been converted to energy.

$$E = \left(0.1586717 \text{ amu} \times \frac{1.661 \times 10^{-24} \text{ g}}{\text{amu}} \times \frac{10^{-3} \text{ kg}}{\text{g}}\right)\left(2.998 \times 10^8 \frac{\text{m}}{\text{S}}\right)^2$$

$$E = 2.3688 \times 10^{-11} \frac{\text{kg m}^2}{\text{s}^2} = 2.3688 \times 10^{-11} \text{ J}$$

While 10^{-11} J seems like a negligible amount of energy, converted to a per mole basis it is enormous.

$$2.3688 \times 10^{-11} \frac{\text{J}}{\text{atom}} \times \frac{6.0221 \times 10^{23} \text{ atoms}}{\text{mol}} = 1.4265 \times 10^{13} \frac{\text{J}}{\text{mol}}$$

Units of MeV (megaelectron volts) also are common for energy.

$$96{,}485 \frac{\text{J}}{\text{mol}} = 1 \frac{\text{MeV}}{\text{atom}}$$

A convenient conversion factor to remember is: 1 amu converts to 931.5 MeV of energy.

The calculation of binding energy results in increasingly larger values as the atomic mass increases. A more informative value than simple binding energy is binding energy *per nucleon*. Figure A3 in the text shows a plot of this energy vs. atomic mass. The most stable nuclei are those with the largest binding energy per nucleon. This occurs for elements with a mass near 60 amu (Fe, Co, Ni, etc.). These stable nuclei are found in large quantities in the earth's crust and in meteorites.

Besides large binding energy per nucleon values, there are several other factors that stable nuclei have in common with each other. Stable nuclei are far more likely to contain even numbers of protons and neutrons than odd. Nuclei which contain "magic numbers" of nucleons (2, 8, 20, 28, 50, 822, 126) are also especially stable.

Example 19.5. Calculate the change in mass that must accompany the following conventional (non-nuclear) reaction at 298 K.

$$C_{(\text{graphite})} + O_{2(g)} \rightarrow CO_{2(g)} \qquad \Delta H° = -393.5 \text{ kJ}$$

$$E = mc^2$$

Chapter 19   The Nucleus

$$3.935 \times 10^5 \text{ J} = \text{mass}(2.998 \times 10^8 \text{ms}^{-1})^2$$

$$4.378 \times 10^{-12} \text{ kg} = \text{mass lost per mole of product}$$

$$\text{or} \quad 4.378 \times 10^{-1} \text{ amu lost per molecule of product}$$

19.2. Radioactivity

A plot of Z vs. N, figure 19.4 in the text, shows that for nuclei of Z<20, a stable configuration is achieved when N = Z. Larger nuclei, Z⩾20, need more neutrons than protons, presumably for buffering the coulombic repulsions, so N>Z. Any nucleus outside this zone of stability represented in figure 19.4 will spontaneously undergo *radioactive decay*. The mode of decay depends on whether the nucleus has too many protons or neutrons, or both.

Source of Instability	Possible Decay Mode	Example
too many neutrons, too few protons	β^- emission	$_6^{14}\text{C} \rightarrow {_7^{14}}\text{N} + {_{-1}^{0}}\text{B}$
too many protons, too few neutrons	positron emission or electron capture	$_8^{15}\text{O} \xrightarrow{\text{EC}} {_7^{15}}\text{N} + {_1^0}\beta$ $_{80}^{197}\text{Hg} \rightarrow {_{79}^{197}}\text{Au}$
too many protons and neutrons	α emission	$_{84}^{210}\text{Po} \rightarrow {_{82}^{206}}\text{Pb} + {_2^4}\alpha$

Alpha emission occurs in large nuclei, Z>80. The nucleus emits an α particle which is two protons and two neutrons. An α particle is a helium nucleus.

$$_2^4\alpha = {_2^4}\text{He}$$

Example 19.6. The radioisotope ^{226}Ra decays by alpha emission. Write a balanced equation indicating the products.

$$_{88}^{226}\text{Ra} \rightarrow {_2^4}\alpha + {_{86}^{222}}\text{Rn}$$

Example 19.7. Given the masses of ^{226}Ra, ^{222}Rn, and ^{4}He below, calculate the energy that accompanies the reaction in example 19.6.

$$^{226}_{88}\text{Ra} \rightarrow {}^{4}_{2}\alpha + {}^{222}_{86}\text{Rn}$$

| mass (in amu) | 226.0254 | 4.0026 | 222.0175 |

The Ra atom has 88 protons, 88 electrons, and 138 neutrons. The Rn atom will have 86 protons, 136 neutrons, and *the same 88 electrons* as the Ra. These two extra electrons are not included in the given weight of Rn but are included in the α particle mass. Although an α particle contains only two protons and two neutrons, by using the mass of an ^{4}He atom we can include the two extra electrons that initially reside on the Rn atom.

$$\Delta \text{ mass} = [4.0026 + 222.0175] - [226.0254] = -0.0053 \text{ amu}$$

The mass decreases, therefore energy is released.

$$E = 0.0053 \text{ amu} \left(931.5 \frac{\text{MeV}}{\text{amu}}\right)$$

$$E = 4.94 \text{ MeV}$$

The energy emitted may be imparted as kinetic energy to the product particles or may be emitted as radiation. Gamma (γ) radiation frequently accompanies α emission.

Example 19.8. What is the wavelength of 0.05 MeV γ rays?

$$E = \frac{hc}{\lambda}$$

$$0.05 \text{MeV} \times \frac{10^6 \text{eV}}{1 \text{MeV}} \times \frac{1.602 \times 10^{-19} \text{ J}}{1 \text{eV}} = \frac{6.626 \times 10^{-34} \text{J} \cdot \text{s} \times 2.998 \times 10^8 \text{ms}^{-1}}{\lambda}$$

$$\lambda = 2.48 \times 10^{-11} \text{m} = 0.0248 \text{ nm}$$

A beta particle is an electron that is emitted from a nucleus that has too many neutrons and not enough protons. Several symbols refer to this particle.

$$^{0}_{-1}e = {}^{0}_{-1}\beta = \beta^{-1} = \beta \text{ particle}$$

Chapter 19   The Nucleus

When a nucleus undergoes beta decay, a neutron becomes a proton and an electron. The proton remains in the nucleus and the electron is emitted. So there is no change in mass number but the atomic number increases by one. Gamma rays or neutrinos (ν) are frequently co-emitted with beta particles. A neutrino is an uncharged particle with almost no mass that carries off excess energy.

Example 19.9. ^{60}Co is a radioactive element used for the radiation treatment of some types of cancer. ^{60}Co is a beta emitter. Cells, both normal and cancerous, are killed by the beta particles. Write the nuclear reaction.

$$^{60}_{27}\text{Co} \rightarrow \, ^{0}_{-1}\text{e} + \, ^{60}_{28}\text{Ni}$$

Both positron emission and electron capture are possible for nuclei with too may protons and not enough neutrons. A positron is a positively charged electron, an antiparticle.

$$^{0}_{1}\text{e} = \, ^{0}_{1}\beta = \beta^{+} = \text{positron}$$

In positron emission, a proton is converted to a neutron and a positron is emitted from the nucleus along with a neutrino. While positron emission occurs for some man-made isotopes, it is not common for natural isotopes due to the energy restrictions (see explanation in text).

Example 19.10. ^{18}F decays by positron emission. Write the reaction.

$$^{18}_{9}\text{F} \rightarrow \, ^{0}_{+1}\text{e} + \, ^{18}_{8}\text{O}$$

Example 19.11. Given the following mass values, calculate the energy involved in the ^{1}F positron decay.

	$^{18}_{9}\text{F}$	$\rightarrow$	$^{0}_{+1}\text{e}$	+	$^{18}_{8}\text{O}$
mass (amu)	18.000950		0.0005486		17.999159
	9p, 9n, 9e$^-$				8p, 10n, 9e$^-$

$$\Delta \text{ mass} = [^{0}_{+1}\text{e} + \, ^{18}\text{O} + \text{extra e}^-] - [^{18}\text{F}]$$

$$\Delta \text{ mass} = [0.0005486 + 17.999159 + 0.0005486] - [18.000950]$$

$$\Delta \text{ mass} = -0.0006938 \text{ amu}$$

$$E = (0.0006938 \text{ amu})\left(931.5 \frac{\text{MeV}}{\text{amu}}\right)$$

$$E = 0.6463 \text{ MeV released}$$

Electron capture is yet another type of beta decay. The nucleus with too many protons can capture an electron from the first quantum shell. This electron combines with a proton to form a neutron. Electron capture is not detectable by emitted particles but when outer electrons fall to the first quantum shell to replace the electron captured, γ or x-rays are emitted.

Some large radioisotopes undergo several decay steps before becoming a stable nucleus. This can involve more than one method of decay.

Example 19.12. ^{232}Th decays to the stable ^{208}Pb isotope in 10 steps. The last three steps are β^-, β^-, and α emissions in that order, starting with ^{212}Pb. Write the three reactions.

$$^{212}_{82}\text{Pb} \rightarrow \,^{0}_{-1}\text{e} + \,^{212}_{83}\text{Bi}$$

$$^{212}_{83}\text{Bi} \rightarrow \,^{0}_{-1}\text{e} + \,^{212}_{84}\text{Po}$$

$$^{212}_{84}\text{Po} \rightarrow \,^{4}_{2}\alpha + \,^{208}_{82}\text{Pb}$$

19.3. Radioactive Decay Rates

Radioactive decay rates are always first order. The equations for first-order processes all apply to the disintegration of radioisotopes, though some of the symbols are different.

$$\ln \frac{N}{N_0} = -\lambda t$$

where N = number of nuclei
N_0 = number of nuclei at time zero
λ = rate constant
t = time

$$\text{rate} = -\frac{dN}{dt} = \lambda N \quad \text{and} \quad t_{\frac{1}{2}} = \frac{0.693}{\lambda}$$

Chapter 19 The Nucleus

The main difference between chemical kinetics and nuclear kinetics is what we measure. For nuclear decays it is easiest to measure the absolute number of atoms decaying per unit time, while in conventional chemical processes we measure loss of reactants with time or formation of products.

Example 19.13. Living tissues have 15.3 disintegrations per minute per gram of carbon, due to ^{14}C decaying via beta emission. The half-life of ^{14}C is 5730 yrs. What is the product of ^{14}C decay? How old is a wooden object that has only 9.1 disintegrations per minute per gram carbon?

$$^{14}_{6}\text{C} \rightarrow {}^{0}_{-1}\text{e} + {}^{14}_{7}\text{N}$$

$$\lambda = \frac{0.693}{t_{\frac{1}{2}}} = \frac{0.693}{5730} = 1.209 \times 10^{-4} \text{ yrs}^{-1}$$

$$\ln \frac{N}{N_0} = -\lambda t$$

$$\ln \frac{9.1}{15.3} = -(1.209 \times 10^{-4} \text{yrs}^{-1})t$$

$$t = 4{,}300 \text{ yrs}$$

Example 19.14. In July 1945, the first man-made atomic explosion took place in a New Mexico desert. Approximately how much of the ^{90}Sr ($t_{\frac{1}{2}} = 29$ yrs) that was originally produced in that explosion still remains?

Since the amount of ^{90}Sr that remains keeps changing, use July 1987 for the calculation.

$$\lambda = \frac{0.693}{t_{\frac{1}{2}}} = 0.02390 \text{ yr}^{-1}$$

$$\ln \frac{N}{N_0} = -t$$

$$\ln \frac{N}{1} = -(0.02390 \text{ yr}^{-1})(42 \text{ yr}) = -1.0038$$

$$N = 0.3665$$

This means that in July 1987 there remains 36.65% of the original ^{90}Sr.

19.4 Nuclear Reactions

The nuclear decay reactions we've discussed thus far have emitted only small particles like the α, β, or positron particles with concurrent small changes in atomic number or mass. There are other reactions that involve reacting two particles to make a new element. All the elements heavier than uranium have been made this way. Curium for example can be synthesized by bombarding a ^{232}Th with ^{12}C particles.

$$^{232}_{90}\text{Th} + {}^{12}_{6}\text{C} \rightarrow {}^{240}_{96}\text{Cm} + 4\,{}^{1}_{0}\text{n}$$

The trans-uranium elements are radioactive. Besides the scientific interest these man-made elements generate, some of them have been put to practical use. ^{241}Am is found in many smoke detectors and ^{99}Tc is used in the medical field to detect damaged heart tissue. The synthesis of man-made elements requires that the colliding particles have sufficient kinetic energy to overcome the nuclear repulsions. Particle accelerators were developed for this purpose.

There are two other types of nuclear reactions, *fission* and *fusion*. During a fission reaction a parent nuclide breaks apart to give two or more daughter nuclides of approximately equal size. Fission of uranium is initiated by a neutron (uranium isotopes undergo α emission in the absence of neutrons). One example of fission is shown below.

$$^{235}_{92}\text{U} + {}^{1}_{0}\text{n} \rightarrow {}^{94}_{36}\text{Kr} + {}^{139}_{56}\text{Ba} + 3\,{}^{1}_{0}\text{n}$$

But just like throwing a baseball at a plate-glass window, where 10 identical throws at 10 windows results in 10 different patterns of breakage, a uranium nucleus can split many different ways. Dozens of elements and scores of isotopes are all formed from ^{235}U fission. The neutron that initiates the reaction must be thermal, or slow-moving. Once started, the fission can be a chain reaction provided the product neutrons are slowed down. If the mass of ^{235}U is too small, the neutrons will escape before reacting and the chain is not sustained. A certain critical mass of ^{235}U is needed to keep the chain going as in conventional nuclear reactors. The energy released by the fission is used to generate electricity. If a supercritical mass of ^{235}U is present (this is never found in reactors) the chain reaction proceeds explosively as in the atomic bomb dropped on Hiroshima in 1945.

Naturally occurring uranium contains less than 1% ^{235}U so enrichment is necessary before the uranium can be used as nuclear fuel. The ^{235}U/^{238}U ratio is increased several fold by the gas phase diffusion of UF_6. The $^{235}UF_6$ is lighter and diffuses faster than $^{238}UF_6$ (see Chapter 2).

Fusion is the combining of light nuclei to form a heavier nucleus. Fusion occurs in stars where helium and hydrogen nuclei interact. Fusion reactions are exothermic and could be used for the generation of energy here on Earth, but there are still practical difficulties to be overcome before it is commercially feasible. Activation energies are so high that temperatures of $>10^8$ K are necessary, where matter is neither solid, liquid, nor gas, but is completely ionized to a *plasma* (a stream of nuclei and electrons). Magnetic fields must be used to contain and direct the plasma. On the plus side, the fuel (deuterium) is abundant and cheap, and the reaction produces no radioactive products.

19.5. Applications of Isotopes

There are a great number of consumer products that rely on radioisotopes, ranging from watches with luminous dials to smoke detectors and static electricity removers for records. The medical field relies heavily on the use of isotopes, from ^{60}Co therapy to positron emission tomography which images the brain. Geologists and archaeologists use ^{14}C, ^{40}K, ^{41}Ar, and many other isotopes to date objects on Earth and even the Earth itself. Chemists and biochemists use radioisotopes to follow the path of reactions: ^{18}O, ^{32}P, and ^{131}I are just a few that are important.

Example 19.15. The following rearrangement is known to take place.

$$-\text{O}-\text{CH}_2\text{CH}=\text{CH}_2 \xrightarrow{\text{heat}} \begin{array}{c} \text{OH} \\ / \\ \text{CH}_2\text{CH}=\text{CH}_2 \end{array}$$

An important clue as to the mechanism comes from the use of ^{14}C experiments. If the terminal C is labelled with ^{14}C, it turns up in the product, to be directly attached to the benzene ring.

$$-\text{O}-\text{CH}_2\text{CH}=*\text{CH}_2 \rightarrow \begin{array}{c} \text{OH} \\ | \\ *\text{CH}_2\text{CH}=\text{CH}_2 \end{array}$$

What can be inferred about the mechanism from this result?

It would seem that the hydrocarbon chain must swing around so that the terminal C can attach to the benzene ring before the O—C bond is broken. If the O—C bond were broken first, then there would be a 50/50 chance for the labelled C to end up at the ring.

Problems

1. What is the density of a ^{12}C nucleus?

2. The sun in our solar system radiates 10^{23} kJ of energy into space every second. How much mass does the sun lose every second?

3. When the stable ^{19}F isotope absorbs a neutron, the unstable ^{20}F isotope is produced. What is the most probable method of decay for ^{20}F?

The Nucleus   Chapter 19

4. Which of the following nuclei do you expect to be radioactive? What is the probable decay mode?

    a) ^{22}Na   d) ^{232}Th

    b) ^{7}Li   e) ^{24}Mg

    c) ^{206}Pb   f) ^{251}Cf

5. Write equations to describe each of the following reactions.

    a) electron capture by ^{207}Po

    b) ^{120}Sb decays to ^{120}Sn

    c) ^{208}Fr emits an α particle

    d) ^{83}Sr emits a positron

    e) ^{87}Rb emits a β particle

6. What is the binding energy per nucleon of an ^{4}He atom? (mass = 4.0026 amu)

7. Which of the following atoms is expected to have the largest binding energy per nucleon?

    ^{1}H, ^{4}He, ^{206}Pb, ^{55}Fe, ^{40}Ca

8. Which of the nuclear decay modes (β^-, β^+, α emissions, or EC) result in an increase of the atomic number?

9. A piece of steel pipe that came out of a nuclear reactor has 950 curies of ^{59}Fe activity, which has a half-life of 45 days. How long will it be before the activity falls to 10 curies?

10. What is the half-life of a radioisotope that loses $\frac{3}{4}$ of its activity in 12.0 minutes?

11. Nuclear decay reactions are unimolecular and independent of temperature. What can you say about the activation energies of such reactions?

12. A biochemist has just received a special shipment of nucleotides containing ^{32}P, which has a $t_{\frac{1}{2}}$ of 14.3 days. How may days does she have to work with the nucleotides before the radioactivity falls to $\frac{1}{16}$ of the original value?

13. How old is a rock that has a ^{238}U/^{206}Pb of 4.15? ($t_{\frac{1}{2}}$ for ^{238}U is 4.5×10^9 yrs).

14. Tritium, ^{3}H, is a β^- emitter with a half-life of 12.3 yrs. If fresh wine has a tritium activity of 1.25 counts per hour per liter, how old is a bottle of wine which has 0.680 counts per hour per liter when opened?

Chapter 19　　　　　　　　　　　　　　　　　　　　　　　　　　　　　　　The Nucleus

15. Write equations for the following reactions.

    a)　　when an ^{27}Al nucleus is hit by an α particle, 3 protons are produced
    b)　　^{222}Rn undergoes two successive α emissions followed by two successive β-emissions
    c)　　^{7}Li absorbs a proton to yield two identical particles

16. Calculate the energy released by the following fusion reaction.

$$2\ {}^{1}_{1}H \rightarrow {}^{2}_{1}H + {}^{0}_{+1}\beta$$

17. Calculate the energy released by the fission of one pound of ^{235}U.

	$^{235}_{92}$U	+	$^{1}_{0}$n	→	$^{141}_{56}$Ba	+	$^{88}_{36}$Kr	+	7 $^{1}_{0}$n
mass (amu)	235.0439		1.008665		140.9137		87.9142		1.008665

Compare this with the energy released by the combustion of one pound of natural gas, (CH_4).

18. When a positron and an electron combine, all the mass is converted to energy and two γ rays are emitted. What is the wavelength of the γ rays?

CHAPTER 20
THE PROPERTIES OF SOLIDS

20.1. Macroscopic Properties of Solids

One way to subdivide solids into types is by crystalline behavior or lack thereof. *Crystalline* solids have a regular order of atoms or ions to form a lattice. *Amorphous* solids may have small pockets of ordering, but are not organized into a large-scale lattice. Both solid types are hard and resist flow. SiO_2, if cooled slowly, will crystallize as quartz. When cooled rapidly, SiO_2 forms an amorphous solid we know as glass. Crystalline solids usually have sharp melting points and are anisotropic, meaning the orientation of the crystal is important to its properties. Amorphous solids melt over a broad temperature range. The randomness of an amorphous solid leads to isotropy since all orientations are the same.

Microscopic properties of crystals are frequently reflected in their macroscopic behavior. Mica, for example, is a mineral composed of silicate sheets, stacked with K^+ ions between the layers.

20.2. Types of Solids

Crystalline solids can be further subdivided into ionic, molecular, covalent networks, and metallic crystals. There is some overlap in characteristics but generalizations concerning the bonding can be made for each crystal type. All types consist of a lattice.

Ionic crystals have lattice points occupied by alternating anions and cations. The crystal is held together by coulombic forces. These strong electrostatic interactions result in a crystal that is solid at room temperature. Ionic crystals are rigid and hard to break. Correspondingly the melting and boiling points are high. When molten or dissolved in a polar solvent the ions are mobile resulting in good electrical conductivity. Salts are examples of ionic crystals.

Molecular crystals have lattice points occupied by neutral molecules. The crystal is held together by weak van der Waals' forces or dipole/dipole interactions. Many compounds that form molecular solids are gases or liquids at room temperature because the intermolecular forces are weak. As solids these molecules are soft and have little electrical conductivity. The bonding is localized and the electrons have no intermolecular mobility. Examples of molecular solids are $H_2O_{(s)}$, $Br_{2(s)}$, and $CCl_{4(s)}$.

Covalent network solids contain an extended lattice with points occupied by neutral atoms. Each atom shares electrons with adjacent atoms. The large array of covalent bonding results in rigid structures with high melting and boiling points. The localization of the valence electrons makes covalent network solids poor conductors when pure. Carborundum (SiC), diamond and quartz (SiO_2) are good examples of covalent network solids.

Metallic crystals have lattice points occupied by metal cations. The valence electrons can be thought of as a free-moving sea surrounding the lattice points. Because the electrons are mobile, metals conduct electricity well.

In the solid state, only molecular solids consist of discrete molecules. Ionic, metallic and network solids do not have individual molecules. A one carat diamond is a single, large carbon "molecule."

Chapter 20   The Properties of Solids

Example 20.1. Predict the type of crystalline solid each of the following substances will form.

$$CH_4 \, , \; GaAs \, , \; MgO \, , \; Al \, , \; NH_3$$

Both CH_4 and NH_3 will form molecular solids. GaAs is isoelectronic with germanium and is expected to be a covalent network solid. Aluminum forms metallic crystals and MgO is an ionic solid.

20.3. X-Rays and Crystal Structure

In earlier chapters we saw that electromagnetic radiation, or light, can be described both as particles and waves. The wave nature of light can be used to determine the geometry of a crystal. Light waves scatter or diffract at the edges of an object. Visible light diffracted by a prism separates into its components; red, orange, yellow, green, blue, indigo, and violet. If x-rays with only one wavelength are aimed at a crystal, the diffraction pattern of the x-rays is a direct consequence of the crystal array. The angle of incidence and reflection of the x-ray at the crystal surface is θ. Knowing θ and the wavelength of the x-ray, the Bragg equation can be used to calculate d, the distance between successive layers of the crystal. The value of n is an integer.

$$n\lambda = 2 \, d \sin \theta$$

The x-ray diffraction pattern also permits the electron density of a crystal to be mapped. The complex patterns can be analyzed by a computer to obtain a "picture" of the crystal, including bond lengths and angles. X-ray crystallography has become one of the most valuable tools available to chemists for determining molecular structure.

20.4. Crystal Lattice.

Crystal lattices may be two-dimensional like a tile floor, or three-dimensional. The unit cell or smallest repeating unit of a tile floor is one tile. The unit cell of a three-dimensional crystal is one of seven crystal systems shown in table 20.2 of the text.

20.5. Common Crystal Structures

All atoms and many spherical molecules such as H_2, O_2, HCl, Fe, Ar, etc., can be thought of as solid spheres that in the solid phase are neatly stacked in order to occupy space in a crystal efficiently. Such packing gives a closest-packed structure. The three general types of closest-packing lead to: face-centered cubic lattice, hexagonal lattice and a body-centered cubic lattice. The unit cells of these structures can be seen in figures 20.24-20.26 in the text. All atoms or molecules have 12 nearest neighbors in the closest-packed structures. There are two kinds of holes or interstices available in the crystal; the tetrahedral sites and the octahedral sites. In ionic crystals these sites may be filled with the cation. Smaller cations occupy tetrahedral sites. Larger cations occupy octahedral sites. The ratio of ionic radii is a good indicator of the cation site. When $r_+/ r_- \geqslant 0.414$ the cation will probably fill an octahedral site. When r_+/ r_- is between 0.414 and 0.225, a tetrahedral site is needed.

Many crystalline solids have geometries that closely match these close-packed conditions.

Example 20.2. Predict the local ionic geometry for NaBr ($r_+/r_- = 0.52$) and CsCl ($r_+/r_- = 0.94$).

NaBr is expected to have a structure similar to NaCl with the Na^+ occupying an octahedral site in a closest-packed cubic Br^- structure. But in the case of CsCl where the size of the cation and anion are nearly identical, we might expect that the cations and the anions are both in alternating cubic arrays. Another way to view this: each Cs^+ is surrounded by 8 Cl^- ions at the corners of a cube, and each Cl^- is surrounded by 8 Cs^+ ions at the corners of a cube.

20.6. Defects in Solid Structures

Many solid properties can be understood in terms of crystal structures. In ionic solids for example, each ion is bound to 4, 6, or 8 oppositely charged ions. The coulombic attractions are maximized and movement in any direction is not energetically favored. This accounts for the rigidity of salts.

Other properties are only understood in terms of defects in the crystal. Insulators can be converted to semi-conductors by incorporation of foreign atoms or ions into a crystal. Pure silicon is doped by adding small amounts of P (which has one more valence electron than Si) or B (which has one less valence electron than Si). The defective crystal sites permit electrons to be transported more readily than in a perfect Si crystal. The doping of silicon is an example of point defects. Other examples are gemstones. The replacement of small amounts of Al^{3+} with Cr^{3+} in Al_2O_3 results in a ruby. If Mn^{3+} replaces some Al^{3+} the gem is known as an amethyst.

Other varieties of defects involve more than an occasional site substitution. If foreign atoms are of much different size than the host atoms, the crystal structure may shift to accommodate the foreign atom. This may lead to edge or screw dislocations.

20.7. Thermal Properties of Solids

It can be readily shown that for a solid consisting of simple atoms the total energy of a mole is given by the following equation.

$$E = 3RT$$

Since C_v, the heat capacity at constant volume, multiplied by a change in temperature equals ΔE:

$$C_v \Delta T = \Delta E$$

$$C_v = \frac{\Delta E}{\Delta T} = \frac{3R \Delta T}{\Delta T} = 3R$$

This means that heat capacities for solids should be constant and near 25 J/mol-K. In fact, heat capacities do vary with temperature. Some of the assumptions used by Einstein to derive the relationships above, are not valid, when $h/kT \ll 1$, but as an approximation we can appreciate its power. Heat capacities, C_p or C_v, are easily measured for solids and can yield a rough molecular weight value.

20.8. Lattice Energy of Ionic Crystals

When we discussed lattice energy in an earlier chapter the correlation between lattice energy, U, and size and charge of the ions was easily shown. But the geometry of the crystal structure also influences U. Using the Born equation given in the text, crystalline NaCl is shown to be 75% more stable than a single Na^+Cl^- pair. This means the lattice energy is 1.75 times lower (more negative) in the crystal than in the ion pair. The factor of 1.75 is called the Madelung constant for NaCl, and arises from the multitude of ionic interactions in the crystal. You should recall that coulombic attractions are felt over large distances while repulsions are important only at very small distances. So the net effect of crystalline packing is an attractive one. Each crystal type has a corresponding value of the Madelung constant associated with it. The Madelung constant as well as the ion charges and ionic radii determine the overall lattice energy. Reactions involving ionic substances may be strongly influenced by the lattice energy so the value of U can be useful when correlating ionic properties.

20.9. Metallic Bonding

The physical properties of metals are familiar to us; metals are lustrous, conduct heat and electricity, are malleable and ductile, and are solids at 25°C (except for Hg). A good model of metallic bonding should be able to explain these characteristics.

The sea of electrons model, though simplistic, accounts for a great deal. The metal may be thought of as positively charged ions surrounded by a common sea of valence electrons. The mobile electrons explain heat and electrical conductivity account for the mallebility of metals: as a metal piece is pounded or shaped, the electrons buffer the movements of the positively charged centers past one another.

A more sophisticated and theoretically satisfying model is *band theory*. Band theory is simply molecular orbital theory applied to metals. All metals have low ionization energies and electron affinities. There is little attraction for their own valence electrons as well as extra electrons. Simple electron-pair covalent bonds are weak between metal atoms, but multiple interractions are more stable. The valence atomic orbitals of many atoms can be combined to form an equal number of molecular orbitals. In the case of one mole (23 g) of Na this corresponds to 6×10^{23} 3s orbitals forming 6×10^{23} molecular orbitals (3×10^{23} σ and 3×10^{23} σ^*). These molecular orbitals are so close together energetically as to form a continuous band occupied by valence electrons. This band is delocalized over the entire metal piece. This valence band may also have contributions from p and d valence atomic orbitals. In this continuous band of molecular orbitals only an infinitesimal quantity of energy is needed to promote an electron from an occupied to an unoccupied orbital where it can move freely over the entire metal crystal. This is conduction. Due to the close spacing of the molecular orbitals all wavelengths of light will be absorbed and emitted. That's why a metal surface is shiny.

Problems

1. Which of the four crystalline solid types: ionic, molecular, network covalent, or metallic, is each substance expected to form?

    a) $P_{4(s)}$

    b) $HCl_{(s)}$

    c) $In_{(s)}$

    d) $Si_{(s)}$

The Properties of Solids   Chapter 20

        e) $Cl_{2(s)}$ f) $Xe_{(s)}$

                                                                  g) $NH_4NO_{3(s)}$

2. What type of bonds hold the atoms in SiO_2 together? Why is SiO_2 a solid and CO_2 a gas at 25°C?

3. Predict whether each cation will occupy an octahedral, tetrahedral, or cubic site in the crystal structure.

    a)   ZnS ($r_+/r_- = 0.41$)

    b)   MnO ($r_+/r_- = 0.59$)

    c)   CsI ($r_+/r_- = 0.79$)

4. Most metals crystallize as body-centered cubes, face-centered cubes or in a hexagonal closest-packed structure. What are the coordination numbers for most metals in the crystalline state?

5. NaCl has the structure shown in figure 20.31 of the text. The internuclear distance of an Na—Cl pair is 282 pm from x-ray diffraction data. This means that each unit cell has sides 564 pm in length. Given this length and the fact that the density of NaCl is 2.16 g/cm^3, calculate Avagadro's number. (Remember to calculate how many ion pairs are in each unit cell).

6. Can you explain why metals conduct heat, in terms of band theory?

APPENDIX A
MATHEMATICAL REVIEW

Scientific Notation and Exponents

A great many of the numbers we will encounter in chemistry are either very small ($<<<1$) or very large ($>>>1$). Scientific notation allows us to conveniently express numbers as exponents of ten. If you have never worked with scientific notation before, this review will probably not provide a sufficient introduction to the topic. It is meant only to give a brief reminder of how exponents are mathematically manipulated.

The purpose of scientific notation is to save time and space. Numbers are represented only by their significant digits and the appropriate power of ten. All non-significant figures, including the necessary zeros to place the decimal point, can therefore be eliminated.

$$153 = 1.53 \times 10^2$$
$$4,500,000,000 = 4.5 \times 10^9$$
$$0.00006790 = 6.790 \times 10^{-5}$$

In the second example above there is no indication that the zeros following the 5 have any significance other than to place the decimal point. Consequently they are not included in the scientific notation. However, in the last example the zero that follows the 9 is significant by virtue of its inclusion while the preceding zeros to the 6 are not significant.

A negative power of ten also can be expressed as given below.

$$10^{-x} = \frac{1}{10^x}$$

Although it is common practice when coverting a number to scientific notation to bring the non-exponential portion to a number between 1 and 10, sometimes it is more convenient not to do so. Consequently there are many ways to express the same number in scientific notation. For example:

$$0.003078 = 3.078 \times 10^{-3} = 30.78 \times 10^{-4} = 3078 \times 10^{-6}$$

This technique comes in handy when adding or subtracting numbers written in scientific notation where the exponents must be the same, as shown below.

$(6.2 \times 10^{-4}) + (5 \times 10^{-5})$ can be calculated only if both numbers have the same exponent.

$$(6.2 \times 10^{-4}) + (5 \times 10^{-5}) = (6.2 \times 10^{-4}) + (0.5 \times 10^{-4}) = 6.7 \times 10^{-4}$$

or

$$(6.2 \times 10^{-4}) + (5 \times 10^{-5}) = (62 \times 10^{-5}) + (5 \times 10^{-5}) = 67 \times 10^{-5}$$

When multiplying numbers expressed in scientific notation, the pre-exponential terms are multiplied while the exponents are added.

$$(A \times 10^a) \times (B \times 10^b) = (A \times B) \times 10^{(a+b)}$$

Mathematical Review   Appendix A

For example:

$$(1.6 \times 10^{12}) \times (5.4 \times 10^{-6}) = 8.6 \times 10^{-18}$$

$$(3 \times 10^6) \times (4 \times 10^{-2}) = 12 \times 10^4 = 1.2 \times 10^5$$

When dividing numbers expressed in scientific notation, the pre-exponential terms are divided while the exponents are subtracted.

$$(A \times 10^a) \div (B \times 10^b) = \frac{(A)}{(B)} \times a0^{(1-b)}$$

For example:

$$\frac{(4.6 \times 10^{-16})}{(20 \times 10^{-8})} = \frac{(4.6)}{(2.0)} \times (10^{(-16-(-8))}) = 2.3 \times 10^{-8}$$

$$\frac{(3.8 \times 10^{-5})}{(9.1 \times 10^3)} = \frac{(3.8)}{(9.1)} \times 10^{(-5-3)} = 0.42 \times 10^{-8} = 4.2 \times 10^{-9}$$

In order to raise a number expressed in scientific notation to a power, the pre-exponential term is raised to that power and the exponent is multiplied by that power.

$$(A \times 10^a)^n = A^n \times 10^{an}$$

For example:

$$(4 \times 10^3)^5 = (4)^5 \times 10^{15} = 1{,}024 \times 10^{15} = 1 \times 10^{18}$$

The reverse procedure is needed to take the root of a number expressed in scientific notation; the root of the pre-exponential term is taken, while the exponent is divided by the root needed.

$$\sqrt[n]{A \times 10^a} = (A \times 10^a)^{1/n} = A^{1/n} \times 10^{a/n}$$

For example:

$$\sqrt[4]{5.0 \times 10^8} = 5^{\frac{1}{4}} \times 10^{\frac{8}{4}} = 1.5 \times 10^2$$

Logarithms

A *logarithm* is an exponent. If the number - base is 10, it is denoted by *log*. If the number - base is e, where e = 2.71828..., it is denoted by *ln*. A logarithm is the power to which the base must be raised in order to equal a given number.

Appendix A — Mathematical Review

$$\begin{array}{cc} \textit{Base 10} & \textit{Base e} \\ n = 10^x & m = e^y \\ \log n = x & \ln m = y \end{array}$$

Logarithms in base 10 and base e are related to each other.

$$\ln x = 2.303 \log x$$

Although a log can easily be found from standard tables, the more direct ln calculations have gained in popularity since calculators have become ubiquitous.

Some manipulations of logarithms may be necessary at some point in problem solving. The following equations show some of the these useful simplifications that apply for both log and ln.

$$\log(ab) = \log a + \log b$$

$$\log \frac{(a)}{(b)} = \log a - \log b$$

$$\log a^n = n \log a$$

$$\log \frac{(1)}{(a)} = -\log a$$

An antilogarithm (or inverse logarithm) is the number whose logarithm is given. For instance, in the equations below,

$$n = 10^x, \log n = x$$

n is the antilog of x. This can again be calculated using standard log tables or by using the inverse log keys on a calculator.

Quadratic Equations

A *quadratic equation* is one in which a variable is raised to a power of 2, but no higher. The standard form of a quadratic equation is:

$$ax^2 + bx + c = 0$$

where a, b and c are numbers. There are two solutions for x given by the *quadratic formula* below

$$x = -b \pm \frac{\sqrt{b^2 - 4ac}}{2a}$$

It is, of course, necessary to have the equation in standard form first, which may require considerable algebraic manipulation, as in the case of equilibrium problems. Usually in chemistry problems only one of the solutions for x will make sense in the context provided.

APPENDIX B
CALCULUS REVIEW

Often it is possible to put the equation that relates two variable x and y into the form

$$y = f(x),$$

which is read: y is is some function (f) of x. The quantity x is called the independent variable, and y is the *dependent variable*, inasmuch as the equation expresses the act that y depends in some fashion on x.

It is frequently of interest to learn what change Δy, in y, is produced by a given change Δx, in x. The equation

$$y + \Delta y = f(x + \Delta x)$$

states that a change Δx in x does produce some change Δy in y. The question to be answered is, what is the relation between Δy and Δx?

When y is a linear function of x, as in

$$y = f(x) = a + bx,$$

where a and b are constants, the answer is simple. We write

$$y + \Delta y = a + b(x + \Delta x),$$

$$y = a + bx,$$

and subtract the second equation from the first we get

$$\Delta y = b(x + \Delta x) - bx,$$

$$\Delta y = b\Delta x.$$

Thus the quantity

$$\Delta y / \Delta x = b,$$

which is the change in y per unit change in x, is for this function a constant, and is equal to the slope of the straight line represented by $y = a + bx$. Because the equation represents a straight line, the slope $b = \Delta y/\Delta x$ is a constant, independent of x and y.

Appendix B — Calculus Review

Suppose we apply the same operation to the function

$$y = f(x) = x^2 ,$$

which represents a parabola. Then we have

$$y + \Delta y = (x + \Delta x)^2$$

The method by which we found dy/dx for the function $y = x^2$ is called the delta process. Application of the delta process shows that for

$$y = x^3 , \quad \frac{dy}{dx} = 3x^2 ,$$

and, in general, for

$$y = x^n , \quad \frac{dy}{dx} = nx^{n-1} ,$$

Some important derivatives are listed in Table B.1.

Table B.1. Some Important Derivatives

$\frac{d}{dx}(x) = 1$	$\frac{d}{dx}(e^u) = e^u \frac{du}{dx}$
$\frac{d}{dx}(au) = a\frac{du}{dx}$	$\frac{d}{dx}(\sin u) = \cos u \frac{du}{dx}$
$\frac{d}{dx}(u^n) = v^{n-1}\frac{du}{dx}$	$\frac{d}{dx}(\cos x) = -\sin u \frac{du}{dx}$
$\frac{d}{dx}(\ln u) = \frac{1}{u}\frac{du}{dx}$	$\frac{d}{dx}(uv) = u\frac{dv}{dx} + v\frac{du}{dx}$

u and v denote functions of x; a is a constant

The following are noteworthy properties of derivatives. If c is any constant, and

$$y = cx^n , \quad \text{then} \frac{dy}{dx} = (c)nx^{n-1} .$$

If we have y not a function of x, as in

$$y = c \quad , \quad \text{then} \frac{dy}{dx} = 0 \quad .$$

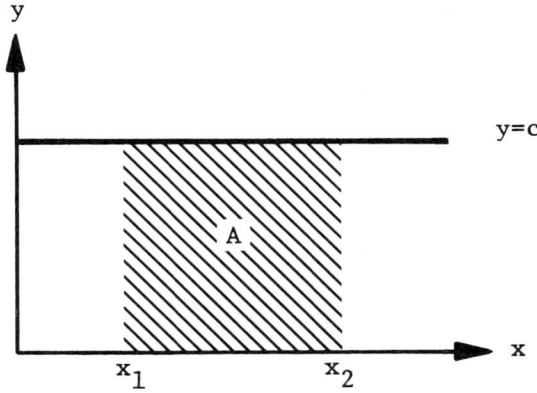

Figure B.1

It is often necessary to find the area under a curve $y = f(x)$ between two values of the independent variable x. When f(x) has one of two simple forms, this presents no difficulty. As shown by Fig. B.1, if $y = c$, a constant, then the area under $y = c$ between x_1 and x_2 is simply $A = c(x_2 - x_1)$. Also, if we have the expression $y = cx$, which represents a straight line through the origin, then the area A between $X = x_1 = 0$ and x_2 is $A = \frac{1}{2}cx_2$, as Fig. B.2 suggests. For situations in which $y = f(x)$ represents a curve, a different approach is required in order to find the required area.

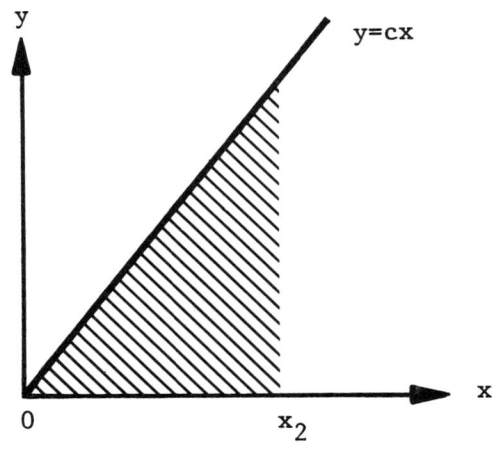

Figure B.2

Appendix B   Calculus Review

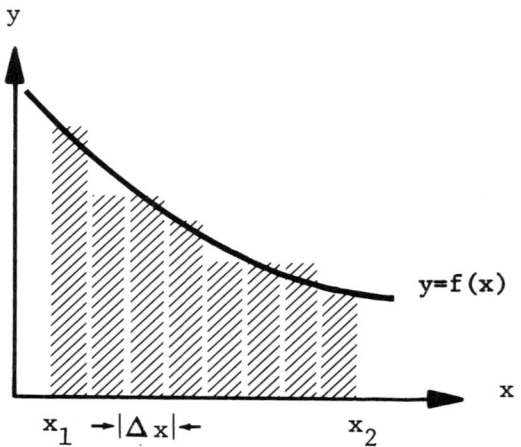

Figure B.3

This type of problem and its possible solution is sketched in Fig. B.3. The area under the curve is approximated by a series of rectangles of equal width Δx and height y_i, for the ith rectangle. Thus if ΔA_i is the area of the ith rectangle,

$$A \simeq \sum_{i=1}^{n} \Delta A_i = \sum_{i=1}^{n} y_i \Delta x \quad .$$

It seems intuitively obvious that the approximation should improve if we were to increase the number of rectangles and make each one narrower. The area of a rectangle of height y is

$$\Delta A = y \Delta x = f(x) \Delta x \quad ;$$

and since

$$\lim_{\Delta x \to 0} \frac{\Delta A}{\Delta A} \equiv \frac{dA}{dx} = f(x) \quad ,$$

we can write

$$dA = f(x) dx \quad .$$

The interpretation of dA is that it is the area of a rectangle of height $f(x)$ and infinitesimal width dx. Consequently, we can express the total area as the sum (or integral) of the infinitesimal contributions

$$A = \lim_{\Delta x \to 0} \sum \Delta A_i = \lim_{\Delta x \to 0} \sum y_i \Delta x \quad ,$$

-290-

Calculus Review   Appendix B

$$A = \int dA = \int_{x_1}^{x_2} f(x)\, dx \quad .$$

That is, the integral is the sum of small areas between x_1 and x_2 in the *limit* where Δx approaches zero.

To evaluate the integral, we note from above the following relation:

$$A = \int dA \quad .$$

Thus, since dA is the differential of A, and the integral sign restores dA to A, the integral sign must stand for the operation of reverse differentiation or antidifferentiation. With this in mind, we can evaluate

$$\int f(x)\, dx \quad .$$

If the function f(x) were itself the derivative with respect to x of another function F(x), that is, if

$$f(x) = \frac{d}{dx} F(x) \quad ,$$

then we could make the substitution

$$\int f(x)\, dx = \int \frac{dF}{dx}\, dx = \int dF \quad .$$

The last form shows that the integral sought is just F, so we might write

$$A = \int f(x)\, dx = F \quad ,$$

where

$$f(x) = \frac{dF}{dx} \quad .$$

We conclude that in order to evaluate the integral of f(x), we must find the function F whose derivative dF/dx is equal to f(x). To check this we start with our answer

$$A = F(x), \quad f(x) = dF/dx \quad ,$$

and differentiate

$$\frac{dA}{dx} = \frac{dF}{dx} \quad ,$$

Appendix B   Calculus Review

$$f(x) = f(x) \quad .$$

There is one very important problem remaining, however. Besides the answer already found, the result

$$A(x) = F(x) + c \quad ,$$

where c is any constant, is apparently equally valid, for differentiation gives us

$$\frac{dA}{dx} = \frac{dF}{dx} + 0 \quad ,$$

$$f(x) = f(x) \quad .$$

Thus the integral we have found is *indefinite* in that any constant may be added to it.

This problem has arisen because we have ignored the fact that what we are seeking is the area under the curve $y = f(x)$ *between* two limits x_1 and x_2. If we regard $A(x_2)$, that is, $A(x)$ evaluated at $x = x_2$, as the area under the curve up to the limit, x_2 plus the constant c, and $A(x_1)$ as the area up to x_1 plus the same constant, then we can write

$$A(x_2) - A(x_1) = A = F(x_2) + c - F(x_1) - c \quad ,$$

$$A = F(x_2) - F(x_1) \quad .$$

That is, the area A under $y = f(x)$ between x_2 and x_1 is just the difference between the integral evaluated at the "upper limit" x_2 and the "lower limit" x_1. This is compactly denoted by

$$\int_0^A dA = \int_{x_1}^{x_2} f(x)\, dx = \int_{x_1}^{x_2} dF \quad ,$$

$$A \Big[_0^A = A - 0 = F(x) \Big]_{x_1}^{x_2} = F(x_2) - F(x_1) \quad .$$

Thus, when the limits of integration are specified, the numerical result is not arbitrary, and we have a *definite* integral.

From the table of derivatives, we can deduce a few specific formulas for the evaluation of integrals. Since

$$\frac{d}{dx} x^n = nx^{n-1} \quad ,$$

Calculus Review Appendix B

we must have

$$\int_{x_1}^{x_2} x^m \, dx = \frac{x^{m+1}}{m+1}\bigg|_{x_1}^{x_2} .$$

Also from

$$\frac{d(\sin x)}{dx} = \cos x ,$$

$$\frac{d(\cos x)}{dx} = -\sin x$$

we deduce that

$$\int_{x_1}^{x_2} \sin x \, dx = -\cos x \bigg|_{x_1}^{x_2} ,$$

$$\int_{x_1}^{x_2} \cos x \, dx = \sin x \bigg|_{x_1}^{x_2} .$$

Because

$$\frac{d \ln x}{dx} = \frac{1}{x} ,$$

$$\int_{x_1}^{x_2} \frac{dx}{x} = \int_{x_1}^{x_2} d \ln x = \ln x \bigg|_{x_1}^{x_2} .$$

Finally, we have the very important relation

$$\frac{d}{dx} e^{ax} = a e^{ax} ,$$

$$\int_{x_1}^{x_2} e^{ax} \, dx = \frac{1}{a} \int_{x_1}^{x_2} d(e^{ax}) = \frac{1}{a} e^{ax} \bigg|_{x_1}^{x_2} .$$

Appendix B Calculus Review

Some other important integrals are given in Table B.2.

Table B.2. Some Important Integrals

$$\int df(x) = f(x) + c \qquad \int \frac{du}{u} = \ln u + c$$

$$\int af(x)\,dx = a\int f(x)\,dx \qquad \int e^u\,du = e^u + c$$

$$\int (u \pm v)\,dx = \int u\,dx \pm \int v\,dx \qquad \int \sin u\,du = -\cos u + c$$

$$\int u^n\,du = \frac{u^{n+1}}{n+1} + c, \ n \neq -1 \qquad \int \cos u\,du = \sin u + c$$

$$\int_a^b f(x)\,dx = -\int_b^a f(x)\,dx \qquad \int_a^c f(x)\,dx = \int_a^{be} f(x)\,dx + \int_b^c f(x)\,dx$$

u and f(x) are functions of x; a, b, and c are constants.

APPENDIX C
PROBLEM ANSWERS

Fundamentals

1.
- a) tetraphosphorus decoxide
- b) oxygen difluoride
- c) sulfur hexafluoride
- d) dinitrogen trioxide
- e) phosphorus pentafluoride
- f) hydrogen bromide

2.
- a) SiC_4
- b) NH_3
- c) N_2O_4
- d) SO_3
- e) ClF_3
- f) P_4I_6

3.
- a) hydrogen cyanide
- b) sodium bicarbonate or sodium hydrogen carbonate
- c) lithium hydride
- d) ammonium nitrate
- e) potassium permanganate
- f) silver chromate
- g) calcium phosphate
- h) sodium hypochlorite
- i) zinc oxide
- j) iron sulfide

4.
- a) $MgSO_4$
- b) NH_4CH_3COO
- c) K_2O_2
- d) $HgCl_2$
- e) Hg_2Cl_2
- f) $NaClO_2$
- g) SnS_2
- h) Al_2S_3

5.
- a) perchloric acid
- b) nitrous acid
- c) acetic acid
- d) hydrocyanic acid

Appendix C   Problem Answers

- e) chromic acid
- f) hydroiodic acid

6.
- a) $HClO_2$
- b) H_3PO_4
- c) H_2CO_3
- d) $HMnO_4$
- e) $H_2C_2O_4$

Chapter 1

1. 8.3×10^{-15} mol of people

2.
   - a) 132.16 g/mol
   - b) 291.07 g/mol

3. 62.04 g

4. 40.0% C, 6.7% H, 53.3% O

5. 0.316 mol S

6. 5×10^{16} C atoms

7. 4.41×10^{23} C atoms

8. 388 moles of Na^+

9. NO_2

10. Nb_2O_5

11. C_5H_{10}

12. $C_5H_{10}O_5$

13. H = 6.25 g/mol, C = 750 g/mol, U = 1,488 g/mol

14. 1.67×10^{-24} g/H atom, 3.95×10^{-22} g/U atom

15.
   - a) $2C_2H_2 + 5O_2 \rightarrow 4CO_2 + 2H_2O$
   - b) $K_2CO_3 + H_2SO_4 \rightarrow K_2SO_4 + H_2O + CO_2$
   - c) $2NH_3 + 4O_2 \rightarrow N_2O_5 + 3H_2O$
   - d) $C_2H_6O + 3O_2 \rightarrow 2CO_2 + 3H_2O$

Problem Answers Appendix C

- e) $2Cr(OH)_2 \rightarrow H_2 + H_2O + Cr_2O_3$
- f) $3NaH_2PO_4 \rightarrow Na_3P_3O_9 + 3H_2O$

16.
- a) $3CO + Fe_2O_3 \rightarrow 3CO_2 + 2Fe$
- b) 2 mol Fe/mol ore
- c) 2.8×10^3 kg Fe

17.
- a) $4HCl + MnO_2 \rightarrow 2H_2O + Cl_2 + MnCl_2$
- b) 8.16 g Cl_2
- c) 2.58 L @ STP
- d) 1.73 g $MnCl_2$

18. 79.9065 g/mol

19. 7.50% ^6Li, 92.50% ^7Li

20. 72.64 g/mol

21. element M is Ce

22. 50.6% C_2H_6, 49.4% C_3H_8 by weight

Chapter 2

1. 16.0/10.7

2. 117 atm, 52.6 atm

3. 6.75 L

4. 221 K

5. a loss of 2.14 moles

6. 122 g/mol

7.
- a) density of O_2 sample is greater than N_2
- b) grams of O_2 are greater than N_2
- c) moles are equal
- d) molecules are equal
- e) velocity of N_2 molecules greater than O_2
- f) average kinetic energies are equal

Appendix C   Problem Answers

8. C_5H_5

9. 47 L

10. 16.3 L

11. N_2: X = 0.366, P = 345 k Pa
    O_2: X = 0.504, P = 475 k Pa
    CO: X = 0.130, P = 123 k Pa

12.
    a) 3.87 kJ/mol   c) 1390 m/s
    b) 6.43 × 10^{-24} kJ/atom

13. a) P_T = 3.36 atm   b) X_{H_2} = 0.173, X_{N_2} = 0.827

14. 45 g/mol

15. 4368 K

16. 4.66 atm

17. N_2H_4

18. 33.5%

19. V_{rms} = 517 m/s for N_2. This is faster than the speed of sound

20. 6.95 × 10^{-6} cm, 2.55 × 10^{-5} cm, 6.95 × 10^{-8} cm

21. gas X

Chapter 3

1. 358 K

2. $\Delta H°$ = 38.9 kJ/mol, T_{bp} = 404 K

3. 364 K or 91 °C

4. The critical temperature

5. 2.27 molal, X_{CH_3OH} = 0.0393

6. 2.34 molal, 2.06 molar, $X_{C_2H_5OH}$ = 0.0404

7. 46 mL

-298-

Problem Answers   Appendix C

8. 200 mL

9. 6 L

10. 4.12 L

11. 0.33 M

12. 1.76 L

13. 23.5 torr

14. 23.0 torr

15. $X_B = 0.524$, $X_H = 0.476$, $P_T = 184$ torr

16. 79 g/mole

17. -2.38 °C

18. $LiCl > CH_3OH > BaBr_2 > C_6H_{12}O_6$

19. 17.7 g

20. 101.08 °C

21. 0.30 M glucose, 0.15 M NaCl

22. 15.8 atm

23. 27,500 g/mol

24. The mole fraction of CO_2 in champagne is high. When opened, the solution attempts to reach an equilibrium vapor pressure of CO_2. This requires all the CO_2 to bubble away.

25.
   a) $BaSO_{4(s)} + 2H_2O$

   b) $AgCl_{(s)}$

   c) $HOAc_{(aq)} + PbS_{(s)}$

   d) $Fe(OH)_{2(s)}$

26. $\Delta H° = -$, hot packs

Appendix C   Problem Answers

Chapter 4

1.
   a) $K = \dfrac{(P_{NO_2})^4}{(P_{N_2O})^2(P_{O_2})^3}$

   b) $K = \dfrac{[Mn^{2+}][Cl_2]}{[Cl^-]^2[[H^+]^4}$

   c) $K = \dfrac{(P_{SO_3})^2}{(P_{SO_2})^2 P_{O_2}}$

   d) $K = \dfrac{P_{SO_3}}{P_{SO_2}(P_{O_2})^{\frac{1}{2}}}$

   e) $K = \dfrac{P_{SO_2}(P_{O_2})^{\frac{1}{2}}}{P_{SO_3}}$

   f) $K = \dfrac{(P_{Cl_2})^6}{(P_{PCl_3})^4}$

   g) $K = (P_{H_2O})^3$

   $K = P_{Br_2}$

2. $K_c = 1.22 \times 10^3$

3. 5.13

4. #2

5. $K_{C_1} = 0.0082$, $K_{C_2} = 1.0 \times 10^{14}$, $K_{C_3} = 3.2 \times 10^{-17}$

6. $\Delta H° = -$

7.
   a) reaction shifts right
   b) no change
   c) reaction shifts right
   d) reaction shifts left
   e) reaction shifts right

8. $K_p = 1.4$

9. $K_c = 1.2 \times 10^{-3}$, $K_p = 0.098$

10. $P_I = 0.250$ atm, $P_{I_2} = 20.4$ atm

11. 1.69

12. 1.05

13. $P_{I_2} = P_{Br_2} = 0.0198$ atm, $P_{IBr} = 0.0395$

Problem Answers   Appendix C

14.
- a) $P_{CO_2} = 0.30$ atm, $K_p = 1.35 \times 10^{-2}$
- b) $P_{H_2O} = 0.125$

15. reaction shifts left

16.
- a) 2.2×10^{-16} atm
- b) 0.0081 atm
- c) endothermic

17. $P_{CO} = 1.196$ atm, $P_{H_2O} = 0.026$ atm, $P_{CO_2} = 0.904$ atm, $P_{H_2} = 1.174$ atm

Chapter 5

1. 3.7×10^{-11}

2. 2.1×10^{-6} g/L

3. 1.8×10^{-10} M

4. 7.3×10^{-9}

5. 6.05×10^{-4} M

6. Both salts precipitate at nearly the same $[SO_4^{2-}]$, but $PbSO_4$ should appear first.

7. 0.70

8. 10.38

9. 6.84

10. b < c < e < d < a

11. $OCl^- > NO_2^- > NO_3^- > ClO_4^-$

12. 20 mL

13. a < b < e < d < c

14. 0.27

15. 0.01%, $K_a = 1 \times 10^{-8}$

16. 1.99

17. 5.08
18. 2.1×10^{-4}
19. 6.12
20. 11.8 g
21. 18.2 mL
22. 427 mL
23. 12.27
24. a) 2.79 b) 4.89 c) 6.21 d) 8.76 e) 11.29
25. pH = pKa at midpoint of titration. $K_a = 3.16 \times 10^{-6}$.
26. pH = 2.19, $[HSO_3^-]$ = 0.0065 M, $[H_2SO_3]$ = 0.0035 M, $[SO_3^{2-}]$ = 6.6×10^{-8} M.
27. 0.057 M
28. 8.48
29. For a pH = 2 buffer: $[NaH_2PO_4]/[H_3PO_4]$ should be 0.708. For a pH = 7.2 buffer: $[Na_2HPO_4]/[NaH_2PO_4]$ = 1.0. For a pH = 12.0 buffer: $[Na_3PO_4]/[Na_2HPO_4]$ = 0.45.
30. 4.8×10^{-6} M, 1.3×10^{-15} M

Chapter 6

1.
   a) :N≡N:
   b) :Ï - Ï:
   c) H—C≡N:
   d) Na^+, $[:\ddot{C}l:]^-$
   e) Ba^{2+}, $[:\ddot{O}:]^{2-}$
   f) H—B̈r:

2.
   a) H^+
   b) $[H:]^-$
   c) $[:\ddot{O}-H]^-$
   d) $[:\ddot{O}:]^{2-}$
   e) Ca^{2+}
   f) Rb^+

3.
   a) 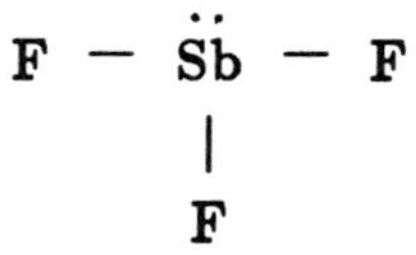 F 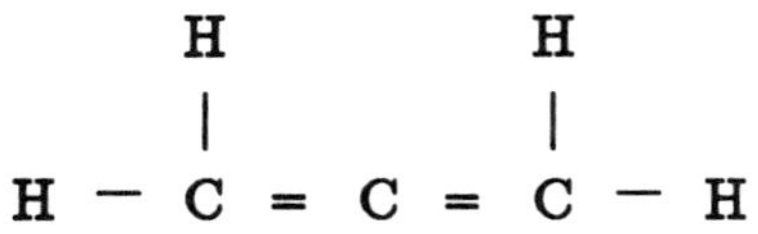 — S̈b — F
          |
          F

   f) H   H
       | |
       H — C = C = C — H

Problem Answers   Appendix C

b) $:\ddot{\underset{..}{O}}-\underset{..}{\ddot{Cl}}-\ddot{\underset{..}{O}}-H$

g)
$$H-\underset{\underset{F}{|}}{\overset{\overset{F}{|}}{C}}-F$$

c) $H-\ddot{\underset{..}{O}}-\ddot{N}=\ddot{\underset{..}{O}}$ (2 resonance forms)

h)
$$Cl-\underset{\underset{Cl}{|}}{\overset{\overset{Cl}{|}}{Si}}-Cl$$

d) $:\ddot{\underset{..}{O}}-\ddot{S}=\ddot{\underset{..}{O}}$ (2 resonance forms)

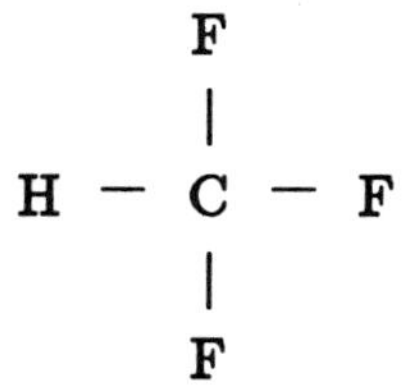

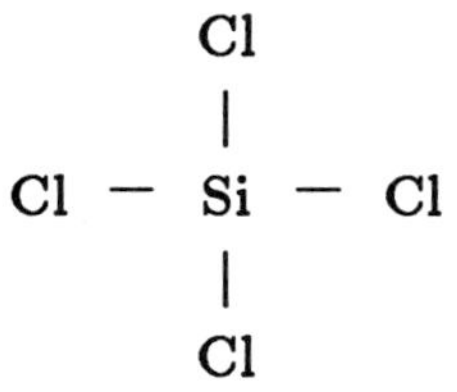

i) $Cl-\ddot{\underset{..}{O}}-Cl$

e) $H-C\equiv C-H$

j) 
$$F-\underset{\underset{F}{|}}{\ddot{N}}-\underset{\underset{F}{|}}{\ddot{N}}-F$$

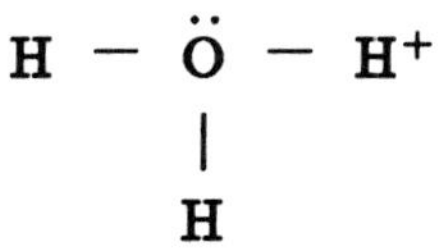

4.

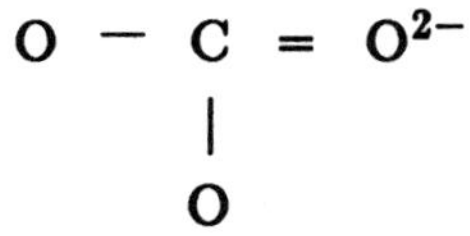

a) 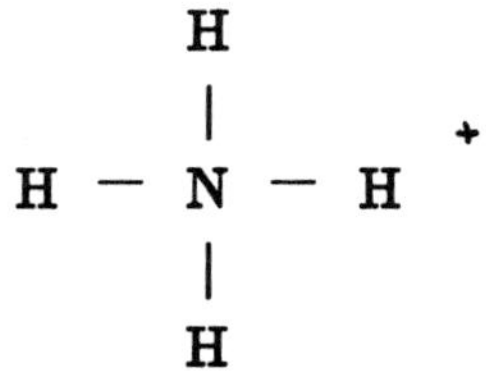
$$H-\underset{\underset{H}{|}}{\overset{\overset{H}{|}}{N}}-H\quad +$$

e) $\ddot{\underset{..}{O}}=N=\ddot{\underset{..}{O}}^+$

b) $:N\equiv O:\ ^+$

f) $:\ddot{\underset{..}{O}}-\ddot{N}=\ddot{\underset{..}{O}}^-$

c)
$$\underset{\underset{H}{|}}{\overset{H-\ddot{O}-H^+}{}}$$

g) $\ddot{N}=N=\ddot{\underset{..}{N}}^-$ (3 resonance forms)

d)
$$\underset{\underset{O}{|}}{\overset{O-C=O}{}}{}^{2-}\quad \text{(3 resonance forms)}$$

5.

a) 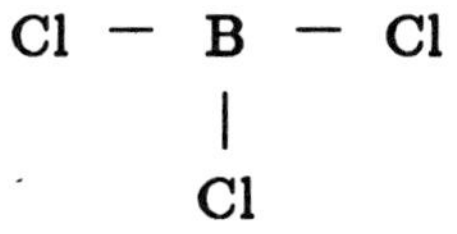
$$\underset{\underset{Cl}{|}}{\overset{Cl-B-Cl}{}}$$

f)
$$\underset{\underset{O}{|}}{\overset{\overset{O}{|}}{O-Xe-O}}$$

b) $H-Be-H$

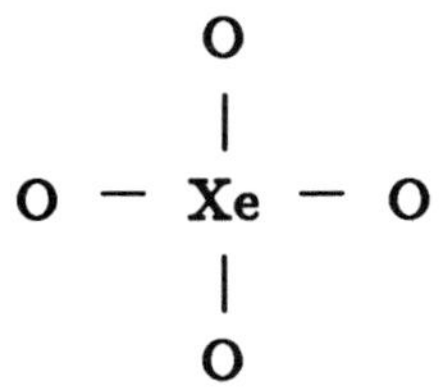

g) $:\ddot{\underset{..}{O}}-\dot{N}=\ddot{\underset{..}{O}}$ (2 resonance forms)

-303-

Appendix C   Problem Answers

c) :Ö—B=Ö⁻ (2 resonance forms)   h) :N≡O:

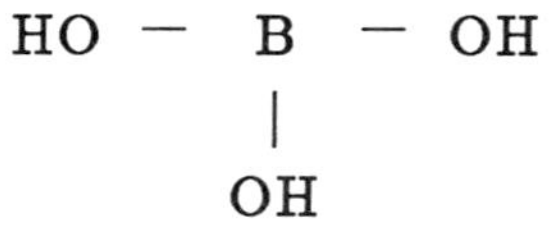

d) 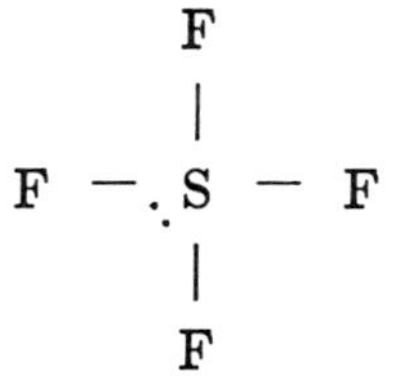 F — S — F   with F above and F below
                                         i) HO — B — OH
                                                   |
                                                   OH

e) F—Xe—F   j) 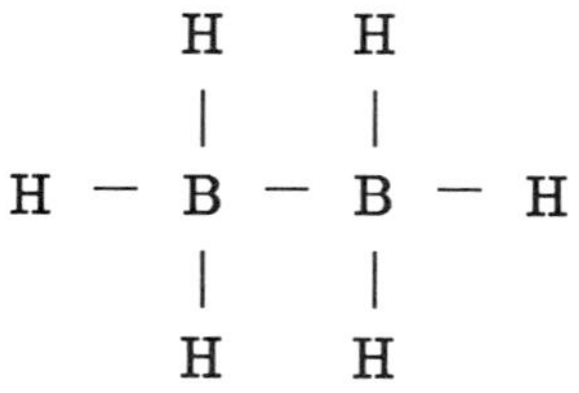 H — B — B — H   with H's above and below each B

6. C̈=N=Ö⁻, :C≡N—Ö:⁻, :C̈—N≡O:⁻
    The first two forms have the lowest formal charges

7. 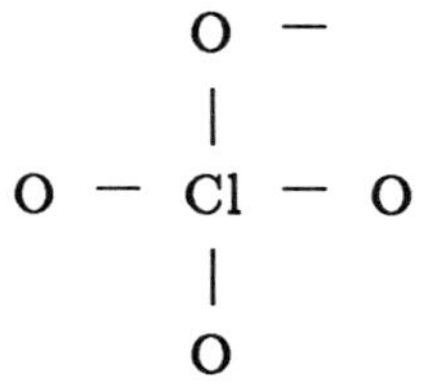
```
        O⁻
        |
  O — Cl — O
        |
        O
```
Cl has a +3 formal charge and each O has a -1

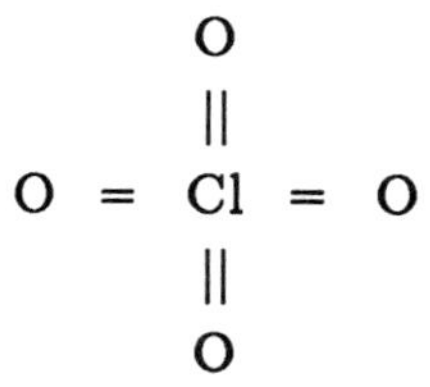

```
        O
        ||
  O = Cl = O
        ||
        O
```
Cl has a -1 formal charge and each O has none

8.
a) yes   e) yes
b) no   f) no
c) yes   g) no
d) yes   h) no

9.
a) tetrahedral, 109°
b) octahedral, 90°
c) trigonal bipyramidal, 90° and 120°
d) trigonal, 120°
e) tetrahedral, 109°
f) bent, 109° or less
g) linear, 180°
h) square pyramid, 90°
i) linear, 180°

-304-

Problem Answers Appendix C

j) bent, 120°
k) square, 90°
l) T-shaped, 90°

10. NH_2^- is bent, with bond angles slightly less than 109°. NH_3 is pyramidal with bond angles slightly less than 109°. NH_4^+ is tetrahedral with 109° angles.

11. SO_2^{2-} is bent, with angles slightly less than 109°. SO_3^{2-} is pyramidal with angles slightly less than 109°. SO_4^{2-} is tetrahedral with 109° angles.

12. $2\ C\equiv O\ +\ O=O \rightarrow 2\ O=C=O$

    $\Delta H° = -566$ kJ. This is hard to rationalize simply based on bond orders.

13. N_2 has a triple bond, which is the shortest. The longest bonds are the single bonds if F_2, Br_2 and Cl_2. Of these, the Br—Br is longest because the Br atoms are largest.

Chapter 7

1.
   a) N^{2-}, H^+
   b) K^+, Br^{7+}, O^{2-}
   c) Li^+, H^-
   d) Ga^{3+}, S^{2-}
   e) Xe^{4+}, O^{2-}
   f) Fe^{2+}, Fe^{3+}, O^{2-}
   g) H^+, O^{2-}, C has an average oxidation state of zero

2.
   a) Fe^{6+}, O^{2-}
   b) V^{4+}, O^{2-}
   c) S^{3+}, O^{2-}
   d) Al^{3+}, H^-
   e) P^{5+}, O^{2-}

3.
   a) $2H^+ + H_2O_2 + Cu \rightarrow 2H_2O + Cu^{2+}$. The oxidizing agent is H_2O_2 and the reducing agent is copper.

   b) $2Cr^{3+} + 3H_2O + 3HNO_2 \rightarrow 2Cr + 3NO_3^- + 9H^+$. The oxidizing agent is Cr^{3+} and the reducing agent is nitrous acid.

   c) $5H^+ + Cr_2O_7^{2-} + 3HClO_2 \rightarrow 2Cr^{3+} + 4H_2O + 2ClO_3^-$. The oxidizing agent is the dichromate ion. The reducing agent is chlorous acid.

   d) $8H_2O + 20HNO_3 + 3As_4 \rightarrow 20NO + 12H_3AsO_4$. The oxidizing agent is nitric acid. The reducing agent is arsenic.

Appendix C   Problem Answers

    e) $2Mn^{2+} + 14H^+ + 5NaBiO_3 \rightarrow 2MnO_4^- + 5Na^+ + 5Bi^{3+} + 7H_2O$. The oxidizing agent is sodium bismuthate. The reducing agent is Mn^{2+}.

    f) $4H^+ + 4NCS^- + 13O_2 \rightarrow 4NO_2 + 4CO_2 + 4SO_2 + 2H_2O$. The oxidizing agent is oxygen. The reducing agent is the thiocyanide ion.

4.
    a) $H_2O + Be_2O_3^{2-} + ClO_2^- \rightarrow 2Be + 2OH^- + ClO_4^-$. The oxidizing agent is $Be_2O_3^{2-}$. The reducing agent is the chlorite ion.

    b) $24OH^- + 3P_4 + 10IO_3^- \rightarrow 12HPO_4^{2-} + 6H_2O + 10I^-$. The oxidizing agent is the iodate ion. The reducing agent is phosphorus.

    c) $2NH_3 + HO_2^- \rightarrow N_2H_4 + H_2O + OH^-$. The oxidizing agent is HO_2^-, the reducing agent is ammonia.

    d) $2C_8H_{18} + 25O_2 + 32OH^- \rightarrow 16CO_3^{2-} + 34H_2O$. The oxidizing agent is oxygen and the reducing agent is octane.

5. $Na^+ < Zn^{2+} < H^+ < Cl_2 < MnO_4^-$

6. $F^- < Mn^{2+} < Cl^- < I^- < H_2 < Na$

7.   Zn, Mn and Fe will all dissolve in 1 M acid.

8.   Any reducing agent between Ag and Mn, such as Cu, Fe, Zn or H_2.

9.   Nitrate ion is a stronger oxidizing agent than chloride ion.

10. Aluminum metal can reduce Ag^+ to Ag.

11.
    a) 0.003 volts   c) 2.45 volts
    b) 2.122 volts   d) 2.20 volts

12.
    a) $\epsilon = \epsilon^\circ$ c) $\epsilon < \epsilon^\circ$
    b) $\epsilon < \epsilon^\circ$ d) 0

13. 0.32 volts

14. 0.12 volts

15. $\epsilon = \epsilon^\circ - \dfrac{RT}{nF} \ln \dfrac{1}{[H^+]^4} = 0.414$ when $[H^+] = 10^{-14}$

16. 0.064 volts

17. $K = 3.5 \times 10^{41}$, $\epsilon^\circ = 0.62$ volts

Problem Answers   Appendix C

18. -277 kJ

19. -0.152 volts

20. 0.005 volts. Sn^{2+} will not disproportionate under standard conditions.

21.
    a) -2.1 volts
    b) -0.14 volts
    c) 0.31 volts

22. 0.243 g

23. 0.0219 M

24. 108 g/mol or multiples thereof. The most likely metal and oxidation state is Ag^+.

25. 17.9 hours

26. 279 hours

Chapter 8

1. -289 J or 69.2 cal

2. $\Delta E = -1197$ J, w = 798 J, q = -1995 J

3. w = 0, $\Delta E = 0$, q = 0

4. w = -1100 J

5. 5.49×10^9 J

6. 676 J

7. -1670 kJ/mol

8. 81.6 J/mol-K

9. $\Delta E = -627.0$ kJ, $\Delta H = -629.5$ kJ

10. $\Delta H^0_{298} = -1134$ kJ, $\Delta H^0_{500} = -1134$ kJ

11. -2552 kJ/mol

12. -220 kJ/mol

Appendix C   Problem Answers

13.
- a) entropy decreases
- b) entropy decreases
- c) no change
- d) entropy increases

14. $\Delta S^0_{298} = 180.41$ J/mol–K, $\Delta S^0_{400} = 189.6$ J/mol–K

15. $\Delta S = 109$ J/mol-L

16. 349.5 K

17. 385 K

18. -959.4 kJ

19. 10^{168}

20. 27.1 kJ

21. -3.88 kJ

22. 8.0×10^{-4} atm

23. 4.69×10^{-6} atm

24.
- a) more soluble at 50 °C
- b) $\Delta G^0_{18} = 56.0$ kJ, $\Delta G^0_{50} = 60.0$ kJ
- c) endothermic

25. T can never be negative so the reaction can never reach equilibrium such that all pressures are 1.0 atm.

26.
- a) $\Delta G° = -162$ kJ
- b) $K = 2.62 \times 10^{28}$
- c) $\epsilon = -0.25$ volts, $\Delta G = 146$ kJ

27. $\Delta H^0_{vap} = 33.85$ kJ/mol, $T_b = 351$ K

$K_b = 2.36$ °C/m (calculated)

Chapter 9

1.
- a) 1.2×10^{-2} Ms^{-1}
- b) $M^{-2}s^{-2}$
- c) 2nd order in NO, 1st order in H_2
- d) $\frac{3}{4} \times$ original rate

-308-

Problem Answers Appendix C

2.
- a) 2.25×10^{-3} M min$^-$
- b) 6.17×10^{-4} M min^{-1}
- c) 2nd order plot, 1/[A] vs t give a straight line

3.
- a) rate = $k[H_3AsO_4]-I^-][H^+]^2$
- b) 4.67×10^{-3} M^{-3} min$^-$ or 7.78×10^{-5} M^{-3} s^{-1} or 0.280 M^{-3} hr$^-$

4. 2.3×10^{-4} M s^{-1}

5. $2 \times$ original rate

6. 0.0688 M

7. 31,500 sec or 525 min or 8.75 hrs

8. 3.56 hr

9. 211 min, 422 min

10. $k_{obs} = 9.24 \times 10^{-3}$ s^{-1}, $k_{real} = 3.93 \times 10^{-3}$ M^{-1} s^{-1}

11.
- a) $CH_3CH_2COOCH_3 + H_2O \rightarrow CH_3CH_2COOH + CH_3OH$
- b) H^+
- c) $CH_3CH_2C(OH)_2^+$, $CH_3CH_2COHOCH_3^+$
- d) rate = $k_2k_1[H^+][CH_3CH_2COOCH_3]$

12.
- a) rate = $k_2K_1[Br^-][BrO_3^-][H^+]^2$
- b) rate = $k_1[BrO_3^-][H^+]^2$

13. mechanism 4

14. $k_{-1} = 1.80 \times 10^{-3}$ M^{-1} s^{-1}

15. $\text{rate}_b = \dfrac{k_{-1}[I_3^-]}{[I^-]^2[H^+]^2}$, $\text{rate}_b = \dfrac{k_{-2}[I_3^-]}{[I^-]^2[H^+]}$

16. Nearly every collision is effective so $E_a \sim 0$ and $\rho \sim 1$.

Appendix C   Problem Answers

17. 275.3 kJ/mol

18. 41.7 kJ

19. 324 K

20. Vary the surface area of the vessel without changing the concentrations of reactants, *ie.* use a 50 mL vessel and compare with a 2.0 L vessel.

Chapter 10

1. Only droplets with charge multiples of 1.60×10^{-19} C are possible. There's no such thing as $1\frac{1}{2}$ electrons.

2. 1.67×10^{-27} kg

3. 2.73×10^{-23} cm^3/atom

4. $\lambda = 2.87$ m, E=6.92×10^{-26} J/quanta = 0.0417 J/mol

5. 434 nm

6. n = 9

7. 1.8×10^{-38} m

8. 1.10×10^{-32} kg

9. a 3p orbital

10. $n = 2, l = 1, m_l = -1, m_s = +\frac{1}{2}$

    $n = 2, l = 1, m_l = -1, m_s = -\frac{1}{2}$

    $n = 2, l = 1, m_l = 0, m_s = +\frac{1}{2}$

    $n = 2, l = 1, m_l = 0, m_s = -\frac{1}{2}$

    $n = 2, l = 1, m_l = 1, m_s = +\frac{1}{2}$

    $n = 2, l = 1, m_l = 1, m_s = -\frac{1}{2}$

11. $n = 5, l = 0, m_l = 0, m_s = +\frac{1}{2}$ or $-\frac{1}{2}$

12. n = 5 shell

13. 32

Problem Answers   Appendix C

14.
- a) $[Kr]5s^2$
- b) $[Kr]5s^2 4d^{10}$
- c) $1s^2 2s^2 2p^2$
- d) $[Kr]5s^2 4d^{10} 5p^1$
- e) $[Ar]4s^2 3d^3$

15.
- a) [Kr]
- b) $[Ar]3d^2$
- c) $[Ar]3d^9$
- d) [Ar]
- e) [Ar]

16. Cu predicted $[Ar]\,4s^2 3d^9$

    Cu^{+1} predicted $[Ar]4s^1 3d^9$

    But the actual configurations, $4s^1 3d^{10}$ and $3d^{10}$, are more stable because of the closed d subshell.

17. They are isoelectronic (have the same electron configuration).

18. Fe^{2+}, Co^{3+}

19. 4

20.
- a) paramagnetic
- b) paramagnetic
- c) paramagnetic
- d) paramagnetic
- e) paramagnetic

21. A ns^1 B $ns^2 np^4$

22. Na < Al < Mg

23. As already has a half-filled shell, and Se has a half-filled shell only by losing one electron.

24. Li

Appendix C   Problem Answers

25.
   a) Sn < As < S < F
   b) As < Sb < Bi

Chapter 11

1. $F^- < Cl^- < Br^- < I^-$

2. $H^+ < H < H^-$

3. $Sc^{3+} < Ca^{2+} < K^+ < Ar < Cl^-$

4. MgO should release more heat, $\Delta H°_{hyd}$ is therefore less. This is because O^{2-} should release more heat than $2F^-$ because O^{2-} has a higher charge to size ratio.

5. V^{3+} releaes the most heat because it is the smallest.

6. -398.7 kJ/mol

7. When an electron is added to H_2, it goes into a σ^* orbital raising the overall energy. When an electron is removed from H_2, it takes energy to do so.

8. He_2 has a $\sigma_{1s}^2 \sigma_{1s}^{*2}$ configuration with no net bonding, the σ^* electrons destabilize He_2 approximately the same amount that σ electrons stabilize. The net result is no bond. He_2^+ has a $\sigma_{1s}^2 \sigma_{1s}^{*1}$ configuration with a net stabilization because the bonding electrons outnumber the antibonding.

9. H_2^{2-} should have the same MO configuration as He_2 and so, not exist.

10. The bonding orbital is lower in energy than the separate atomic orbitals, and the antibonding is higher in energy.

11. 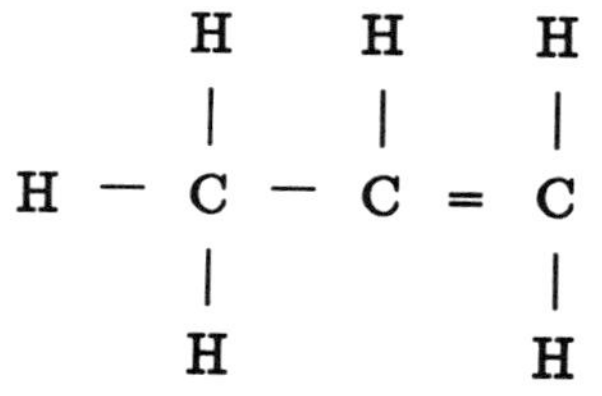
```
        H   H   H
        |   |   |
    H - C - C = C
        |       |
        H       H
```

   a) 8 single bonds, 1 double bond
   b) from left to right the carbon hybridizations are sp^3, sp^2, sp^2.

12.
   a) sp^2 d) sp^2
   b) sp^3 e) sp^2
   c) sp^3 f) sp^2

-312-

Problem Answers   Appendix C

g) sp^2 i) sp^2

h) sp^3 j) sp

13.
   a) sp^3d d) sp^3d^2

   b) sp^3d^2 e) sp^3d^2

   c) sp^3d

14.

$$H-\underset{\underset{H}{|}}{\overset{\overset{H}{|}}{C}}-\underset{\underset{O}{\|}}{C}-\ddot{\underset{..}{O}}-\ddot{\underset{..}{O}}-N\overset{\displaystyle O}{\underset{\displaystyle O}{}}$$

         sp^3 sp^2 sp^2

Chapter 12

1. One α and two π bonds

2.
   a) O_2 $\sigma_s^2(\sigma_s^*)^2\sigma_p^2\,\pi_{px}^2\pi_{py}^2(\pi_{px}^*)^1(\pi_{py}^*)^1$

      O_2^+ $\sigma_s^2(\sigma_s^*)^2\,\sigma_p^2\,\pi_{px}^2\pi_{py}^2(\pi_{px}^*)^1$

      O_2^- $\sigma_s^2(\sigma_s^*)^2\,\sigma_p^2\,\pi_{px}^2\pi_{py}^2(\pi_{px}^*)^2(\pi_{py}^*)^1$

      O_2^{2-} $\sigma_s^2\,(\sigma_s^*)^2\,\sigma_p^{22}\,\pi_{px}^2\pi_{py}^2(\pi_{px}^*)^2(\pi_{py}^*)^2$

   b) O_2^{2-} has the longest bond

   c) O_2^+ has the shortest bond

   d) O_2, O_2^+ and O_2^- are paramagnetic

3. bond order = $\frac{1}{2}$

4. HOMO = π_p^*, LUMO = σ_p^*

5.
   a) N_2 (bond order = 3) has a shorter and stronger bond than O_2 (bond order = 2)

Appendix C   Problem Answers

b) Both Cl_2 and I_2 have bond orders = 1, but Cl is a smaller atom so Cl_2 has a shorter, stronger bond.

6. 1 unpaired electron in the σ_p orbital.

7. The bond length increases and bond strength decreases in N_2^{2-}.

8. A π_p orbital

9.
   a) bond order = $2\frac{1}{2}$
   b) paramagnetic
   c) the unpaired electron resides in a π^* orbital

10. bond order = $1\frac{1}{2}$

11.
    a) bent
    b) linear
    c) linear

12.
    a) linear
    b) bent
    c) bent

Chapter 13

1. Fluorine

2. Radium

3. Bi < Au < W $\simeq$ Mo < Cs

4. Germanium and selenium

5. F > S < As > Sn > Tl

6. B—Cl polar covalent, Mg—Cl ionic, Na—Cl ionic, C—Cl polar covalent, O—Cl nonpolar covalent

   O—Cl < C—Cl < B—Cl < Mg—Cl < Na—Cl

7. Group VIA, the chalcogens

Problem Answers Appendix C

8. Electronegativity is defined as attraction for electrons *in a bond*. Since most noble gases do not readily form compounds, this property has little meaning. But based on the ionization energies and electron affinities, the noble gases should have high X values.

9. Groups IIIA, IIIB or VA, VB

10. Cr can have positive oxidation states up to +6. The +6 state ([Ar]) and +3 state ($3d^5$) should be especially stable.

11. Cl^-, Cl^{+1}, Cl^{+3}, Cl^{+5}, Cl^{+7}. Chlorine will only have a positive oxidation state when combined with more electronegative elements. Only O and F are more electronegative than Cl.

12. Na_2O, MgO, Al_2O_3, SiO_2, P_2O_3, P_2O_5 (or P_4O_{10}), SO_2, SO_3, Cl_2O, ClO, Cl_2O_3, ClO_2, Cl_2O_5, Cl_2O_7.

13. AsH_3 is a covalent hydride.

14. $4H_2O + LiAlH_4 \rightarrow 4H_2 + Al(OH)_3 + Li(OH)$. H^- has been oxidized, H^+ has been reduced.

15. $MgO < Al_2O_3 < P_4O_{10} < SO_3$

16.
   a) $H_2C = CH_2 + H_2 \xrightarrow{\text{Ni or Pt}} H_3C - CH_3$

   b) $ZnO + 2HCl \rightarrow Zn^{2+} + 2H_2O + 2Cl^-$

   c) $NaH + H_2O \rightarrow H_2 + Na^+ + OH^-$

   d) $N_2O_5 + H_2O \rightarrow 2HNO_3$

   e) $ZnO + 2NaOH + H_2O \rightarrow Zn(OH)_4^{2-}$

17. Cl_2O_5

18. PH_3, AsH_3, and SbH_3 should show an increase in boiling points, but NH_3 can hydrogen bond and so will be much higher than expected (just like H_2O, H_2S, H_2Se).

19. The ΔH_{vap}^0 trend is the same as the boiling point trend.

20. Carbon in CH_4 has no lone valence electron pairs to hydrogen-bond. Also, the molecule has no dipole moment.

Chapter 14

1. I values are strictly for gases while ϵ° values incorporate heats of atomization and hydration energies.

Appendix C   Problem Answers

2.
   a) $2K + 2H_2O \rightarrow 2KOH + H_2$

   b) $NaH + H_2O \rightarrow NaOH + H_2$

   c) $Na_2O + H_2O \rightarrow 2NaOH$

   d) $Li_2O_2 + CO_2 \rightarrow Li_2CO_3 + \frac{1}{2}O_2$

3. In water, H^+ is reduced to H_2 more easily than Na^+ to Na.

4.
   a) $CaH_2 + 2H_2O \rightarrow Ca(OH)_2 + 2H_2$

   b) $SrO + H_2O \rightarrow Sr(OH)_2$

   c) $Mg(OH)_2 + 2HCl \rightarrow 2H_2O + MgCl_2$

5. $CaCO_3 + CO_2 + H_2O \rightarrow Ca^{2+} + 2HCO_3^-$

6. The group IA elements have a much lower ionization energy.

7. $Al^{3+} + 3OH^- \rightarrow Al(OH)_3 \xrightarrow{OH^-} Al(OH)_4^-$

8. B_2H_6 is the Lewis acid, H^- is the Lewis base.

9. $B(OH)_3 + H_2O \rightarrow B(OH)_4^- + H^+$

   $B(OH)_3 + OH^- \rightarrow B(OH)_4^-$

10. Much more energy is required to reduce Al^{3+} to Al than for Fe^{3+} to Fe (see $\epsilon\,°$ values).

11.
    a) $CaC_2 + 2H_2O \rightarrow Ca(OH)_2 + C_2H_2$

    b) $Sn + 2HCl \rightarrow H_2 + SnCl_2$

12. The chlorides of the nonmetals and semimetals (C, Si, Ge) should be covalent and the chlorides of the metals (Sn, Pb) should be ionic.

13. $Sn(OH)_2 + 2H^+ \rightarrow Sn^{2+} + 2H_2O$

    $Sn(OH)_2 + 2OH^- \rightarrow Sn(OH)_4^{2-}$

14. d (diamond) > d (graphite)

15. $Pb + PbO_2 + 2H_2SO_4 \rightarrow 2PbSO_4 + 2H_2O$

Problem Answers   Appendix C

Chapter 15

1. N—N bonds are considerably weaker than C—C bonds, possibly because the N atoms are smaller and there is lone electron-pair repulsion between the N atoms.

2. $2NH_3 + ClO^- \rightarrow N_2H_4 + Cl^- + H_2O$

3. $N_2O_3 + H_2O \rightarrow 2HNO_2$

   $NO + NO_2 + H_2O \rightarrow 2HNO_2$

4.
   a)
   $$F-\ddot{N}(F)-\ddot{N}(F)-F$$

   b)
   $$Cl-N=O$$
   $$\quad\; |$$
   $$\quad\; O$$

   c)
   $$Cl-P(Cl)(=O)-Cl$$

5. ClNO has 18 valence electrons and so should be bent.

6. $P_4O_{10} + 6H_2O \rightarrow 4H_3PO_4$

7. NO_3^- is a stronger oxidizing agent than H^+ and Cl^-

8. $2NO_2 \rightarrow N_2O_4$, N_2O_4 is colorless and diamagnetic

9. $HCO_3^- + H_2PO_4^- \rightarrow CO_2 + H_2O + HPO_4^{2-}$

10. $Na_2O_2 + H_2O \rightarrow \frac{1}{2}O_2 + 2NaOH$

11. $\left[\begin{array}{c} O \\ | \\ S-S(O)-O \end{array}\right]^{2-}$

12. O_2^+ has a bond order of $2\frac{1}{2}$, O_2 has a bond order of 2, O_2^- has a bond order of $1\frac{1}{2}$, and O_2^{2-} has a bond order of 1.

13. $3Ag_2S + 2Al \rightarrow Al_2S_3 + 6Ag$

Appendix C   Problem Answers

14. If the S_8 ring were planar, the S—S bond angles would be 140°, but the molecule is bent. The angles are near 105°.

15. The larger SF_6 molecule cannot diffuse out of the ball as quickly as N_2 or O_2, so it will retain its bounce.

16. $2NaX_{(s)} + H_2SO_{4(conc)} \rightarrow 2HX_{(g)} + Na_2SO_{4(s)}$

17.
   a) hypobromous acid
   b) bromous acid
   c) hypoiodous acid
   d) iodic acid

18. $3IF \rightarrow IF_3 + I_2$

19.
   a) linear

   b)

   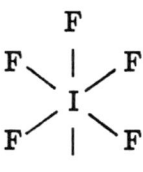

   a sawhorse geometry

   c)

   a square pyramid

20. $K = 6.9 \times 10^2$

21. $\epsilon° = -0.27$ volts

22. linear, square planar, distorted octahedron

23. The noble gases have complete octets and so were not expected to give up, acquire, or share any valence electrons.

Problem Answers Appendix C

Chapter 16

1. $Mn[Ar]4s^2 3d^5$, $Mn^{2+}[Ar]3d^5$, $Mn^{4+}[Ar]3d^3$, $Mn^{7+}[Ar]$

2. $Cu[Ar]4s^1 3d^{10}$

3. $[Cr(H_2O)_4Cl_2]^+ Cl^-$

4.
   a) tetrachloroethylenediamineplatinum (IV)
   b) potassium tetracyanonickelate (II)
   c) diamminetetracyanocobaltate (III) ion
   d) hexafluoroferrate (III) ion
   e) pentacarbonyliron (0)
   f) dichlorodihydroxodiiodoferrate (III) ion

5.
   a) $Cu(NH_3)_4^{2+}$
   b) $[Ru(NH_3)_4Br_2]NO_3$
   c) $[Rh(en)_2Cl(H_2O)]Cl_2$
   d) $K[Co(C_2O_4)_2(NH_3)_2]$
   e) $Ca_2[Fe(CN)_6]$

6. 4c) and 4f) have cis/trans isomers

7.
   a) no   b) yes   c) no optical iosomers, but geometric isomers

8. $2[Co(H_2O)_6]Cl_2 \xrightarrow{heat} Co[CoCl_4] + 12H_2O$

9. 4 unpaired electrons

10.
   a) $t_{2g}^5 e_g^2$
   b) $t_{2g}^5 e_g^0$
   c) $t_{2g}^5 e_g^0$

11. V^{2+}, Cr^{3+}, Mn^{4+}

12.
   a) $Co(NH_3)_6^{3+}$
   b) $Fe(H_2O)_6^{3+}$
   c) $CoCl_6^{4-}$

13. 464 nm

14. The t_{2g} and e_g orbitals are filled so there is no low energy transition that can occur.

15. Sc and Ti

Appendix C Problem Answers

16. Co^{3+} oxidizes H_2O

17. $2Cu + H_2O + CO_2 + O_2 \rightarrow Cu(OH)_2 + CuCO_3$

18. Co^{1+}, s^1d^7 or more likely d^8 would probably prefer square planar or a distorted octahedron.

19.
   a) MnO_4^-
   b) $K_2Cr_2O_7$
   c) Ni^{3+}

Chapter 17

1.

a) 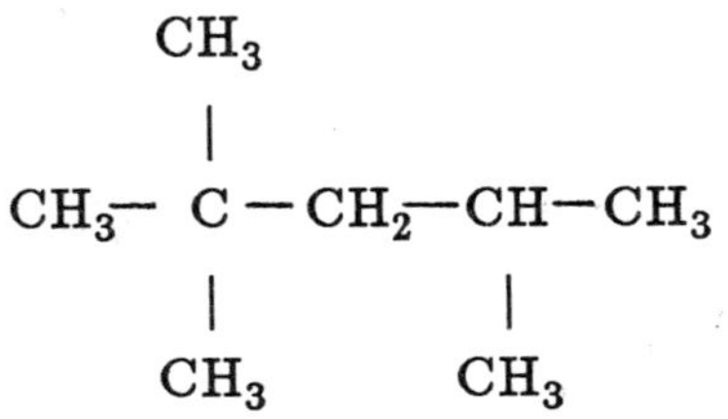
```
        CH3
         |
CH3 — C — CH2 — CH — CH3
         |            |
        CH3          CH3
```

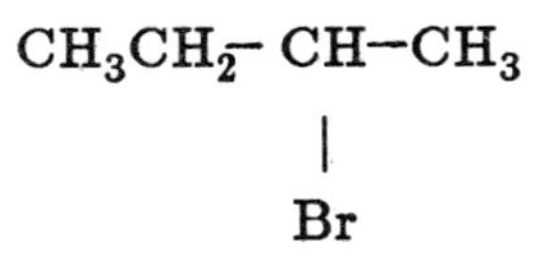

b)
```
CH3CH2 — CH — CH3
          |
          Br
```

c)
```
              CH3        CH3
               |          |
CH3 — CH2 — CH  —  C  —  CH2CH2CH3
                          |
                         CH2CH3
```

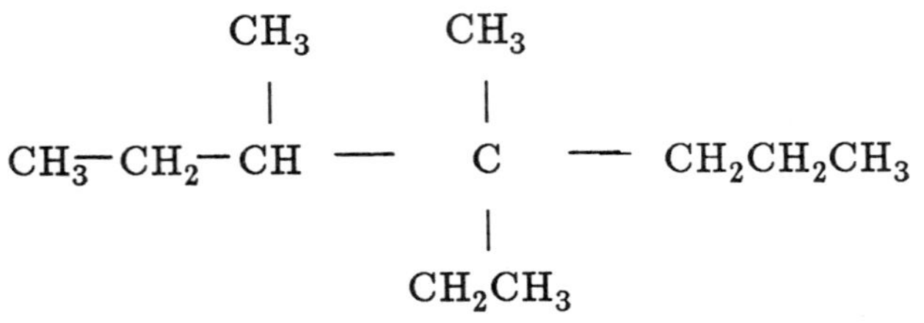

d) 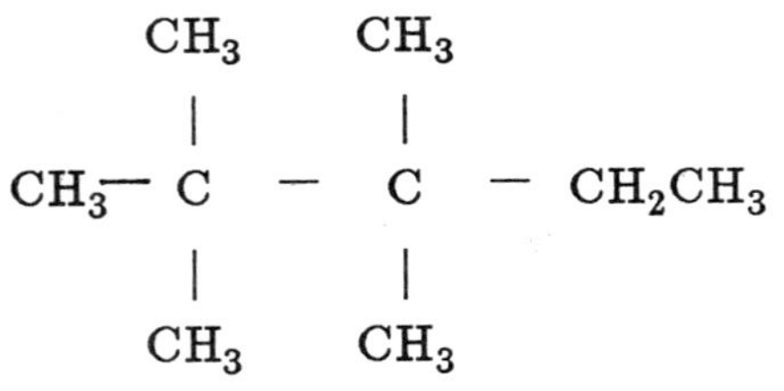
```
        CH3       CH3
         |         |
CH3 — C   —   C — CH2CH3
         |         |
        CH3       CH3
```

2. butane and 2-methylpropane

3. 1-pentene, 2-pentene, or cyclopentane

4.
   a) ethyne (or acetylene)
   b) 7-methyl-2,6-octadienol
   c) 2-methylpentene

5.
   a) none
   b) one pair of cis/trans isomers
   c) none

6. 1-butanol, 2-butanol, 2-methylpropanol, 1,1-dimethylethanol

7. 2-isopropoxypropane

8.
   a) 2-butenal
   b) 4-methyl-2-pentanone

Problem Answers   Appendix C

    c) 3-hydroxy-pentanol
    d) ethyl ethanoate or ethyl acetate
    e) trichloroethanoic acid or trichloroacetic acid
    f) ammonium methanoate or ammonium formate
    g) calcium propanoate

9.
    a) 1,2-dichloropentane
    b) ethane
    c) 1-butanol
    d) 1-chloropropane

10. $CH_3OH + \frac{3}{2}O_2 \rightarrow CO_2 + 2H_2O$

    ($C^{2-} \rightarrow C^{4+}$, $O_2 \rightarrow 2O^{2-}$)

11.
    a) $CH_2=CH_2$

    b) no reaction

    c) $(CH_2CH_3)_3NH^+Cl^-$

    d) CH_3CHO or CH_3COOH

    e) $CH_3SH_2-CH-CH_2OH$
                   |
                   CH_3

    f) $CH_2=CHCH_2COOCH_3$

    g) $(CH_3)_2-C-CH_2CH_2CH_2CH_3$
                |
               OH

12. $3CH_2=CH_2 + 2H_2O + 2H^+ + 2KMnO_4 \rightarrow CH_2CH_2 + 2K^+ + 2MnO_2$
                                                                              | |
                                                                                OH OH

Chapter 18

1. $K = 1.67 \times 10^5$

2. histidine

3. 2

Appendix C   Problem Answers

4. $CH_3-\underset{\underset{O}{\|}}{C}-COO^- \rightarrow CH_3-\underset{\underset{NH_2}{|}}{CH}-COO^-$

5.
   a) +2870 kJ/mol
   b) 41.5 L

6. $(CH_3)_2CHCH_2-\underset{\underset{NH_2}{|}}{CH}-COONH-\underset{\underset{CH_3}{|}}{CH}-COOH$

7. Fats are more reduced than carbohydrates, and they are hydrophobic so there are no waters of hydration.

8. $Na^+\ ^-OOCCH_2CH_2-\underset{\underset{NH_3^+}{|}}{CH}-COO^-$

9. ser, cys, thr, asp, glu, tyr, asn, gln, his, lys, arg

10. 1:1

Chapter 19

1. 1.7×10^{14} g/cm³

2. 10^6 kg/s

3. β^- decay

4.
   a) β^+ decay or EC
   b) stable
   c) stable
   d) α emission
   e) stable
   f) α emission

5.
   a) $^{207}_{84}Po \xrightarrow{EC} {}^{207}_{83}Bi$

   b) $^{120}_{51}Sb \rightarrow {}^{120}_{50}Sn + {}^{0}_{1}\beta$

   c) $^{208}_{87}Fr \rightarrow {}^{4}_{2}\alpha + {}^{204}_{85}At$

   d) $^{83}_{38}Sr \rightarrow {}^{0}_{1}\beta + {}^{83}_{37}Rb$

   e) $^{87}_{37}Rb \rightarrow {}^{0}_{-1}\beta + {}^{87}_{38}Sr$

Problem Answers   Appendix C

6. 7.07 MeV/nucleon

7. ^{55}Fe

8. β^- only

9. 297 days

10. 6.0 min

11. $E_a \simeq 0$

12. 57.2 days

13. 1.6×10^9 yrs

14. 11.4 yrs

15.
  a) $^{27}_{13}Al + ^{4}_{2}\alpha \rightarrow 3^{1}_{0}n + ^{28}_{15}P$

  b) $^{22}_{86}Rn \rightarrow ^{4}_{2}\alpha + ^{218}_{84}Po \rightarrow ^{4}_{2}\alpha + ^{214}_{82}Pb \rightarrow ^{0}_{-1}e + ^{214}_{83}Bi \rightarrow ^{0}_{-1}e + ^{214}_{84}Po$

  c) $^{7}_{3}Li + ^{1}_{1}p \rightarrow 2^{4}_{2}\alpha$

16. 9×10^7 kJ/mol

17. 2.8×10^{10} kJ/pound of ^{235}U, 2.5×10^4 kJ/ pound of CH_4

18. 0.0024 nm

Chapter 20

1.
  a) molecular
  b) molecular
  c) metallic
  d) network
  e) molecular
  f) molecular
  g) ionic

2. SiO_2 forms a covalent network crystal where all Si—O bonds are covalent. There is no discrete SiO_2 molecule. These strong, rigid interactions can only be broken at high temperatures.

3.
  a) tetrahedral site
  b) octahedral site
  c) cubic or octahedral site

Appendix C   Problem Answers

4. 8, 12, 12 respectively

5. 6.04×10^{23}

6. Similar to electrical conductivity, small amounts of heat can promote an electron to an unoccupied molecular orbital in the band. There it is free to move over the entire crystal. When it falls back to a lower energy orbital it re-emits the heat.